IET MATERIALS, CIRCUITS AND DEVICES 32

Oscillator Circuits

Other volumes in this series:

Volume 2	**Analogue IC Design: The current-mode approach** C. Toumazou, F.J. Lidgey and D.G. Haigh (Editors)
Volume 3	**Analogue–Digital ASICs: Circuit techniques, design tools and applications** R.S. Soin, F. Maloberti and J. France (Editors)
Volume 4	**Algorithmic and Knowledge-based CAD for VLSI** G.E. Taylor and G. Russell (Editors)
Volume 5	**Switched Currents: An analogue technique for digital technology** C. Toumazou, J.B.C. Hughes and N.C. Battersby (Editors)
Volume 6	**High-frequency Circuit Engineering** F. Nibler *et al.*
Volume 8	**Low-power High-frequency Microelectronics: A unified approach** G. Machado (Editor)
Volume 9	**VLSI Testing: Digital and mixed analogue/digital techniques** S.L. Hurst
Volume 10	**Distributed Feedback Semiconductor Lasers** J.E. Carroll, J.E.A. Whiteaway and R.G.S. Plumb
Volume 11	**Selected Topics in Advanced Solid State and Fibre Optic Sensors** S.M. Vaezi-Nejad (Editor)
Volume 12	**Strained Silicon Heterostructures: Materials and devices** C.K. Maiti, N.B. Chakrabarti and S.K. Ray
Volume 13	**RFIC and MMIC Design and Technology** I.D. Robertson and S. Lucyzyn (Editors)
Volume 14	**Design of High Frequency Integrated Analogue Filters** Y. Sun (Editor)
Volume 15	**Foundations of Digital Signal Processing: Theory, algorithms and hardware design** P. Gaydecki
Volume 16	**Wireless Communications Circuits and Systems** Y. Sun (Editor)
Volume 17	**The Switching Function: Analysis of power electronic circuits** C. Marouchos
Volume 18	**System on Chip: Next generation electronics** B. Al-Hashimi (Editor)
Volume 19	**Test and Diagnosis of Analogue, Mixed-signal and RF Integrated Circuits: The system on chip approach** Y. Sun (Editor)
Volume 20	**Low Power and Low Voltage Circuit Design with the FGMOS Transistor** E. Rodriguez-Villegas
Volume 21	**Technology Computer Aided Design for Si, SiGe and GaAs Integrated Circuits** C.K. Maiti and G.A. Armstrong
Volume 22	**Nanotechnologies** M. Wautelet *et al.*
Volume 23	**Understandable Electric Circuits** M. Wang
Volume 24	**Fundamentals of Electromagnetic Levitation: Engineering sustainability through efficiency** A.J. Sangster
Volume 25	**Optical MEMS for Chemical Analysis and Biomedicine** H. Jiang (Editor)
Volume 26	**High Speed Data Converters** Ahmed M.A. Ali
Volume 27	**Nano-Scaled Semiconductor Devices** E. A. Gutiérrez-D (Editor)
Volume 29	**Nano-CMOS and Post-CMOS Electronics: Devices and Modelling** Saraju P. Mohanty and Ashok Srivastava
Volume 30	**Nano-CMOS and Post-CMOS Electronics: Circuits and Design** Saraju P. Mohanty and Ashok Srivastava

Oscillator Circuits

Frontiers in Design, Analysis and Applications

Edited by
Yoshifumi Nishio

The Institution of Engineering and Technology

Published by The Institution of Engineering and Technology, London, United Kingdom

The Institution of Engineering and Technology is registered as a Charity in England & Wales (no. 211014) and Scotland (no. SC038698).

First published 2016

The Institution of Engineering and Technology
Michael Faraday House
Six Hills Way, Stevenage
Herts, SG1 2AY, United Kingdom

www.theiet.org

British Library Cataloguing in Publication Data
A catalogue record for this product is available from the British Library

ISBN 978-1-78561-057-8 (hardback)
ISBN 978-1-78561-058-5 (PDF)

Typeset in India by MPS Limited

Contents

Chapter 1

Introduction

Yoshifumi Nishio[1]

Oscillator circuit has been an absorbing theme of great interest for long time [1]. The researchers of oscillator circuit have been working in order to develop various electrical engineering systems from radio communication circuits to clock generators for microprocessors.

At the same time, synchronization of oscillator circuits has attracted attentions of many researchers not only in the electrical engineering field but also in the fields of mathematics, physics, chemistry, biology, medical science, neuroscience, social science, and so on. Oscillation waves (periodic or non-periodic) are the most basic and the most important signals in natural and artificial systems, and hence developing new oscillators and investigating synchronization and related phenomena in coupled oscillators are essential topics for many researchers in various fields. Oscillator circuits are good models of various systems generating oscillations in the sense that they are real physical systems realized easily and handled easily. This is the reason why oscillator circuits have been a common subject for researches in diverse fields.

This book is a collection of recent researches on oscillator circuits. Design, analysis, and applications of oscillator circuits are described by different authors.

In Chapter 2, bifurcation analysis of oscillator circuits is described with some example circuits. It is important to know what kinds of bifurcations exist in the circuits and how the oscillation states change as circuit parameters in order to understand the features of the circuits and to use the circuits for real applications. The authors of this chapter classify the bifurcations and describe how to calculate the parameters at which the bifurcations occur.

Because of the recent development of circuit devices and nonlinear theory, several researchers have been proposing novel oscillator circuits. In this book, three types of oscillator circuits are introduced by different authors.

In Chapter 3, fractional-order oscillator circuits are described. Fractional-order calculus is a mathematical theory dealing with non-integer order differentiation [2]. The authors describe how to realize oscillator circuits with fractional-order circuit elements and their advantages with a possible application in biomaterial characterization.

[1]Department of Electrical and Electronic Engineering, Tokushima University, Tokushima, Japan

In Chapter 4, oscillator circuits including memristive devices are described. After the first nanoscale device was reported in 2008 [3], many researchers have been trying to produce different kinds of circuits including memristive devices. In this chapter, the authors concentrate on the floating memristor emulator, which extends the application possibility of memristive devices, and utilize it to realize oscillator circuits.

In Chapter 5, piecewise-constant oscillator circuits are described. The circuits are extremely simple and are also realized very easily. The most important advantage of the circuits is that rigorous analysis using exact solutions becomes possible. In spite of their simplicities, the circuits exhibit a rich nonlinear phenomena. The authors show two example oscillator circuits with theoretical analysis using exact solutions.

Coupled systems of oscillator circuits have been an important subject in relatively wide range of research field. Basic coupled oscillator circuits were investigated in 1960s [4,5] and they are extended to larger systems in 1970s [6–8]. Since then, many different types of coupled oscillator circuits have been reported and investigated.

In Chapter 6, a new type of master–slave synchronization method is described. Namely, common non-periodic pulse is injected to oscillator circuits as an external force. Complete and stochastic synchronizations are achieved according to the features of the external force. This method may offer a simple control method to attain synchronization of engineering systems such as power-grid networks, communication networks, and so on. Also, the investigation of the synchronization mechanism may help to understand the behavior of a group of small animals.

Oscillator circuit containing a nonlinear resistor whose voltage–current characteristics are described by a fifth-power nonlinear polynomial equation is known to exhibit hard excitation [9,10]. Namely, the origin is asymptomatically stable and proper initial values are necessary to generate oscillation. Such an oscillator circuit is often called as a hard oscillator or said to have a hard nonlinearity. In Chapter 7, coupled oscillator circuits with hard nonlinearities are investigated. Multimode oscillation, one of typical oscillation phenomena observed in such circuits, is described in detail. The authors extend their circuits to one-dimensional ring structure and report propagating waves.

In Chapter 8, the authors report the wave propagation of phase difference between adjacent circuits in one-dimensional and two-dimensional arrays of inductively coupled oscillator circuits. They describe the physical mechanism of the wave using the relationship between the instantaneous frequency and the phase difference.

In Chapter 9, a concept of frustration in coupled oscillator circuits is introduced, which is a kind of difficult situation for circuits to generate apparent stable synchronization patterns. The authors investigate coupled polygonal oscillator networks as an example of oscillator circuit networks with frustration and describe the calculation method of synchronization states by using power consumption under some assumptions.

Synchronization of complex networks has been intensively investigated in these two decades. In Chapter 10, the authors describe the method to achieve global synchronization of complex networks composed of coupled oscillator circuits using the topological information of the network. The method is explained to be especially useful for networks with growing topologies. Successful practical applications are expected in near future.

In Chapter 11, a lot of interesting nonlinear phenomena observed from oscillator networks, more specifically coupled Chua's circuits which are the most famous chaotic oscillator circuits, are reported. It should be emphasized that almost all results in this chapter are obtained from real circuit experiments, which are not easy for large-size networks.

There have been many different engineering applications of oscillator circuits. In this book, two different applications are introduced.

In Chapter 12, an application to power electronics is described. The author has developed a design method of tuned power oscillators satisfying the class-E conditions, which achieves high-power conversion efficiency at high frequencies. This chapter describes how to design the class-E tuned power oscillators.

In Chapters 13 and 14, an application of ring oscillators to true random number generators is described, which is one of key technologies in data security systems. In Chapter 13, the authors introduce different types of ring oscillators and how to construct true random number generators with those ring oscillators. The characteristics of different true random number generators are explained with their advantages and disadvantages in detail. In Chapter 14, a possible attacking on a true random number generator based on a ring oscillator is described. Improving and attacking are a kind of two wheels of one car to develop a robust security system. In the next step, the authors may add an innovative idea to ring oscillators to make true random number generators robust against electromagnetic field.

References

[1] B. van der Pol, "On relaxation-oscillations," *The London, Edinburgh, and Dublin Philosophical Magazine and Journal of Science: Series 7*, vol. 2, no. 11, pp. 978–992, 1926.

[2] K.B. Oldham and J. Spanier, *The Fractional Calculus: Theory and Applications of Differentiation and Integration to Arbitrary Order*. Amsterdam: Elsevier Science, Jun. 1974.

[3] D.B. Strukov, G.S. Snider, D.R. Stewart, and R.S. Williams, "The missing memristor found," *Nature*, vol. 453, pp. 80–83, May 2008.

[4] B.R. Nag, "Characteristics of a system of two capacitively coupled tuned anode oscillators," *Transactions on AIEE Part I: Communication and Electronics*, vol. 82, no. 5, pp. 584–588, Nov. 1963.

[5] T. Suezaki and S. Mori, "Mutual synchronization of two oscillators," *Transactions on IECE*, vol. 48, no. 9, pp. 1551–1557, Sep. 1965.

[6] A.C. Scott, "Distributed multimode oscillators of one and two spatial dimensions," *IEEE Transactions on Circuit Theory*, vol. 17, no. 1, pp. 55–60, Feb. 1970.

[7] T. Endo and S. Mori, "Mode analysis of a multimode ladder oscillator," *IEEE Transactions on Circuits and Systems*, vol. 23, no. 2, pp. 100–113, Feb. 1976.

[8] T. Endo and S. Mori, "Mode analysis of a ring of a large number of mutually coupled van der Pol oscillators," *IEEE Transactions on Circuits and Systems*, vol. 25, no. 1, pp. 7–18, Jan. 1978.

[9] V.I. Arnold, *Geometrical Methods in the Theory of Ordinary Differential Equations*. Berlin: Springer-Verlag, pp. 267–269, 1983.

[10] C. Hayashi, *Nonlinear Oscillations in Physical Systems*. Princeton, NJ: Princeton University Press, p. 367, 1984.

Chapter 2

Analysis of bifurcations in oscillatory circuits

Hiroyuki Asahara[1], *Takuji Kousaka*[2] *and Tetsushi Ueta*[3]

In this chapter, we investigate the bifurcation phenomena observed in oscillatory circuits. The stability and bifurcation phenomena in autonomous systems are introduced by focusing on the equilibrium point and the fixed point. The characteristics and conditions of the saddle-node bifurcation, Hopf bifurcation, and pitchfork bifurcation are discussed for the equilibrium point. Likewise, the characteristics and conditions of the saddle-node bifurcation, period-doubling bifurcation, Neimark–Sacker bifurcation, and pitchfork bifurcation are introduced for the fixed point. The method for computing the bifurcation points of the equilibrium point and the periodic points is also introduced, and an example of an application is presented.

2.1 Introduction

It is impossible to realize many industrial devices without oscillators. Their basic, but vital function of providing a fixed period and amplitude is widely used. Typical applications include reference clocks for synchronization, audio synthesizer components, or carrier signals for modulators.

The simplest oscillator is a harmonic oscillator, which we learned about in physics or mathematics in high school. An example of a real harmonic oscillator is a small-displacement pendulum, and the corresponding differential equation is:

$$\frac{d^2x}{dt^2} + k\frac{dx}{dt} + \omega^2 x = 0, \tag{2.1}$$

where $x \in R$ is an angle, and k and ω are parameters. Throughout this chapter, we assume $k = 0$. Since the characteristic multipliers of (2.1) are complex conjugates, the general solution is oscillatory motion, as expressed by

$$x(t) = C_1 \cos \omega t + C_2 \sin \omega t, \tag{2.2}$$

[1]Department of Electrical and Electronic Engineering, Okayama University of Science, Okayama, Japan
[2]Department of Mechanical and Energy Systems Engineering, Oita University, Oita, Japan
[3]Center for Administration of Information Technology, Tokushima University, Tokushima, Japan

where C_1 and C_2 are arbitrary constants. For such oscillating motion, notice that ω is a constant that governs the frequency, but the amplitude depends on the initial value $(x(0), dx(0)/dt)$. Obtaining the same amplitude of oscillation for any initial condition is simple. Moreover, in practice, k should be taken as positive because of the ubiquitous friction losses; in other words, any oscillation must decay to zero amplitude over time for $k > 0$. Equation (2.1) is the fundamental equation and it affords an intuitive description of the oscillation; however, it is useless for practical applications unless some appropriate regulations or compensations are applied to the initial values and friction loss.

In electronic circuits, (2.1) corresponds to an LC tank circuit. The capacitor current $i = Cdv/dt$ and the inductor voltage $v = -Ldi/dt$ yield $d^2v/dt^2 + v/(LC) = 0$, which are equivalent to (2.1). However, this oscillator is not suitable for practical use because a tiny resistance R is always present in real circuits, and it is not easy to regulate the initial voltage of the tank circuit. In other words, no mechanism is built into the dynamics to retain a desirable oscillation. This begs the questions: How do we retain the oscillation? How can we become independent of the initial values?

The Poincaré–Bendixon theorem puts forward the existence of limit cycles as periodic orbits in planar autonomous systems. With harmonic oscillations, in contrast, a limit cycle is attractive (i.e., the cycle can absorb nearby trajectories). Given any initial value, the system always settles to a fixed-period, fixed-amplitude oscillation. This property surely matches the requirements described above for engineering applications. Since the planar and linear differential equation (2.1) only generates periodic orbits when $k = 0$ (otherwise, a stable origin or divergence is the final state), it must be modified to produce limit cycles. For example, if a nonlinear conductor is modeled as $i = -g_1 v + g_3 v^3$, where g_1 and g_3 are parameters, and it is connected in parallel with an LC tank circuit, the dynamics are described by the following (after suitable variable transformations) equation:

$$\frac{d^2x}{dt^2} - \varepsilon(1 - x^2)\frac{dx}{dt} + x = 0, \tag{2.3}$$

where ε is a parameter. Equation (2.3) describes what is called a "van der Pol oscillator". For $\varepsilon > 0$, a stable limit cycle exists in the large. To give the conductor cubic characteristics would probably require certain nonlinear elements, such as a transistor, field-effect transistor (FET), or operational amplifier. For $x^2 \gg 1$, such a nonlinear conductor behaves like a linear conductor that consumes energy, which corresponds to $k > 0$ in (2.1). Therefore, any solution starting from a large initial value heads toward the origin. For small x, however, the element acts as a negative conductor, which destabilizes the origin (i.e., any solution that starts near the origin oscillates and departs from the origin). A limit cycle is a compromise between these two mutually exclusive mechanisms.

Equation (2.3) is a nonlinear differential equation that does not allow analytical solutions such as (2.2), and the superposition principle is also not valid in this case. In general, a nonlinear oscillator may have multiple equilibria and multiple limit cycles for the same parameter values. In addition, their stability may change for different parameter values. Such topological changes are called bifurcations, and computing actual bifurcations requires numerical methods. In the following sections, we first classify bifurcations and then show how to calculate them.

2.2 Analysis of bifurcations of autonomous systems

2.2.1 Stability of equilibrium point

Consider the following n-dimensional autonomous system:

$$\frac{dx}{dt} = f(x, \lambda), \tag{2.4}$$

where $x \in \boldsymbol{R}^n$ is a state variable, and $\lambda \in \boldsymbol{R}^s$ is a parameter. We assume that f is a smooth function of x and is infinitely differentiable with respect to t and λ. The condition for the equilibrium point is

$$f(x_0, \lambda) = 0, \tag{2.5}$$

where x_0 satisfies (2.5) and is called the equilibrium point. The equilibrium point is a solution of (2.4) and does not change macroscopically. Assume the following perturbation around the equilibrium point:

$$x(t) = x_0 + \xi(t), \tag{2.6}$$

where $\xi(t)$ denotes the variation of x_0. Here, we insert (2.6) into (2.4) to obtain:

$$\frac{dx(t)}{dt} = \frac{dx_0}{dt} + \frac{d\xi}{dt} = f(x_0 + \xi(t), \lambda). \tag{2.7}$$

We assume that the variation $\xi(t)$ is sufficiently small and use a Taylor series expansion as follows:

$$f(x_0 + \xi(t), \lambda) = f(x_0, \lambda) + \left.\frac{\partial f}{\partial x}\right|_{x(t)=x_0} \xi(t) + \cdots, \tag{2.8}$$

where $\cdots$ denotes higher-order terms. Based on (2.6) and (2.8), we obtain:

$$\frac{d\xi}{dt} = J\xi(t), \tag{2.9}$$

where

$$J = \left.\frac{\partial f}{\partial x}\right|_{x(t)=x_0}. \tag{2.10}$$

The $n \times n$ matrix J is the Jacobian matrix. Equation (2.9) is called the variational equation or the linearized equation. The characteristic equation for the variational equation is:

$$\chi(\mu) = \det(\mu I_n - J) = \mu^n + a_1\mu^{n-1} + a_2\mu^{n-2} + \cdots + a_{n-1}\mu + a_n = 0, \tag{2.11}$$

Table 2.1 Relationship between coefficients a_1, a_2 and type of equilibrium point in the two-dimensional characteristic equation

Symbol	Nominal designation	Condition
${}_0O$	Sink	$a_1 > 0,\ a_2 > 0$
${}_1O$	Saddle	$a_2 < 0$
${}_2O$	Source	$a_1 < 0,\ a_2 > 0$

where I_n is the identity matrix of size n, μ is the characteristic multiplier, and a_n is a coefficient. By examining a_n, we can understand the stability of the equilibrium point. In a two-dimensional system, for example, the characteristic equation is:

$$\mu^2 + a_1\mu + a_2 = 0. \tag{2.12}$$

Table 2.1 shows the relationship between the coefficient a_n and the type of equilibrium point in the two-dimensional characteristic equation. The value of the characteristic multiplier μ varies depending on a_1 and a_2. If $|\mu| < 1$ is satisfied, the equilibrium point is stable. Symbols ${}_0O$, ${}_1O$, and ${}_2O$ show hyperbolic equilibrium points, where subscripts show the number of unstable dimension.

2.2.2 *Bifurcation at equilibrium point*

Consider now a bifurcation phenomenon at the equilibrium point of (2.4). Suppose the bifurcation parameter is λ. The current number of equilibrium points changes at the singular point of the Jacobian matrix. In other words, the existing number of equilibrium points changes as we tune the parameter that makes the characteristic multipliers of the Jacobian matrix go to zero. However, the stability of an equilibrium point can change when the characteristic multipliers of the Jacobian matrix are pure imaginary numbers. These two qualitative characteristics of the equilibrium points can be observed when the bifurcation parameter varies and the equilibrium point loses its hyperbolic characteristic. In general, this phenomenon is referred to as "bifurcation" and occurs when the bifurcation parameter takes on the appropriate value (called the "bifurcation-parameter value").

We assume that the bifurcation-parameter value is $\lambda = \lambda^*$, and that the other parameters are fixed. Note that the following condition is satisfied at the hyperbolic equilibrium point:

$$\mathrm{R_e}(\mu_i) \neq 0, \quad i = 1, 2, \ldots, n. \tag{2.13}$$

If a characteristic multiplier does not satisfy (2.13), a bifurcation occurs. Bifurcations at the equilibrium point are classified as one of the three types: saddle-node bifurcations, Hopf bifurcations, or pitchfork bifurcations.

2.2.2.1 Saddle-node bifurcation

Upon tuning the bifurcation-parameter value, a pair of equilibrium points appears or disappears. Here, we assume that the properties of the other characteristic multipliers do not change. Therefore, the pair of equilibrium points satisfies the following condition:

$$ {}_{k}O + {}_{k+1}O \Leftrightarrow \emptyset, \quad k = 0, 1, 2, \ldots, n-1, \tag{2.14} $$

where Ø means that no equilibrium point exists at the given point. The bifurcation that satisfies (2.14) is called a saddle-node bifurcation and is shown conceptually in Figure 2.1(a). Suppose that a saddle-node bifurcation occurs at λ^*. The equilibrium points ${}_{k}O$ and ${}_{k+1}O$ exist when $\lambda < \lambda^*$. The saddle-node bifurcation occurs at $\lambda = \lambda^*$. The equilibrium points disappear when $\lambda > \lambda^*$. By substituting $\mu = 0$ into (2.11), we have:

$$ \chi(0) = \det(-J) = a_n = 0, \tag{2.15} $$

where (2.15) gives the bifurcation condition for the saddle-node bifurcation.

2.2.2.2 Hopf bifurcation

Consider now the situation in which a pair of characteristic multipliers passes over the imaginary axis. In this situation, the limit cycle emerges from or converges to the equilibrium point. This phenomenon is called Hopf bifurcation. Hopf bifurcations satisfy the following conditions:

$$ {}_{k}O \Leftrightarrow {}_{k+2}O + LC({}_{k}D) \tag{2.16} $$

and

$$ {}_{k}O + LC({}_{k+1}D) \Leftrightarrow {}_{k+2}O, \tag{2.17} $$

where $LC({}_{k}D)$ denotes the limit cycle and $k = 0, 1, \ldots, n-2$. Note that ${}_{k}D$ denotes the characteristic of the fixed point that is discussed in Section 2.2.3.

For a two-dimensional system, a sink bifurcates via Hopf bifurcation; this is called a supercritical Hopf bifurcation. Equation (2.16) gives the requisite condition for a supercritical Hopf bifurcation. However, a subcritical Hopf bifurcation satisfies (2.17). Figure 2.1(b) and (c) shows a conceptual diagram of supercritical and subcritical Hopf bifurcations. The condition for a Hopf bifurcation is:

$$ \chi(j\omega) = \det(j\omega I_n - J) = 0, \tag{2.18} $$

where ω denotes the angular frequency of the limit cycle.

2.2.2.3 Pitchfork bifurcation

If the system is symmetric, the equilibrium point diverges for a certain parameter, where one of the characteristic multipliers equals zero; this is called the pitchfork bifurcation. The conditions for the pitchfork bifurcation are:

$$ \begin{aligned} &{}_{k}O \Leftrightarrow {}_{k+1}O + 2 \times {}_{k}O, \\ &{}_{k}O + 2 \times {}_{k+1}O \Leftrightarrow {}_{k+1}O. \end{aligned} \tag{2.19} $$

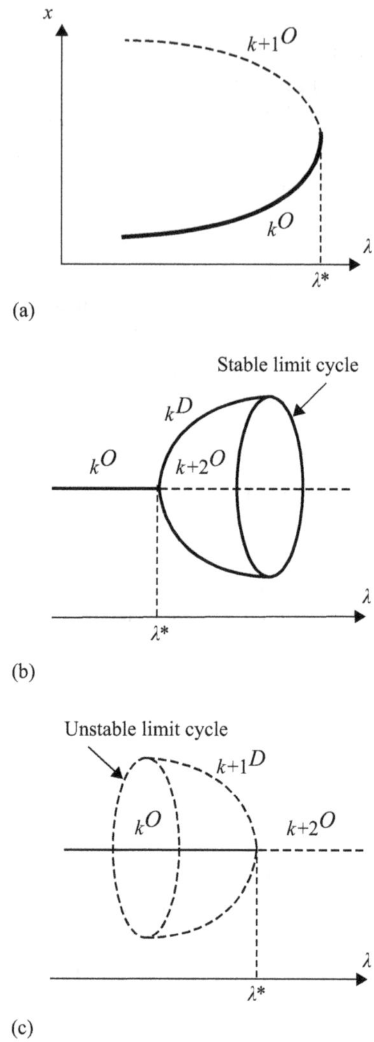

Figure 2.1 *Bifurcation at equilibrium point. (a) Saddle-node bifurcation; (b) supercritical Hopf bifurcation; and (c) subcritical Hopf bifurcation*

2.2.3 Stability of fixed point

Let a solution of (2.4) be:

$$x(t) = \varphi(t, x_0), \tag{2.20}$$

where x_0 denotes an initial value that is defined as:

$$x_0 = x(0) = \varphi(0, x_0). \tag{2.21}$$

Assume that (2.4) has a limit cycle. Consider a cross section Π, in which the orbit of the limit cycle exists; this cross section may be expressed as:

$$\Pi = \{x \in \boldsymbol{R}^n \mid q(x) = 0\}, \tag{2.22}$$

where q is the scalar function

$$q : \boldsymbol{R}^n \to \boldsymbol{R}; \quad x \mapsto q(x). \tag{2.23}$$

Note that the limit cycle intersects the cross section, so the following condition is satisfied:

$$\frac{\partial q(x_0)}{\partial x} \cdot f(x_0) \neq 0 \quad \text{for all} \quad x_0 \in \Pi. \tag{2.24}$$

The cross section that satisfies (2.24) is called the local cross section. We now consider a solution x that exists on the local cross section near the fixed point. By assuming $\Delta = x(L(x_0)) - x^*(L(x_0))$, where x^* denotes the fixed point, the following map P may be derived:

$$\begin{aligned} P:\ & \Delta \to \Pi \\ & x \mapsto \varphi(L(x_0), x_0), \end{aligned} \tag{2.25}$$

where $L(x_0)$ is the return time during which the orbit travels from the initial point $x \in \Pi$ back to Π again. The map P is called the Poincaré map, and the fixed point of such a map is:

$$x^*(L(x_0)) = P(x^*(L(x_0))). \tag{2.26}$$

Consider now the stability of the fixed point. Let a perturbation near the fixed point be ξ:

$$\xi = x(L(x_0)) - x^*(L(x_0)). \tag{2.27}$$

Substituting (2.27) into (2.25) and using a Taylor expansion leads to

$$\xi(k+1) = \frac{\partial P}{\partial x_0}\xi(k). \tag{2.28}$$

The Jacobian matrix of (2.28) can be obtained by calculating the derivative of (2.25) for an initial value x_0, which requires the application of the chain rule. Thus, the Jacobian matrix of (2.28) becomes:

$$\frac{\partial P}{\partial x_0} = \frac{\partial \varphi}{\partial x_0} + \frac{\partial \varphi}{\partial t}\frac{\partial \tau}{\partial x_0} = \frac{\partial \varphi}{\partial x_0} + f\frac{\partial \tau}{\partial x_0}, \tag{2.29}$$

where f denotes $f(\varphi(\tau(x_0)))$. Equation (2.29) is calculated by numerically integrating the following equation from $t = 0$ to $t = \tau(x_0)$:

$$\frac{d}{dt}\frac{\partial \varphi}{\partial x_0} = \frac{\partial f}{\partial x}\frac{\partial \varphi}{\partial x_0}, \tag{2.30}$$

where

$$\left.\frac{\partial \varphi}{\partial x_0}\right|_{t=0} = I_n. \tag{2.31}$$

The characteristic equation for $\frac{\partial \varphi}{\partial x_0}$ is:

$$\xi(\mu) = \det\left(\frac{\partial \varphi}{\partial x_0} - \mu I_n\right) = 0. \tag{2.32}$$

The derivative of $q(x_1) = q(\varphi(\tau(x_0), x_0)) = 0$ with initial value x_0 gives

$$\frac{\partial q}{\partial x}\left(\frac{\partial \varphi}{\partial x_0} + f\frac{\partial \tau}{\partial x_0}\right) = 0, \tag{2.33}$$

where (2.33) is rewritten as follows:

$$\frac{\partial q}{\partial x} f \frac{\partial \tau}{\partial x_0} = -\frac{\partial q}{\partial x}\frac{\partial \varphi}{\partial x_0}. \tag{2.34}$$

Based on the condition given by (2.24), the following equation may be derived:

$$\frac{\partial \tau}{\partial x_0} = -\frac{1}{\frac{\partial q}{\partial x} f}\frac{\partial q}{\partial x}\frac{\partial \varphi}{\partial x_0}. \tag{2.35}$$

Therefore, (2.29) may be rewritten as follows:

$$\frac{\partial P}{\partial x_0} = \frac{\partial \varphi}{\partial x_0} - \frac{1}{\frac{\partial q}{\partial x} f} f \frac{\partial q}{\partial x}\frac{\partial \varphi}{\partial x_0} = \left[I_n - \frac{1}{\frac{\partial q}{\partial x} f} f \frac{\partial q}{\partial x}\right]\frac{\partial \varphi}{\partial x_0}. \tag{2.36}$$

Based on the roots of the following equation, we classify the stability of the fixed points:

$$\frac{\partial P}{\partial x_0} - \mu I_n = 0. \tag{2.37}$$

Table 2.2 shows the types of fixed points in two-dimensional systems. Symbols ${}_0D$, ${}_1D$, ${}_1I$, and ${}_2D$ show type-D or type-I fixed points, where subscripts show the number of unstable dimension.

Note that (2.26) cannot be used as a condition to locate x_0 because the coordinates of the fixed point are not independent. This fact affects the convergence of numerical computations. To overcome this difficulty, we provide an $(n-1)$-dimensional

Table 2.2 Types of fixed points in two-dimensional systems

Symbol	**Nominal designation**	**Condition**
${}_0D$	Completely stable	$\lvert\mu_1\rvert < 1,\ \lvert\mu_2\rvert < 1$
${}_1D$	Directly unstable	$0 < \mu_1 < 1 < \mu_2$
${}_1I$	Inversely unstable	$\mu_1 < -1 < \mu_2 < 0$
${}_2D$	Completely unstable	$\lvert\mu_1\rvert > 1,\ \lvert\mu_2\rvert > 1$

local coordinate frame Σ attached to the cross section. The projection $h : \Pi \to \Sigma$ is defined as:

$$h : \Pi \to \Sigma \subset \boldsymbol{R}^{n-1}; \quad x \mapsto u = h(x). \tag{2.38}$$

The induced embedded map is:

$$h^{-1} : \Sigma \to \Pi \subset \boldsymbol{R}^{n}; \quad x \mapsto x = h^{-1}(u). \tag{2.39}$$

Assume that $u_0 \in \Sigma \subset \boldsymbol{R}^{n-1}$ is located in the local coordinate frame, such that $h(x_0) = u_0$. Also, let $\hat{\Sigma}$ be the neighborhood of $u_0 \in \Sigma$, and let u_1 be a point on $\hat{\Sigma}$ such that $h^{-1}(u_1) = x_1 \in \hat{\Pi}$, and $\varphi(t, x_1)$ is the solution starting from x_1. Moreover, suppose that the orbit $\varphi(t, x_1)$ intersects Π at the point $x_2 \in \Pi$ within the return time $\tau(x_1)$: namely,

$$x_2 = \varphi(\tau(x_1), x_1). \tag{2.40}$$

In this case, we can define the Poincaré map in the local coordinate frame by:

$$\begin{aligned} P_\ell :\ \hat{\Sigma} &\to \Sigma \\ u_1 \mapsto u_2 &= h(\varphi(\tau(h^{-1}(u_1)), h^{-1}(u_1))) \\ &= h \circ P \circ h^{-1}(u_1). \end{aligned} \tag{2.41}$$

Finally, we have a discrete system derived from (2.41):

$$u_{k+1} = P_\ell(u_k). \tag{2.42}$$

The fixed point u_0 is defined as:

$$P_\ell(u_0) = u_0. \tag{2.43}$$

The fixed point is given by solving via Newton's method in the following equation for u, which is unknown:

$$F_1(u) = [P_\ell(u) - u]. \tag{2.44}$$

However, the stability of the fixed point is given by:

$$\chi(\mu) = \det\left(\frac{\partial P_\ell}{\partial u_0} - \mu I_n\right) = 0, \tag{2.45}$$

where:

$$\frac{\partial P_\ell}{\partial u_0} = \frac{\partial h}{\partial x}\frac{\partial P}{\partial x_0}\frac{\partial h^{-1}}{\partial u_0}. \tag{2.46}$$

Therefore, based on (2.35), $\frac{\partial P_\ell}{\partial u_0}$ can be expressed as:

$$\frac{\partial P_\ell}{\partial u_0} = \frac{\partial h}{\partial x}\left[I_n - \frac{1}{\frac{\partial q}{\partial x}f}f\frac{\partial q}{\partial x}\right]\frac{\partial \varphi}{\partial x_0}\frac{\partial h^{-1}}{\partial u_0}. \tag{2.47}$$

The Jacobian matrix of (2.44) is given by:

$$\frac{\partial F_1}{\partial u_0} = \frac{\partial P_\ell}{\partial u_0} - I_{n-1}. \tag{2.48}$$

The bifurcation parameter can be calculated by solving the simultaneous equations (2.44) and (2.45), which are described as:

$$F_2(u, \lambda) = \begin{bmatrix} P_\ell(u) - u \\ \chi(\mu) \end{bmatrix}, \tag{2.49}$$

where the derivation of P_ℓ with λ is:

$$\frac{\partial P_\ell}{\partial \lambda} = \frac{\partial h}{\partial x}\frac{\partial P}{\partial \lambda} = \frac{\partial h}{\partial x}\frac{\partial \varphi}{\partial \lambda}. \tag{2.50}$$

Based on the above, we complete the equations that need to be computed numerically.

2.2.4 Bifurcation of fixed point

The bifurcation phenomena at the fixed point occur when a part of the characteristic multiplier exits the unit circle. Therefore, the absolute value of the characteristic multiplier is unity at the bifurcation-parameter value; that is, the characteristic multiplier exists only on the unit circle. Bifurcation phenomena at the fixed point can be classified into one of the four types: saddle-node bifurcations, period-doubling bifurcations, Neimark–Sacker bifurcations, and pitchfork bifurcations; that of at the periodic points are the same.

2.2.4.1 Saddle-node bifurcation

The saddle-node bifurcation occurs when a pair of fixed points appears or disappears upon modifying the bifurcation parameter. The characteristic multiplier becomes unity at the bifurcation-parameter value. Therefore, the bifurcation condition is:

$$\chi(1) = \det(I_n - J) = 1 + a_1 + \cdots + a_{n-1} + a_n = 0. \tag{2.51}$$

The pairs of fixed points that satisfy the following equation can pass through the saddle-node bifurcation:

$$\begin{aligned} {}_kD + {}_{k+1}D &\Leftrightarrow \phi \quad (k = 0, 1, \ldots, n-1), \\ {}_kI + {}_{k+1}I &\Leftrightarrow \phi \quad (k = 0, 1, \ldots, n-1). \end{aligned} \tag{2.52}$$

2.2.4.2 Period-doubling bifurcation

The period-doubling bifurcation occurs when the characteristic multiplier goes to -1. The stability of the fixed point changes upon passing through the period-doubling bifurcation. The bifurcation condition is:

$$\chi(-1) = \det(-I_n - J) = (-1)^n + a_1(-1)^n + \cdots - a_{n-1} + a_n = 0. \tag{2.53}$$

The period-doubling bifurcation satisfies:

$$\begin{aligned}
{}_kD &\Leftrightarrow {}_{k+1}I + 2 \times {}_kD^2 \quad (k = 0, 1, \ldots, n-2),\\
{}_kD &\Leftrightarrow {}_{k-1}I + 2 \times {}_kD^2 \quad (k = 2, 3, \ldots, n),\\
{}_kI &\Leftrightarrow {}_{k+1}D + 2 \times {}_kD^2 \quad (k = 1, 2, \ldots, n-1),\\
{}_kI &\Leftrightarrow {}_{k-1}D + 2 \times {}_kD^2 \quad (k = 1, 2, \ldots, n-1),
\end{aligned} \tag{2.54}$$

where ${}_kD^2$ denotes the period-two solution of type D whose unstable dimension is k.

2.2.4.3 Neimark–Sacker bifurcation

The Neimark–Sacker bifurcation occurs when a pair of characteristic multipliers that exist on the complex plane exit the unit circle. The bifurcation condition is:

$$\chi(e^{j\theta}) = \det(e^{j\theta}I_n - J) = e^{jn\theta} + a_1 e^{j(n-1)\theta} + \cdots + a_{n-1}e^{j\theta} + a_n = 0. \tag{2.55}$$

The Neimark–Sacker bifurcation satisfies:

$$\begin{aligned}
{}_kD &\Leftrightarrow {}_{k+2}D + ICC \quad (k = 0, 1, \ldots, n-2),\\
{}_kD + ICC &\Leftrightarrow {}_{k+2}D \quad (k = 0, 1, \ldots, n-2),
\end{aligned} \tag{2.56}$$

where ICC denotes the invariant closed curve.

2.2.4.4 Pitchfork bifurcation

The pitchfork bifurcation is a type of co-dimension-two bifurcation. If a system is symmetric, the pitchfork bifurcation appears when a characteristic multiplier becomes unity. The pitchfork bifurcation satisfies:

$$\begin{aligned}
{}_kD &\Leftrightarrow {}_{k+1}D + 2{}_kD \quad (k = 0, 1, \ldots, n-1),\\
{}_kD + 2{}_kD &\Leftrightarrow {}_{k+1}D \quad (k = 0, 1, \ldots, n-1).
\end{aligned} \tag{2.57}$$

2.3 Example of bifurcation analysis applied to an autonomous system

The Bonhöffer-van der Pol (BVP) equation is a simplified model of the Hodgkin–Huxley equation and was introduced by FitzHugh [1] and Nagumo *et al.* [2]. The dynamic behavior of the BVP equation can be investigated experimentally by using simple passive elements and single nonlinear conductor. The BVP equation qualifies as one of the classic nonlinear oscillator models.

The BVP oscillator is a suitable model for understanding mutual synchronization, which occurs in biological systems. The dynamic behavior and qualitative characteristic of the BVP equation have been investigated by many researchers over many years. Compared with the van der Pol equation, a greater variety of nonlinear phenomena have been observed in the BVP model [3,4].

Analyzing mutually coupled oscillators is important for understanding the qualitative characteristics of physical, biological, and chemical systems. For example, the synchronization phenomenon is of interest when analyzing coupled oscillators. Many synchronization phenomena have been clarified by applying the model of coupled BVP oscillators [5,6]. Thus, the BVP model has proven to be a very useful oscillating unit for describing coupled systems, and coupled BVP oscillators have been analyzed in numerous publications [7–10].

In this section, as an example, we apply the method of bifurcation analysis introduced in Section 2.2 to single and coupled BVP oscillators.

2.3.1 Single BVP oscillator

Figure 2.2 shows the circuit model of a single BVP oscillator. The circuit equations describing this oscillator are:

$$\begin{aligned} C\frac{dv}{dt} &= -i - g(v), \\ L\frac{di}{dt} &= v - ri + E. \end{aligned} \tag{2.58}$$

For simplicity, we assume that $g(v)$ is nonlinear, and neglect E. Equation (2.58) is symmetric under the permutation $(v, i) \to (-v, -i)$. Moreover, the DC source E may be used to destroy the symmetry property of the system, thereby avoiding degeneracy. When the nonlinear conductor $g(v)$ is implemented by a 2SK30A FET, the nonlinear characteristic may be approximated by:

$$g(v) = -a \tanh bv. \tag{2.59}$$

Applying the Marquardt–Levenberg method (nonlinear least squares method) to fit this function gives the parameter values $a = 6.89099 \times 10^{-3}$ and $b = 0.352356$. Although $g(v)$ is given by a third-order polynomial in the original BVP equation, here we simply assume the sigmoidal function (2.59) based on the results of experimental measurements of the FET. This assumption is reasonable because the dynamic range of v in the coupled system (2.59) is essentially restricted to the region of the third-power term.

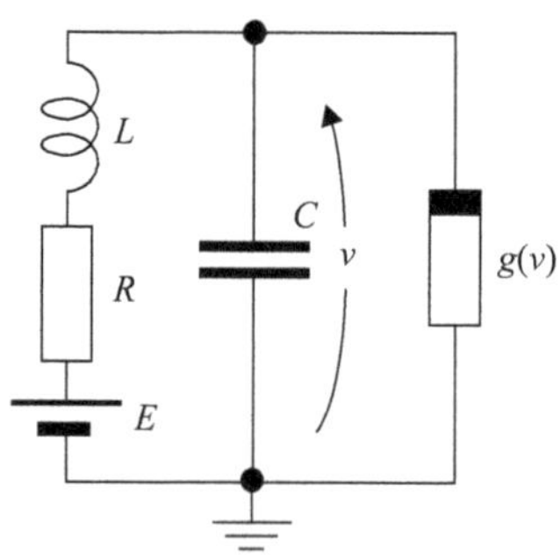

Figure 2.2 Circuit diagram of the BVP oscillator

Let the dimensionless values be:

$$x = \frac{1}{a}\sqrt{\frac{C}{L}}v, \quad y = \frac{i}{a}, \quad \tau = \frac{1}{\sqrt{LC}}t, \quad k = r\sqrt{\frac{C}{L}}, \quad \gamma = ab\sqrt{\frac{L}{C}}. \tag{2.60}$$

Therefore, (2.58) takes the following form:

$$\begin{aligned} \frac{dx}{dt} &= -y + \tanh \gamma x, \\ \frac{dy}{dt} &= x - ky. \end{aligned} \tag{2.61}$$

Figure 2.3 shows the bifurcation diagram of a single BVP oscillator in the γ–k plane. In the bifurcation diagram, pf, h, sn, and H denote the pitchfork bifurcation, Hopf bifurcation, saddle-node bifurcation, and the homoclinic loop of the limit cycle, respectively. This parameter plane contains six distinct areas, which are labeled (I)–(VI). Area (I) includes a sink at the origin O. The sink origin bifurcates to a saddle origin via the pitchfork bifurcation, generating two sinks $C^+ = (x_c, y_c)$ and $C^- = (-x_c, -y_c)$ in area (II), where the pitchfork bifurcation is given by $k = \gamma^{-1}$. However, via the supercritical Hopf bifurcation at the origin, which is given by $k = \gamma$ with $0 \leq \gamma \leq 1$, the sink of area (I) bifurcates into a source, and a stable limit cycle is generated in area (III). By changing the parameters to go from area (III) to area (IV), the source origin observed in area (III) bifurcates into a saddle origin, and two sources C^+ and C^- are generated. Because the parameters change in going from area (IV)

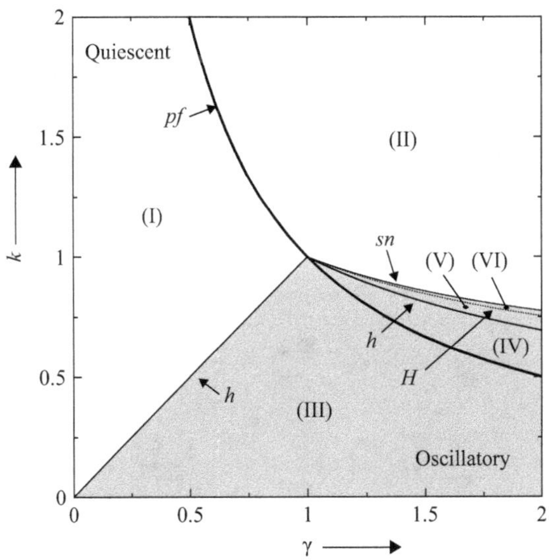

Figure 2.3 Bifurcation diagram of equilibria in (2.61). The labels pf, h, sn, and H denote the pitchfork bifurcation, Hopf bifurcation, saddle-node bifurcation, and the homoclinic loop of the limit cycle, respectively

to area (V), the subcritical Hopf bifurcation occurs, so that sources C^+ and C^- become sinks, and two unstable limit cycles are generated. In area (V), two unstable limit cycles are knit together and form two separatrix loops. Next, the homoclinic loop of the limit cycle occurs and stable and unstable limit cycles are observed in area (VI). Finally, because the parameters change in going from area (VI) to (II), the stable and unstable limit cycles disappear altogether via the saddle-node bifurcation. To summarize, areas (I)–(IV), which are defined by the bifurcation curves, may be classified as follows:

- **(I)** a sink exists;
- **(II)** a saddle and two sinks coexist;
- **(III)** a source and a stable limit cycle coexist;
- **(IV)** two sinks, a saddle, one or two unstable limit cycles, and one stable limit cycle coexist;
- **(V)** two sources, a saddle, a stable limit cycle, and two unstable limit cycles coexist;
- **(VI)** two sources, a saddle, a stable limit cycle, and an unstable limit cycle coexist.

2.3.2 *Coupled BVP oscillators*

We now consider the diffusively coupled unbalanced BVP oscillators shown in Figure 2.4. The circuit equations describing this system of oscillators are (with $G = 1/R$):

$$\begin{aligned}
C\frac{dv_1}{dt} &= -i_1 + a\tanh bv_1 - G(v_1 - v_2),\\
L\frac{di_1}{dt} &= v_1 - ri_1,\\
C\frac{dv_2}{dt} &= -i_2 + a\tanh bv_2 - G(v_2 - v_1),\\
L\frac{di_2}{dt} &= v_2 - ri_2.
\end{aligned} \tag{2.62}$$

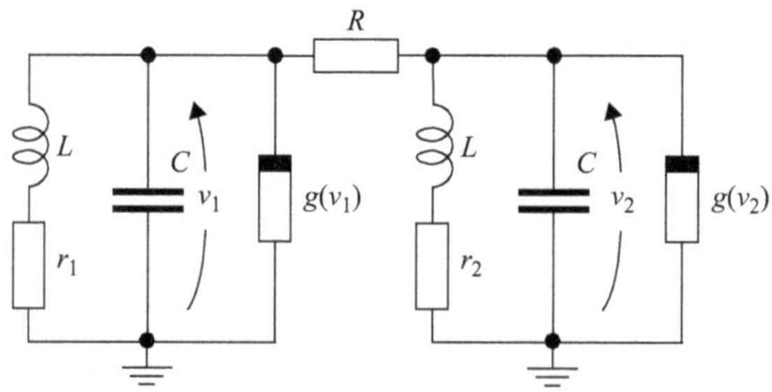

Figure 2.4 Circuit diagrams of coupled BVP oscillators

Let the dimensionless values be:

$$x_j = \frac{1}{a}\sqrt{\frac{C}{L}}v_j, \quad y_j = \frac{i_j}{a}, \quad k_j = r_j\sqrt{\frac{C}{L}}, \tag{2.63}$$

$$\tau = \frac{1}{\sqrt{LC}}t, \quad \gamma = ab\sqrt{\frac{L}{C}} \quad \delta = \sqrt{\frac{L}{C}}G, \tag{2.64}$$

where $j = 1, 2$. Therefore, the normalized equation takes the following form:

$$\begin{aligned} \frac{dx_1}{dt} &= -y_1 + \tanh \gamma x_1 - \delta(x_1 - x_2), \\ \frac{dy_1}{dt} &= x_1 - k_1 y_1, \\ \frac{dx_2}{dt} &= -y_2 + \tanh \gamma x_2 - \delta(x_2 - x_1), \\ \frac{dy_2}{dt} &= x_2 - k_2 y_2. \end{aligned} \tag{2.65}$$

In the following, the coupling coefficient δ is fixed as $\delta = 0.337$ ($R = 2$ kΩ). Figure 2.5 shows a bifurcation diagram at equilibrium. The diagram is symmetrical for k_1 and k_2, which means, for example, that a solution $(x_1^*, y_1^*, x_2^*, y_2^*)$ at (k_1^*, k_2^*)

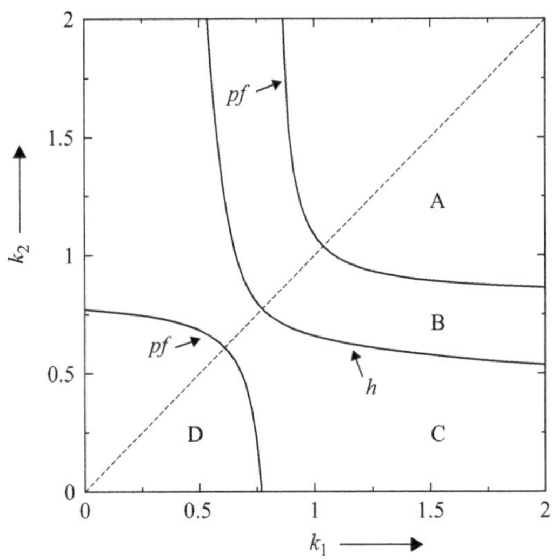

Figure 2.5 Bifurcation of equilibria

implies the existence of another solution $(x_2^*, y_2^*, x_1^*, y_1^*)$ at (k_2^*, k_1^*). Moreover, pitchfork bifurcation curves *pf* at the origin can be expressed analytically as follows:

$$k_2 = \frac{-k_1(\delta - \gamma) - 1}{\delta - \gamma + k_1\gamma(\gamma - 2\delta)}, \tag{2.66}$$

from which they can be plotted on the bifurcation diagram. Note that this equation forms a hyperbolic curve in the k_1–k_2 parameter plane.

Sinks $C^+ = (x_{1c}, y_{1c}, x_{2c}, y_{2c})$ and $C^- = (-x_{1c}, -y_{1c}, -x_{2c}, -y_{2c})$ exist in region A of Figure 2.5, and a two-dimensionally unstable saddle appears at the origin, which changes into a three-dimensionally unstable saddle after crossing the pitchfork bifurcation curve that separates regions B and C, generating two one-dimensionally unstable saddles. Likewise, after again crossing the pitchfork bifurcation curve, one finds a four-dimensionally unstable origin in region D.

Because the parameters change upon going from region B to region C, the sinks C^+ and C^- simultaneously become two-dimensionally unstable saddles via the Hopf bifurcation *h*. Theoretically, this bifurcation generates two stable limit cycles, so we conclude that limit cycles exist in regions C and D. In region D, the origin becomes a completely unstable equilibrium because it again touches the pitchfork bifurcation curve.

Thus, the limit cycles exhibit complicated motion because of the local bifurcation phenomena. In the following analysis, we analyze these bifurcation phenomena. Figure 2.6 shows a global bifurcation diagram for limit cycles in the k_1–k_2 plane with

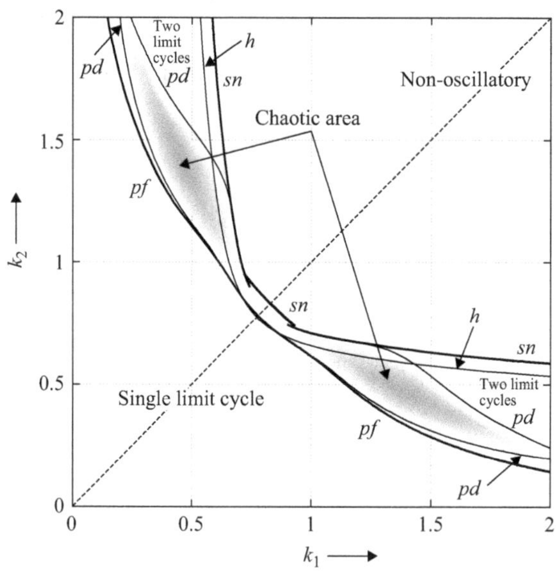

Figure 2.6 Bifurcation of limit cycles with δ = 0.337

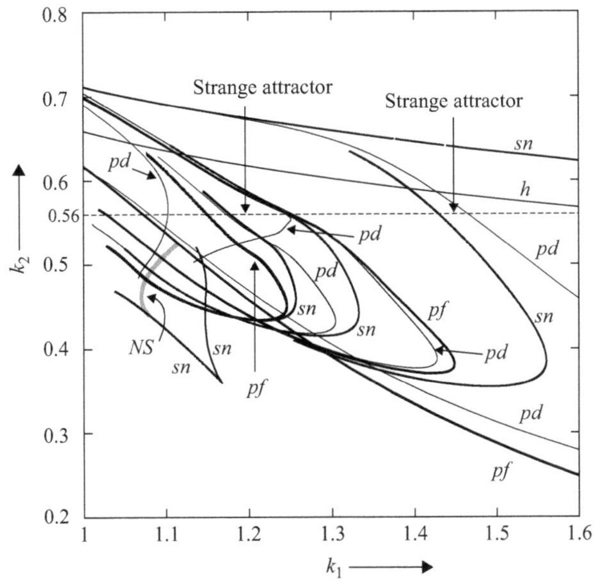

Figure 2.7 Enlarged version of bifurcation diagram from Figure 2.6

$\delta = 0.337$, where *pd* denotes the period-doubling bifurcation. Two parameter regions appear in the k_1–k_2 plane defining oscillatory and nonoscillatory regions, and the saddle-node bifurcation *sn* sits at the border between these regions. In the area encircled by *pd* (i.e., the period-doubling bifurcation), the chaotic attractor like a double scroll attractor [11,12] that is generated by a period-doubling cascade is found.

Now consider Figure 2.7, which shows an enlarged version of Figure 2.6. In the following, the behavior of this circuit is investigated with the parameter k_2 fixed at $k_2 = 0.56$. Figures 2.8 and 2.9 show the phase portraits corresponding to the parameter range shown in Figure 2.7 and with $k_2 = 0.56$. Two sinks exist near $k_1 = 1.5$ that touch the Hopf bifurcation, and two limit cycles are generated. A period-doubling cascade occurs near $k_1 = 1.45$, and the limit cycles bifurcate toward the chaotic attractor. Upon tuning the parameters, the chaotic attractor bifurcates to a strange attractor.

2.4 Conclusions

In this chapter, we investigated the bifurcation phenomena that occur in oscillatory circuits. More specifically, we discussed bifurcation phenomena and stability in autonomous systems, such as single and coupled BVP oscillators. Based on the

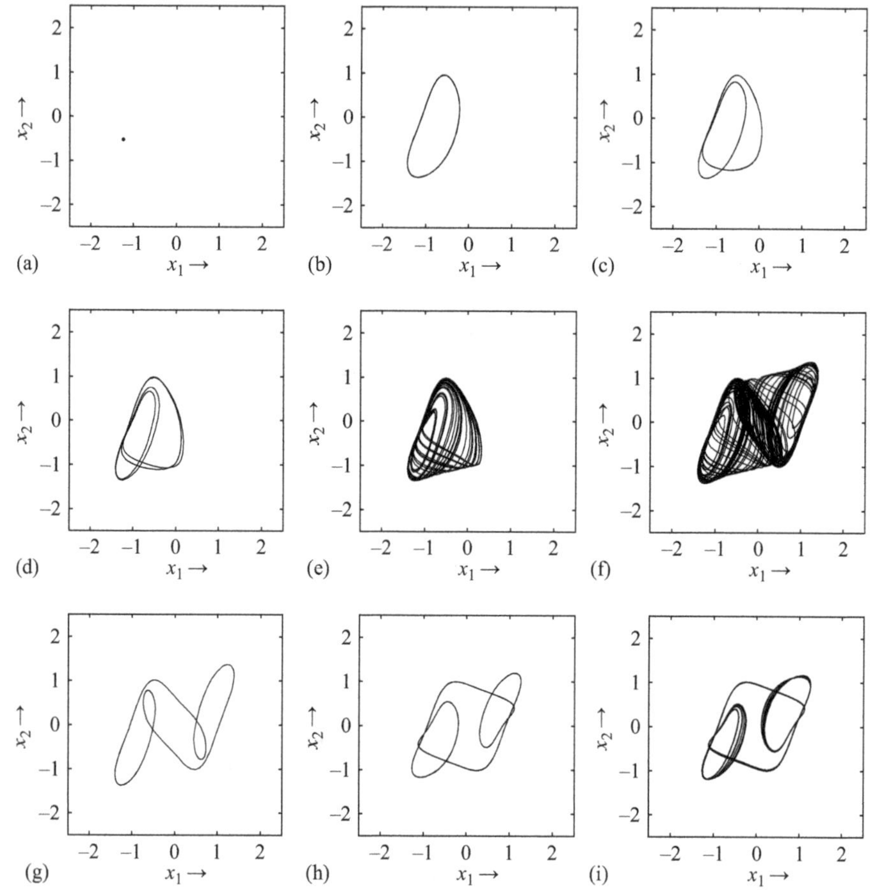

Figure 2.8 Phase portraits for parameter range shown in Figure 2.7 with $k_2 = 0.56$. (a) $k_1 = 1.51$, (b) $k_1 = 1.461$, (c) $k_1 = 1.451$, (d) $k_1 = 1.441$, (e) $k_1 = 1.43$, (f) $k_1 = 1.248$, (g) $k_1 = 1.245$, (h) $k_1 = 1.24$, and (i) $k_1 = 1.112$

characteristic equation, the saddle-node bifurcation, Hopf bifurcation, and the pitchfork bifurcation are classified at the equilibrium point. In addition, the saddle-node bifurcation, period-doubling bifurcation, Neimark–Sacker bifurcation, and the pitchfork bifurcation are discussed at the fixed point, and we introduce the methods used to compute the bifurcation points of the equilibrium point and the periodic points, and bifurcation diagrams are calculated using the methods. Finally, as an example of an application, the bifurcation phenomena for single and coupled BVP oscillators are analyzed.

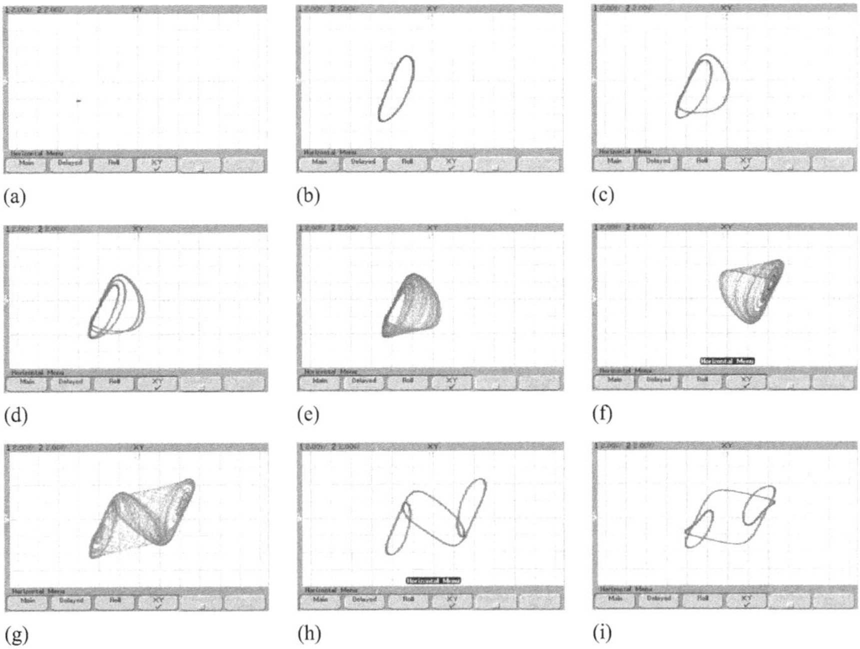

Figure 2.9 Experimental phase portraits for parameter range shown in Figure 2.7 with $k_2 = 0.56$. (a) $k_1 = 1.51$, (b) $k_1 = 1.461$, (c) $k_1 = 1.451$, (d) $k_1 = 1.441$, (e) $k_1 = 1.43$, (f) $k_1 = 1.248$, (g) $k_1 = 1.245$, (h) $k_1 = 1.24$, and (i) $k_1 = 1.112$

Acknowledgments

We gratefully acknowledge the authors of Reference 13 and Professor H. Kawakami for their contributions.

Figures 2.8 and 2.9 are cited by Reference 13, T. Ueta, H. Miyazaki, T. Kousaka and H. Kawakami, "Bifurcation and chaos in coupled BVP oscillators," *International Journal of Bifurcation and Chaos*, vol. 14, no. 4, pp. 1305–1324, 2004.

References

[1] R. FitzHugh, "Impulses and physiological states in theoretical models of nerve membrane," *Biophysical Journal*, vol. 1, pp. 445–466, 1961.

[2] J. Nagumo, S. Arimoto, and S. Yoshizawa, "An active pulse transmission line simulating nerve axon," *Proceedings of the IRE*, vol. 50, no. 10, pp. 2061–2070, 1962.

[3] A. N. Bautin, "Qualitative investigation of a particular nonlinear system," *Journal of Applied Mathematics and Mechanics*, vol. 39, no. 4, pp. 606–615, 1975.

[4] H. Kitajima, Y. Katsuta, and H. Kawakami, "Bifurcations of periodic solutions in a coupled oscillator with voltage ports," *IEICE Transactions on Fundamentals*, vol. E81-A, no. 3, pp. 476–482, 1998.

[5] D. A. Linkens, "Analytical solution of large numbers of mutually coupled nearly sinusoidal oscillators," *IEEE Transactions on Circuits and Systems*, vol. CAS-21, no. 2, pp. 294–300, 1974.

[6] T. Endo and S. Mori, "Mode analysis of a ring of a large number of mutually coupled van der Pol oscillators," *IEEE Transactions on Circuits and Systems*, vol. CAS-25, no. 1, pp. 7–18, 1978.

[7] M. Hoque and H. Kawakami, "Resistively coupled oscillators with hybrid connection," *IEICE Transactions on Fundamentals*, vol. E78-A, no. 9, pp. 1253–1256, 1995.

[8] O. Papy and H. Kawakami, "Symmetrical properties and bifurcations of the periodic solutions for a hybridly coupled oscillator," *IEICE Transactions on Fundamentals*, vol. E78-A, no. 12, pp. 1816–1821, 1995.

[9] O. Papy and H. Kawakami, "Symmetry breaking and recovering in a system of *n* hybridly coupled oscillators," *IEICE Transactions*, vol. E79-A, no. 10, pp. 1581–1586, 1996.

[10] K. Tsumoto, T. Yoshinaga, and H. Kawakami, "Bifurcation of a modified BVP circuit model for neurons generating rectangular waves," *IEICE Transactions on Fundamentals*, vol. E82-A, no. 9, pp. 1729–1736, 1999.

[11] L. O. Chua, M. Komuro, and T. Matsumoto, "The double scroll family," *IEEE Transactions on Circuits and Systems*, vol. CAS-33, pp. 1073–1118, 1986.

[12] L. O. Chua, "Global unfolding of Chua's circuit," *IEICE Transactions*, vol. 76-A, no. 5, pp. 704–734, 1993.

[13] T. Ueta, H. Miyazaki, T. Kousaka, and H. Kawakami, "Bifurcation and chaos in coupled BVP oscillators," *International Journal of Bifurcation and Chaos*, vol. 14, no. 4, pp. 1305–1324, 2004.

Chapter 3
Fractional-order oscillators

Ahmed G. Radwan[1,2], *Brent J. Maundy*[3] *and* *Ahmed S. Elwakil*[2,4]

Fractional-order calculus is the branch of mathematics which deals with non-integer-order differentiation and integration. Fractional calculus has recently found its way to engineering applications; particularly electronic circuits with promising results showing the feasibility of fabricating fractional-order capacitors on silicon. Fractional-order capacitors are lossy non-deal capacitors with an impedance given by $Z_c = (1/j\omega C)^{\alpha}$, where C is the pseudo-capacitance and α is its order ($0 < \alpha \leq 1$). When these fractional-order capacitors are employed within an oscillator (sinusoidal or relaxation) circuit, this oscillator is called a fractional-order oscillator and is described by non-integer-order differential equations. Therefore, an oscillator of order 1.5 or 2.6 is possible to obtain. While the oscillation frequency in integer-order oscillators is related to their RC time constants, fractional-order oscillators have their oscillation frequencies also related to α. This adds more design freedom and enables extremely high or extremely low oscillation frequencies even with large RC time constants. This chapter aims at reviewing the theory of designing fractional-order oscillators accompanied by several design examples. Experimental results are also shown.

3.1 Fractional-order sinusoidal oscillators

Sinusoidal oscillators are the first class of oscillators introduced in this chapter. A sinusoidal oscillator is designed to generate one or more sinusoidal signals with a prescribed oscillation frequency when an oscillation start-up condition is satisfied. The oscillation condition is obtained from stability analysis in the frequency domain. Therefore, in the following subsection, the basic concepts of stability analysis for a linear time-variant fractional-order system are reviewed.

[1]Department of Engineering Mathematics, Faculty of Engineering, Cairo University, Egypt
[2]Nanoelectronics Integrated Systems Center (NISC), Nile University, Cairo, Egypt
[3]Department of Electrical and Computer Engineering, University of Calgary, Alberta, Canada
[4]Department of Electrical and Computer Engineering, University of Sharjah, Sharjah, United Arab Emirates

3.1.1 Fractional-order linear time invariant systems (FLTI)

Linear integer-order circuits are a narrow subset of rational-order circuits which are a subset of fractional-order ones. The general definition of fractional-order continuous linear time invariant (FOC-LTI) system can be described by a fractional-order differential equation of constant coefficients as follows [1,2]:

$$\sum_{i=0}^{N} a_i D^{\alpha_i} y(t) = \sum_{j=0}^{M} b_j D^{\beta_j} x(t) \tag{3.1}$$

where x and y are the input and output signals, respectively, D^{α_i} and D^{β_j} are the fractional-order derivative operators of orders α_i and β_j, respectively, and $[a_i, b_j]$ are the coefficients parameters. Note that the parameters N and M are independent as there is no condition for the number of terms on both sides, however, for causality $\max(\beta_j)$ should not exceed $\max(\alpha_i)$. Since the Laplace transform of the fractional-order operator under zero initial conditions is $L(D^{\alpha} f(t)) = s^{\alpha} F(s)$, the transfer function of the previous equation can be written as follows:

$$H(s) = \frac{Y(s)}{X(s)} = \frac{\sum_{j=0}^{M} b_j s^{\beta_j}}{\sum_{i=0}^{N} a_i s^{\alpha_i}} \tag{3.2}$$

This transfer function can be used in many modeling and design applications such as fractional-order filter design [3–12]. Note that a qth-order fractional-order Laplacian s^q is equivalent to:

$$s^q = (j\omega)^q = \omega^q \left(\cos\left(q\frac{\pi}{2}\right) + j\sin\left(q\frac{\pi}{2}\right)\right) \tag{3.3}$$

3.1.2 Stability analysis

Stability is related to the poles' locations in the s-plane where the system is stable if none of the poles lies in the right-half plane (RHP). The poles' extractions are calculated based on the characteristic equation $Q(s) = \sum_{i=0}^{N} a_i s^{\alpha_i} = 0$. However, in the fractional-order domain, two different planes are used to find these poles.

3.1.2.1 The F-plane and W-plane

If the fractional-order parameters α_i are irrational numbers but multiples of a common number α, then $Q(s)$ can be written as $Q(s) = \sum_{i=0}^{N} a_i s^{k_i \alpha} = \sum_{i=0}^{N} a_i F^{k_i}$ where k_i are integers and $F = s^{\alpha}$. Note that the physical s-plane has an angle $\theta_s \in (-\pi, \pi]$, so the equivalent phase in the F-plane is $\theta_F \in (-\alpha\pi, \alpha\pi]$ as introduced in Reference 1. The general F-plane is shown in Figure 3.1(a). The characteristic equation becomes a polynomial in F and its solution can be obtained easily. However, if the fractional-order parameters α_i are rational numbers with a common denominator, i.e., $\alpha_i = \frac{r_i}{m}$, then $Q(s)$ can be written as $Q(s) = \sum_{i=0}^{N} a_i s^{\frac{r_i}{m}} = \sum_{i=0}^{N} a_i W^{r_i}$, where m, r_i are integers and $W = s^{\frac{1}{m}}$. Note that in the W-plane $\theta_W \in (-\frac{\pi}{m}, \frac{\pi}{m}]$ [1]. The region of the W-plane where $|\theta_W| > \frac{\pi}{m}$ has no physical s-plane equivalent as shown in Figure 3.1(b) for the case $m = 3$.

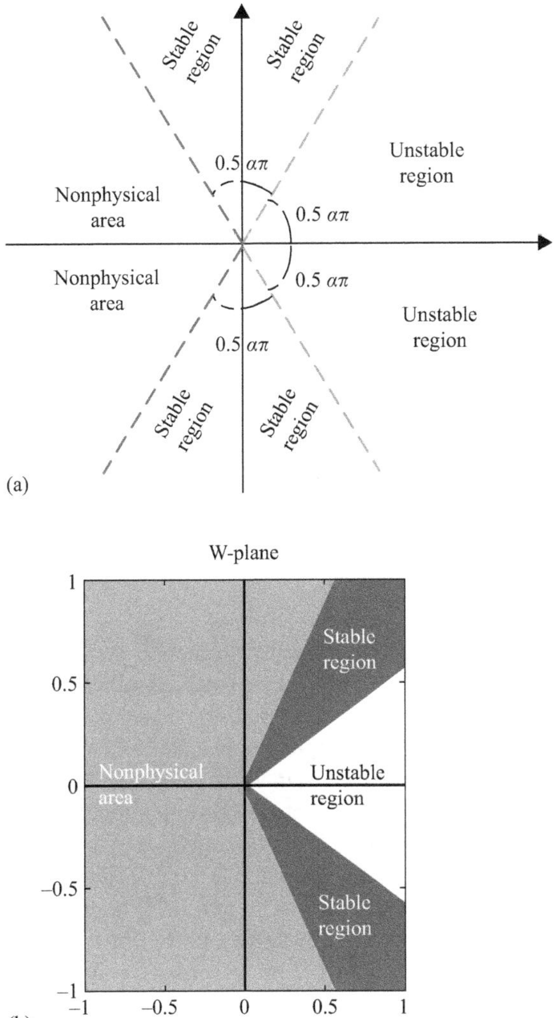

Figure 3.1 *Real versus imaginary values in the (a) F-plane and (b) W-plane at $m = 3$. Nonphysical area means it cannot be mapped into the s-plane*

3.1.3 Stability analysis examples

Two examples are given here to illustrate stability analysis of a fractional-order system:

3.1.3.1 Example 1

Let us study the stability conditions for the fractional-order differential equation:

$$\frac{d^{0.8}y(t)}{dt^{0.8}} + a\frac{d^{0.3}y(t)}{dt^{0.3}} + by(t) = f(t) \tag{3.4}$$

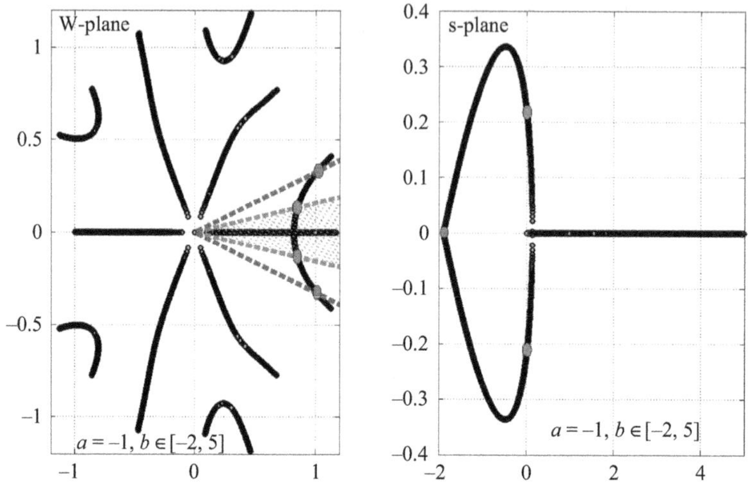

Figure 3.2 Poles of Example 1 in the W-plane and s-plane when a = −1

Defining $W = s^{0.1}$, which is common between 0.3 and 0.8, Figure 3.2 shows the poles in the W-plane and the physical s-plane, respectively, when $a = -1$ and $b \in [-2, 5]$ where the number of physical poles changes as the value of b changes. When $b = -2$, the system has one physical pole at $s = 4.9933$ (unstable); however, when $b = 0$, the number of physical poles increases to two $s = (0, 1)$ (unstable). These two poles approach each other and turn into the complex roots $s \approx 0.14 \pm 0.0034j$ at $b = 0.347$ with an unstable response. Then, the real value of these roots decreases as b increases until they are located on the $j\omega$ axis at ($s = \pm 0.2278j$) (the two inner points in the W-plane; closest to the real axis in Figure 3.2), at $b = b_{osc} = 0.477$ which corresponds to oscillatory behavior. These two poles enter the stable region with complex roots until they merge again (the pure negative real point in the s-plane) at $s \approx -1.8935$ when $b = b_{phy} = 2.06$. For any value $b > 2.06$ all poles will be located in the nonphysical region of the W-plane which means that no-physical poles exist in the s-plane. Therefore, in conclusion, this system is stable if $b > b_{osc}$ and has no equivalent physical poles if $b > b_{phy}$.

Now let us discuss the impulse response of the above system and extract the condition of oscillation which means that $s = j\omega_{osc} = e^{\frac{j\pi}{2}}\omega_{osc}$ is a solution of its characteristic equation. Therefore, by using Euler formula and eliminating ω_{osc}, the condition of oscillation for this fractional-order differential equation is found to be:

$$b_{osc} = \frac{\sin^{0.6}(0.15\pi)}{\sqrt{2}\sin^{1.6}(0.4\pi)}a^{1.6} \approx 0.477a^{1.6} \tag{3.5}$$

which for $a = -1$ would be equal to 0.477, in agreement with Figure 3.2.

3.1.3.2 Example 2

Let us consider the fractional-order $RL_\alpha C_\beta$ circuit shown in Figure 3.3(a) where the governing differential equation of this circuit is [7–10]:

$$Ri(t) + L\frac{d^\alpha i(t)}{dt^\alpha} + \frac{1}{C\Gamma(\beta)}\int_0^t (t-\tau)^{\beta-1} i(\tau)d\tau = E(t) \tag{3.6}$$

where $\Gamma(\beta)$ is the well-known gamma function. Note that L and C are the pseudo-inductance and pseudo-capacitance, respectively, measured in *Henry* and *Farad* only when $\alpha = \beta = 1$. The impulse response ($E(t) = V_o\delta(t)$) of the above circuit in the s-domain under zero initial conditions is thus:

$$I(s) = \frac{V_o}{R + s^\alpha L + \frac{1}{s^\beta C}} = \frac{V_o s^\beta C}{s^{\alpha+\beta}LC + s^\beta RC + 1} \tag{3.7}$$

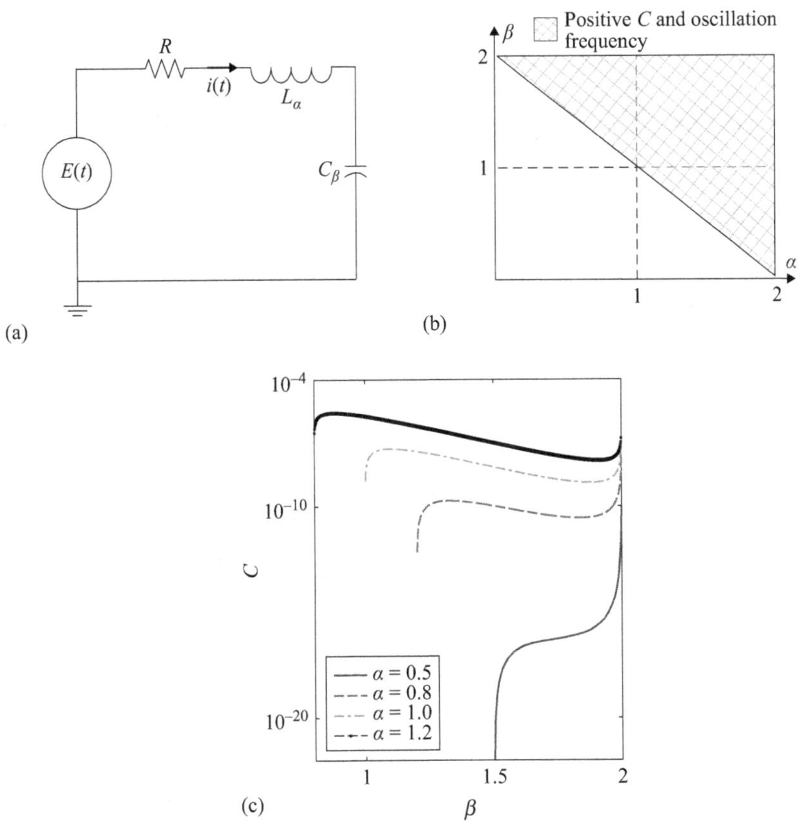

Figure 3.3 (a) $RL_\alpha C_\beta$ circuit, (b) α–β oscillating region, and (c) the condition of oscillation versus β at R = 100 Ω and L = 0.01 H

which shows a band pass filter response. Substituting with $s = j\omega_{osc} = e^{\frac{j\pi}{2}}\omega_{osc}$ in the characteristic equation, it is seen that this network can admit oscillations with a frequency:

$$\omega_{osc}^{\alpha} = \frac{-R\sin\left(\frac{\beta\pi}{2}\right)}{L\sin\left((\alpha+\beta)\frac{\pi}{2}\right)} \tag{3.8}$$

if the value of C satisfies the condition $C = \frac{\sin\left(\frac{\beta\pi}{2}\right)}{R\omega_{osc}^{\beta}\sin\left(\frac{\alpha\pi}{2}\right)}$ and only if $\alpha + \beta > 2$, as shown in Figure 3.3(b). The oscillation $C - \beta$ contours are given in Figure 3.3(c) for different values of α.

3.1.4 *General theorems of fractional-order LTI systems*

This section discusses the condition of oscillation as well as oscillation frequency for any continuous LTI systems with two different fractional-order derivatives.

3.1.4.1 Systems with two fractional-order elements

The state-space representation of two-element fractional-order dynamical system is given by:

$$\begin{pmatrix} D^{\alpha}v_1(t) \\ D^{\beta}v_2(t) \end{pmatrix} = \begin{pmatrix} a_{11} & a_{12} \\ a_{21} & a_{22} \end{pmatrix} \begin{pmatrix} v_1(t) \\ v_2(t) \end{pmatrix} + B\begin{pmatrix} v_1(t) \\ v_2(t) \end{pmatrix} = Av + B(v) \tag{3.9}$$

where A is the linear state matrix while $v_1(t)$ and $v_2(t)$ are the state voltage variables (the voltages across two fractional-order capacitors). B is a nonlinear function which in oscillators are only used to guarantee bounded oscillations. For the linearized system, the relationship between the voltages in the s-domain under zero initial conditions is given by:

$$\frac{V_2(s)}{V_1(s)} = \frac{s^{\alpha} - a_{11}}{a_{12}} = \frac{a_{21}}{s^{\beta} - a_{22}} \tag{3.10}$$

and the characteristic equation is thus:

$$s^{\alpha+\beta} - a_{11}s^{\beta} - a_{22}s^{\alpha} + |A| = 0 \tag{3.11}$$

where $|A|$ is the determinant of the state matrix. Then, the condition of oscillation and the oscillation frequency can be obtained when substituting $s = j\omega_{osc}$ in the characteristic equation leading to the following two equations:

$$\omega_{osc}^{\alpha+\beta}\cos\left(\frac{(\alpha+\beta)\pi}{2}\right) - a_{11}\omega_{osc}^{\beta}\cos\left(\frac{\beta\pi}{2}\right) - a_{22}\omega_{osc}^{\alpha}\cos\left(\frac{\alpha\pi}{2}\right) + |A| = 0 \tag{3.12}$$

$$\omega_{osc}^{\beta}\sin\left(\frac{(\alpha+\beta)\pi}{2}\right) - a_{11}\omega_{osc}^{(\beta-\alpha)}\sin\left(\frac{\beta\pi}{2}\right) - a_{22}\sin\left(\frac{\alpha\pi}{2}\right) = 0 \tag{3.13}$$

Table 3.1 summarizes the oscillation frequency and condition of oscillation for some important cases.

Note that $tr(A)$ is the trace of the state matrix ($tr(A) = a_{11} + a_{22}$). Also, note that the second case in Table 3.1 corresponds to a second-order oscillator while the last case corresponds to a quadrature oscillator.

Table 3.1 Special cases for fractional-order oscillators

Case	Design equations
$\alpha = \beta$	$\omega_{osc} = \left(\frac{tr(A)}{2\cos(0.5\alpha\pi)}\right)^{\frac{1}{\alpha}} = \lvert A\rvert^{\frac{1}{2\alpha}}$
$\alpha = \beta = 1$	$\omega_{osc} = \sqrt{\lvert A\rvert},\ tr(A) = 0$
$a_{11} = a_{22} = 0$	$\omega_{osc} = \sqrt{\lvert A\rvert},\ \alpha + \beta = 2$

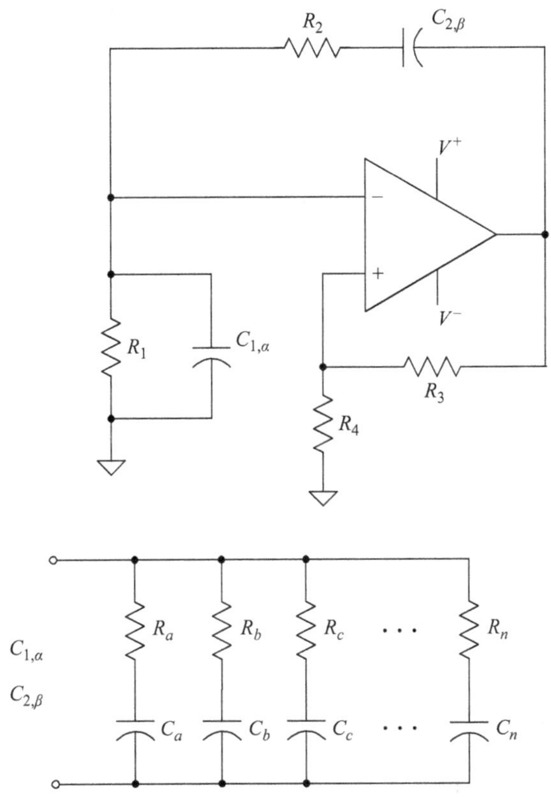

Figure 3.4 Wien-bridge oscillator with two fractional-order capacitors and the RC-tree circuit used to approximate these fractional-order capacitors

3.1.4.2 Example: Wien-bridge oscillator

One of the most famous oscillators is the Wien-bridge oscillator shown in Figure 3.4. Its fractional-order version includes two fractional-order capacitors with pseudo-capacitances C_1, C_2 and fractional orders α, β, respectively [13–15].

The state-space representation of this oscillator is given by:

$$\begin{pmatrix} D^{\alpha} v_{c1}(t) \\ D^{\beta} v_{c2}(t) \end{pmatrix} = \begin{pmatrix} \frac{a-1}{R_2 C_1} - \frac{1}{R_1 C_1} & -\frac{1}{R_2 C_1} \\ \frac{a-1}{R_2 C_2} & -\frac{1}{R_2 C_2} \end{pmatrix} \begin{pmatrix} v_{c1}(t) \\ v_{c2}(t) \end{pmatrix} + \begin{pmatrix} \frac{b}{R_2 C_1} \\ \frac{b}{R_2 C_2} \end{pmatrix} \tag{3.14}$$

$$(a, b) = \begin{cases} (0, V_{sat}) & \text{if } kv_{c1}(t) \geq V_{sat} \\ (k, 0) & \text{if } -V_{sat} < kv_{sat}(t) < V_{sat} \\ (0, -V_{sat}) & \text{if } -V_{sat} \geq kv_{c1}(t) \end{cases} \tag{3.15}$$

where V_{sat} is the saturation voltage of the operational amplifier (opamp) operating as an amplifier with gain $k = 1 + \frac{R_3}{R_4}$. The nonlinearity arising from the opamp saturation will guarantee oscillations with a fixed amplitude. Using the results given above and noting that $|A| = \frac{1}{R_1 R_2 C_1 C_2}$, it can then be shown that the oscillation frequency is obtained by solving:

$$\omega_{osc}^{\alpha+\beta} \sin\left(\frac{\alpha\pi}{2}\right) + \frac{\omega_{osc}^{\alpha}}{R_2 C_2} \sin\left(\frac{(\alpha-\beta)\pi}{2}\right) - \frac{1}{R_1 R_2 C_1 C_2} \sin\left(\frac{\beta\pi}{2}\right) = 0 \tag{3.16}$$

which simplifies to $\omega_{osc} = \frac{1}{R_1 R_2 C_1 C_2}$ as expected when $\alpha = \beta = 1$. The value of amplifier gain k needed to start-up oscillations is then obtained by solving:

$$\frac{\omega_{osc}^{\beta} \sin\left(\frac{(\alpha+\beta)\pi}{2}\right) + \frac{1}{R_2 C_2} \sin\left(\frac{\alpha\pi}{2}\right)}{\omega_{osc}^{\beta-\alpha} \sin\left(\frac{\beta\pi}{2}\right)} + \frac{1}{R_1 C_1} = \frac{k-1}{R_2 C_1} \tag{3.17}$$

which again simplifies to the well-known formula $k = 1 + \frac{1}{R_2 C_1} + \frac{1}{R_1 C_1}$ when $\alpha = \beta = 1$.

In the case that $\alpha = \beta \neq 1$, the above equations simplify to $\omega_{osc} = \left(\frac{1}{R_1 R_2 C_1 C_2}\right)^{\frac{1}{2\alpha}}$ for the oscillation frequency and $k = 1 + \frac{R_2}{R_1} + \frac{C_1}{C_2} + 2\sqrt{\frac{R_2 C_1}{R_1 C_2}} \cos\left(\frac{\alpha\pi}{2}\right)$ for the start-up gain condition. Note that the oscillation frequency is dependent on α and not only on C_1, C_2. This offers an extra degree of design freedom, particularly for very high and very low frequency oscillators. Table 3.2 shows some typical design values assuming $R_1 = R_2 = 10$ kΩ. Frequency scaling can be used to scale the oscillation frequency higher or lower as desired. Note in particular from Table 3.1 that when $\alpha = \beta = 1$, i.e., with ideal capacitors, the start-up gain is $k = 3$ as well known for the Wien-bridge oscillator when $C_1 = C_2$. This is the only case in Table 3.2 where C_1, C_2 are measured in proper *Farads*.

Numerical simulations were carried out using a backward difference integration algorithm based on the Grünwald–Letnikov approximation of a fractional-order derivative [13] with step size $\frac{T}{5{,}000}$ (T is the period of the sinusoid). Figure 3.5 shows the numerical simulation result for $\beta = 2, \alpha = 1$.

In order to perform circuit simulations with SPICE or experimental validation, fractional-order capacitors are approximated using semi-infinite *RC* trees [12–14], as

Table 3.2 Design parameters for the Wien-bridge oscillator assuming $R_1 = R_2 = 10\ k\Omega$

α	β	C_1	C_2	ω	k
0.1	0.1	10^{-4}	10^{-4}	1.0	4.9754
0.3	0.3	10^{-4}	10^{-4}	1.0	4.782
0.4	0.4	3.7×10^{-5}	3.7×10^{-5}	11.9	4.618
0.5	0.5	10^{-5}	10^{-5}	100	4.4142
0.7	0.7	10^{-5}	10^{-5}	26.827	3.908
1.0	1.0	10^{-6}	10^{-6}	100	3
0.3	0.5	10^{-4}	10^{-5}	112.09	10.039
0.5	0.3	10^{-5}	10^{-4}	7.8455	2.8219
0.5	1.0	10^{-5}	10^{-6}	200	3.5
1.0	0.5	10^{-6}	10^{-5}	50	3.5
0.4	1.0	3.7×10^{-5}	10^{-6}	193.46	5.3919
0.4	0.5	3.7×10^{-5}	10^{-5}	72.405	6.8832

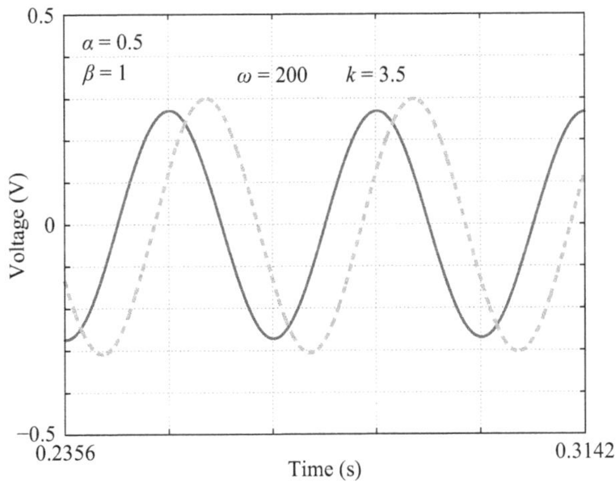

Figure 3.5 Numerical simulation results of the Wien-bridge oscillator equations when $R_1 = R_2 = 10\ k\Omega$, $C_1 = 10\ \mu F$, and $C_2 = 1\ \mu F$

shown in inside of Figure 3.4. The choice of the number of branches is based on the required accuracy of the approximation over a certain frequency range. The values of the resistors ($R_a, R_b, \ldots, R_n$) and capacitors ($C_a, C_b, \ldots, C_n$) are obtained by continued fraction expansion (CFE) [16,17]. A Wien-bridge oscillator was experimentally tested with two fractional-order capacitors having $\alpha = \beta = 0.8$ and $C_1 = C_2 = 10^{-8}$. The resistors were selected as $R_1 = R_2 = 1\ \text{k}\Omega$ and the constructed circuits along with the experimentally observed sinusoids are shown in Figure 3.6. The start-up gain was $k = 2.6$ and the measured oscillation frequency was 314.3 kHz. The experimental waveforms show the voltages across the two capacitors.

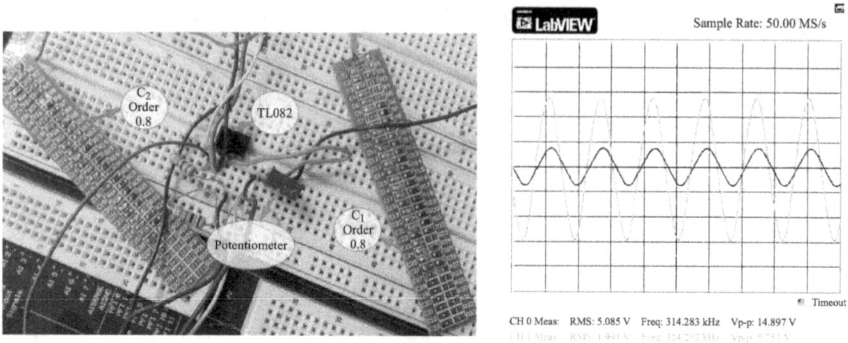

Figure 3.6 Circuit experiment of the fractional-order Wien-bridge oscillator and observed waveforms

3.1.4.3 Systems with three fractional-order elements

The state-space representation of such a linear system is given by:

$$\begin{pmatrix} D^{\alpha} v_1(t) \\ D^{\beta} v_2(t) \\ D^{\gamma} v_3(t) \end{pmatrix} = \begin{pmatrix} a_{11} & a_{12} & a_{13} \\ a_{21} & a_{22} & a_{23} \\ a_{31} & a_{32} & a_{33} \end{pmatrix} \begin{pmatrix} v_1(t) \\ v_2(t) \\ v_3(t) \end{pmatrix} = A \begin{pmatrix} v_1(t) \\ v_2(t) \\ v_3(t) \end{pmatrix} \tag{3.18}$$

and therefore, it has a characteristic equation given by:

$$s^{(\alpha+\beta+\gamma)} - a_{22}s^{(\alpha+\gamma)} - a_{33}s^{(\alpha+\beta)} - a_{11}s^{(\beta+\gamma)} + |A_{\alpha}|s^{\alpha} + |A_{\beta}|s^{\beta} + |A_{\gamma}|s^{\gamma} - |A| = 0 \tag{3.19}$$

where $|A|$ is the determinant of the A matrix, and:

$$|A_{\alpha}| = \begin{vmatrix} a_{22} & a_{23} \\ a_{32} & a_{33} \end{vmatrix}, \quad |A_{\beta}| = \begin{vmatrix} a_{11} & a_{13} \\ a_{31} & a_{33} \end{vmatrix}, \quad |A_{\gamma}| = \begin{vmatrix} a_{11} & a_{12} \\ a_{21} & a_{22} \end{vmatrix} \tag{3.20}$$

The oscillation start-up condition and oscillation frequency are obtained by simultaneously solving the equations [14]:

$$\begin{aligned} &\omega_{osc}^{(\alpha+\beta+\gamma)} \cos(0.5(\alpha+\beta+\gamma)\pi) - a_{22}\omega_{osc}^{(\alpha+\gamma)} \cos(0.5(\alpha+\gamma)\pi) \\ &\quad - a_{33}\omega_{osc}^{(\alpha+\beta)} \cos(0.5(\alpha+\beta)\pi) - a_{11}\omega_{osc}^{(\beta+\gamma)} \cos(0.5(\beta+\gamma)\pi) \\ &\quad + |A_{\alpha}|\omega_{osc}^{\alpha} \cos(0.5\alpha\pi) + |A_{\beta}|\omega_{osc}^{\beta} \cos(0.5\beta\pi) + |A_{\gamma}|s^{\gamma} \cos(0.5\gamma\pi) - |A| = 0 \end{aligned} \tag{3.21}$$

and:

$$\begin{aligned} &\omega_{osc}^{(\alpha+\beta+\gamma)} \sin(0.5(\alpha+\beta+\gamma)\pi) - a_{22}\omega_{osc}^{(\alpha+\gamma)} \sin(0.5(\alpha+\gamma)\pi) \\ &\quad - a_{33}\omega_{osc}^{(\alpha+\beta)} \sin(0.5(\alpha+\beta)\pi) - a_{11}\omega_{osc}^{(\beta+\gamma)} \sin(0.5(\beta+\gamma)\pi) \\ &\quad + |A_{\alpha}|\omega_{osc}^{\alpha} \sin(0.5\alpha\pi) + |A_{\beta}|\omega_{osc}^{\beta} \sin(0.5\beta\pi) + |A_{\gamma}|s^{\gamma} \sin(0.5\gamma\pi) = 0 \end{aligned} \tag{3.22}$$

It can be verified that for $\alpha = \beta = \gamma = 1$ (classical third-order system), we obtain:

$$\omega_{osc} = \sqrt{\frac{|A|}{tr(A)}} = \sqrt{(|A_\alpha| + |A_\beta| + |A_\gamma|)} \tag{3.23}$$

as well known.

3.1.4.4 Example: phase-shift oscillator

The phase-shift oscillator circuit is shown in Figure 3.7 and its state-space matrix is given by

$$A = \frac{-1}{k+1}\begin{pmatrix} \frac{1}{C_1}(\frac{1}{R_1} + \frac{1}{R_2} + \frac{1}{R_3}) & \frac{1}{C_1}(\frac{1}{R_2} + \frac{1}{R_3} - \frac{k}{R_1}) & \frac{1}{C_1}(\frac{1}{R_3} - \frac{k}{R_1} - \frac{k}{R_2}) \\ \frac{1}{C_2}(\frac{1}{R_2} + \frac{1}{R_3}) & \frac{1}{C_2}(\frac{1}{R_2} + \frac{1}{R_3}) & \frac{1}{C_2}(\frac{1}{R_3} - \frac{k}{R_2}) \\ \frac{1}{R_3C_3} & \frac{1}{R_3C_3} & \frac{1}{R_3C_3} \end{pmatrix} \tag{3.24}$$

We consider for simplicity the special case where $\alpha = \beta = \gamma > \frac{2}{3}$, $R_1 = R_2 = R_3 = R$, and $C_1 = C_2 = C_3 = C$. In this case, the oscillation frequency is found to be:

$$\omega_{osc} = \frac{1}{6RC}\left(\sqrt{x^2+6} - 5x\right) \tag{3.25}$$

and the start-up condition is:

$$k+1 = \frac{18\left(10x^2 - 5 - 2x\sqrt{x^2+6}\right)}{(4x^2-1)(13x^2+3-5x\sqrt{x^2+6})} \tag{3.26}$$

where $x = \cos\left(\frac{\alpha\pi}{2}\right)$. If $\alpha = \beta = \gamma = 1$, these equations will simplify to:

$$\omega_{osc} = \frac{1}{\sqrt{6}RC} \tag{3.27}$$

$$k+1 = 30 \tag{3.28}$$

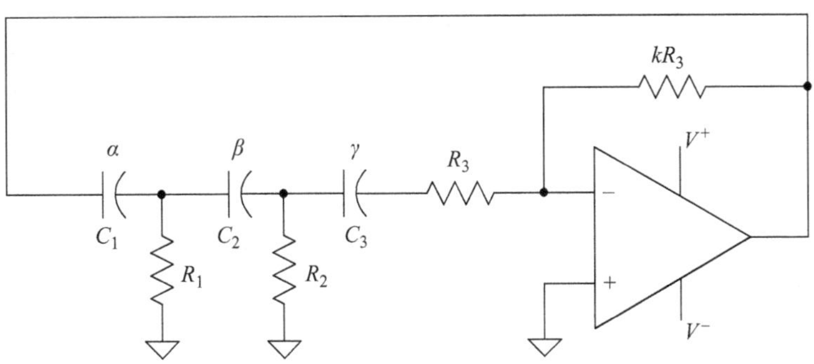

Figure 3.7 The phase-shift oscillator with three fractional-order capacitors

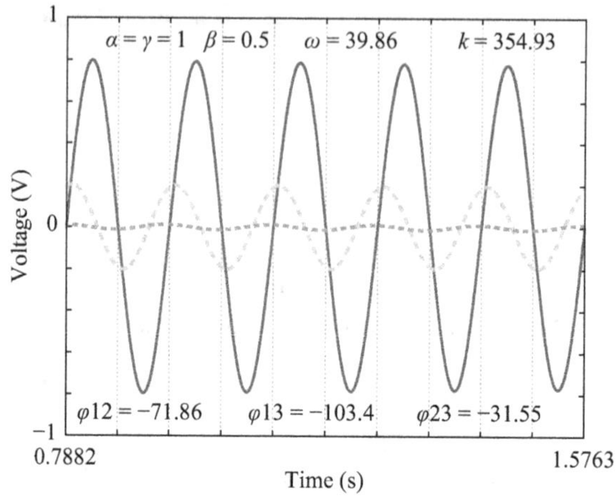

Figure 3.8 Numerical simulations for a phase-shift oscillator designed with $\alpha = \gamma = 2\beta = 1$

as expected for the classical phase-shift oscillator. Figure 3.8 shows a numerical simulation result for this oscillator with $\alpha = \gamma = 2\beta = 1$. The phase shift between the voltage signals across the three capacitors is also given in Figure 3.8. Exact equations for this phase shift can be found in Reference 14.

3.2 Fractional-order relaxation oscillators

Relaxation oscillators find many uses in function generators and voltage-controlled oscillators where they are often used to produce sawtooth, triangular, or square waveforms. Implementation can take many forms, but one of the simplest is via the popular Schmitt trigger which can be implemented using operational amplifiers or inverters. In this section, we consider the effect of using a fractional-order capacitor in place of a normal capacitor in the popular single opamp multivibrator which is configured as a Schmitt trigger. We show that the fractional-order capacitor has the ability to increase the oscillator frequency significantly all the while using reasonable circuit time constants. The results can be easily extended to other forms of multivibrator circuits. Because most multivibrators employ only one capacitor, closed form formulae are possible to derive [16].

3.2.1 The fractional multivibrator

Consider the circuit shown in Figure 3.9 which represents the basic free-running multivibrator without amplitude limiting control. Let C_α be the fractional capacitor whose impedance $Z(s) = 1/C_\alpha s^\alpha$, where $0 < \alpha < 1$. Resistors R_2 and R_3 set the voltage v_z at the non-inverting node, and also determine the oscillation frequency. We assume that the opamp employed here has a gain bandwidth product that far exceeds that of the desired oscillation frequency. For the simple case of $\alpha = 1$, it is well

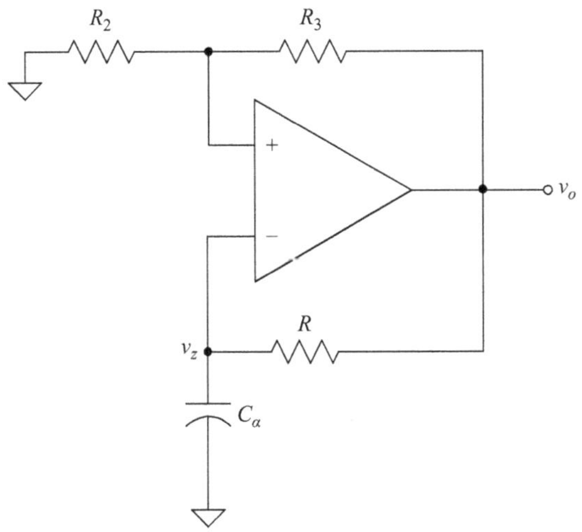

Figure 3.9 The free running multivibrator with fractional capacitor C_α

known that the period of oscillation T of this multivibrator is linearly related to the time constant $\tau = RC_\alpha$ by:

$$\frac{T}{2} = \tau \ln\left(\frac{1+\beta}{1-\beta}\right) \tag{3.29}$$

where $\beta = R_2/(R_2 + R_3)$. For the case when $\alpha < 1$, an examination of the step response of $v_z(t)$, as obtained from (23) of Reference 18 of the fractional RC circuit during half a period, shows that the period T and τ are related by the following closed form expression:

$$\frac{1-\beta}{1+\beta} = \sum_{n=0}^{\infty} \frac{\left(-\frac{1}{\tau}\right)^n \left(\frac{T}{2}\right)^{n\alpha}}{\Gamma(n\alpha + 1)} \tag{3.30}$$

where $\Gamma(\cdot)$ is the gamma function and we have made the substitution $a = \frac{1}{\tau}$, $t = \frac{T}{2}$ and $q = \alpha$ into (10b) of Reference 18 and used it in (23) of Reference 18.

Note the right-hand side of (3.30) is nothing more than the one-term Mittag-Leffler function $E_\alpha[\cdot]$. That is:

$$\frac{1-\beta}{1+\beta} = E_\alpha\left[-\frac{1}{\tau}\left(\frac{T}{2}\right)^\alpha\right] \tag{3.31}$$

The similarities between (3.31) and (3.29) become obvious when we consider $\alpha = 1$. In this case, $E_1\left[-\frac{T}{2\tau}\right] = e^{-\frac{T}{2\tau}}$ and hence (3.31) reduces to (3.29) as expected. For the special case of $\alpha = 0.5$, (3.30) reduces to:

$$\frac{1-\beta}{1+\beta} = e^{\frac{T}{2\tau^2}} \cdot \operatorname{erfc}\left(\frac{1}{\tau}\sqrt{\frac{T}{2}}\right) \tag{3.32}$$

where erfc$(\cdot)$ is the compliment of the error function. A simple unit plot showing the relationship between the period T and τ at two different values of β is shown in Figure 3.10(a). For the case of $\beta = \frac{1}{3}$, the solid line obtained from (3.32) is compared

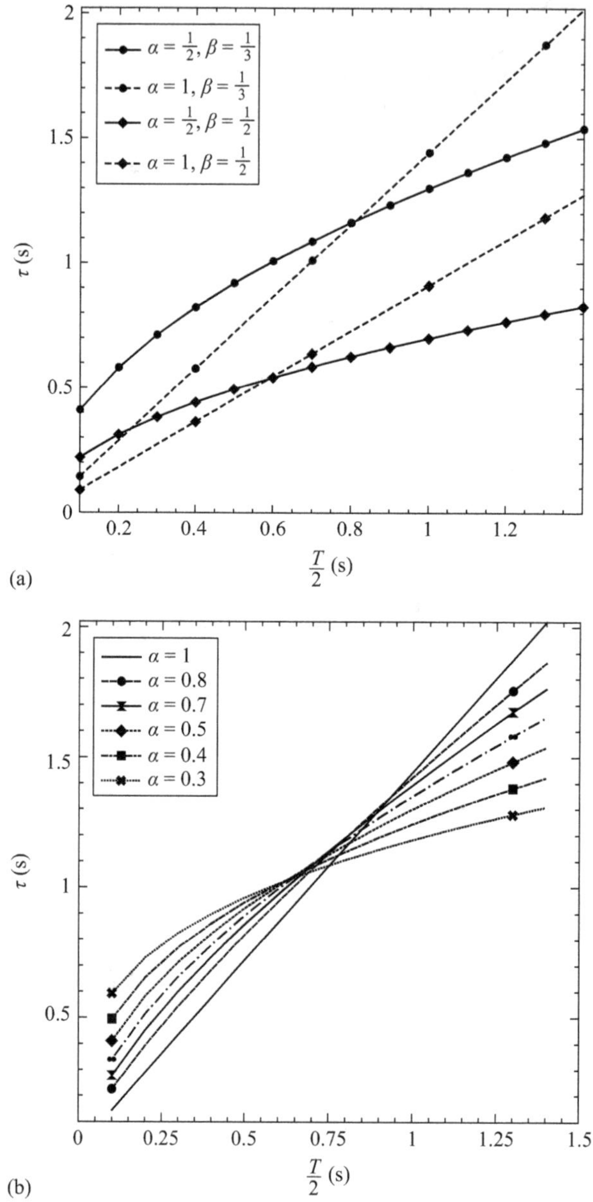

Figure 3.10 (a) Plot of time constant τ versus T/2 to two values of β for the multivibrator, $\beta = \frac{1}{2}$ and $\beta = \frac{1}{3}$. (b) Plot of time constant τ versus T/2 to various values of α with $\beta = \frac{1}{3}$

to the dashed straight line obtained from (3.29). It can be seen that for certain values of τ, referred to as τ_c (in this case $\tau_c < 1.172$ for $\beta = \frac{1}{3}$ and $\tau_c < 0.537$ for $\beta = \frac{1}{2}$), the period of the fractional multivibrator is much less than that of the normal multivibrator. The difference becomes even more apparent when we use a more practical example such as $\tau = 1$ ms, then at $\beta = \frac{1}{2}$ we obtain $f_o^{fractional} = 845.3$ kHz compared to $f_o^{integer} = 721.34$ Hz. This represents an increase of 1 172 times. If β is decreased even further to $\beta = \frac{1}{3}$, then $\frac{f_o^{fractional}}{f_o^{integer}} = 1\,464$ with $f_o^{integer}$ and $f_o^{fractional}$ increasing to 807.7 Hz and 1.12 MHz, respectively. The reason for this is that the step response of a fractional-order *RC* circuit rises much faster than the conventional exponential response. It is thus clear that for low values of β, the fractional multivibrator can produce much higher frequencies than its integer order counterpart. Note that for a given β and α, τ_c can be easily computed from the simultaneous solution of (3.31) and (3.29). For values of $\tau > \tau_c$, $\frac{f_o^{fractional}}{f_o^{integer}} < 1$. Finally, Figure 3.10(b) shows the relationship between τ and $T/2$ to various values of α for the case of $\beta = \frac{1}{3}$. It shows that for a given $\tau < \tau_c$ then as α decreases, T decreases, which means that the frequency of oscillation increases.

3.2.2 Verification via simulation and experimentation

To verify the previous results for the circuit of Figure 3.9, a fractional capacitor of value 1 μF/s$^{1-\alpha}$ with $\alpha = 0.5$ was approximated using Carlson's method [19] to a second order and realized using the partial fraction decomposition circuit shown in Figure 3.11. The opamp used was the LT1364 with a gain bandwidth product of 70 MHz and $\beta = \frac{1}{3}$ was used. Resistor R was set at 1 kΩ which yielded $\tau =$ 1 ms. The SPICE results shown in Figure 3.12(a) revealed the new multivibrator oscillated with a frequency of 1.57 MHz with no output amplitude stabilization. The experimental results shown in Figure 3.12(b) yielded an oscillation frequency of 830.16 kHz which is less than both the SPICE and the theoretical result. This, however,

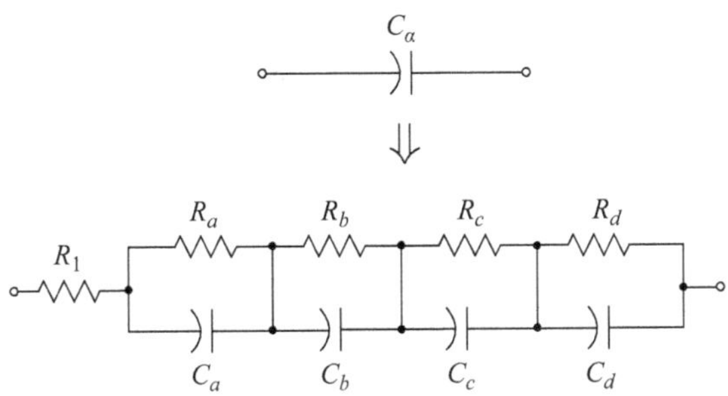

Figure 3.11 Approximation to the fractional capacitor $C_\alpha = 1\ \mu F$ with $\alpha = 0.5$. Values used for the fractional capacitor approximation are $R_1 = 0.111$ kΩ, $R_a = 7.37$ kΩ, $R_b = 0.252$ kΩ, $R_c = 0.379$ kΩ, $R_d = 0.889$ kΩ, and $C_a = 4.364$ nF, $C_b = 0.526$ nF, $C_c = 1.859$ nF, and $C_d = 3.375$ nF

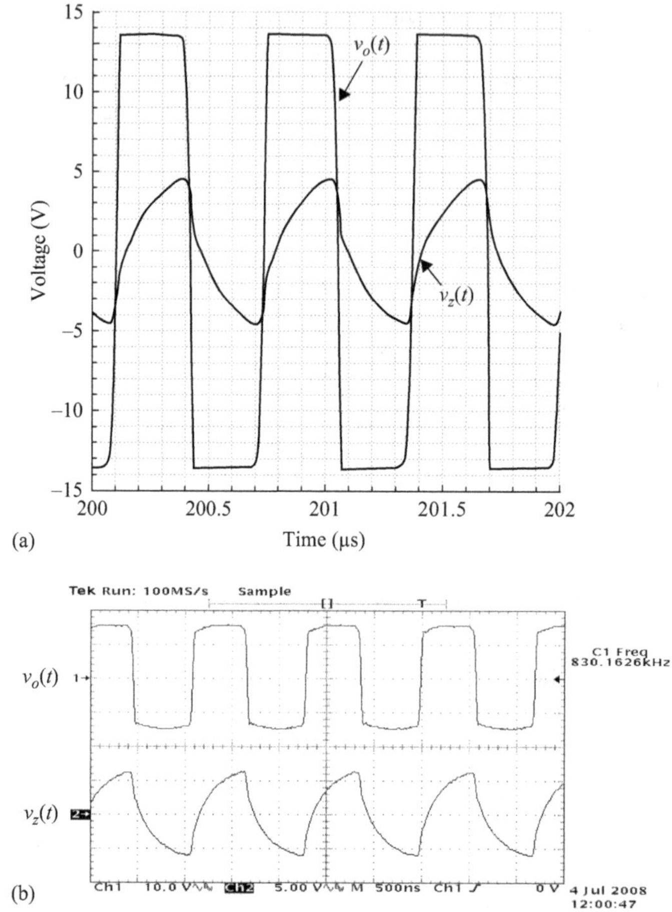

Figure 3.12 (a) SPICE simulation of the multivibrator using an approximation to the fractional capacitor. The observed oscillation frequency was 1.57 MHz. (b) Experimental results of the multivibrator using an approximation to the fractional capacitor. The measured oscillation frequency was 830.16 kHz. The fractional capacitor was made using the following up 10% tolerance resistors and 20% tolerance capacitors: $R_1 = 111\ \Omega$, $R_a = 7.36\ k\Omega$, $R_b = 251\ \Omega$, $R_c = 378\ \Omega$, $R_d = 888\ \Omega$ and $C_a = 0.0047\ \mu F$, $C_b = 0.55\ nF$, $C_c = 1.82\ nF$, and $C_d = 3.3\ nF$.

is due to the fact that exact values to the approximation could not be obtained using discrete components and the low tolerance of the discrete components available. If a normal 1 μF capacitor was used, the oscillation frequency obtained in SPICE was 731.5 Hz as shown in Figure 3.13(a) and in the experimental result shown in Figure 3.13(b) it was 819.68 Hz. This shows the usefulness of employing the fractional

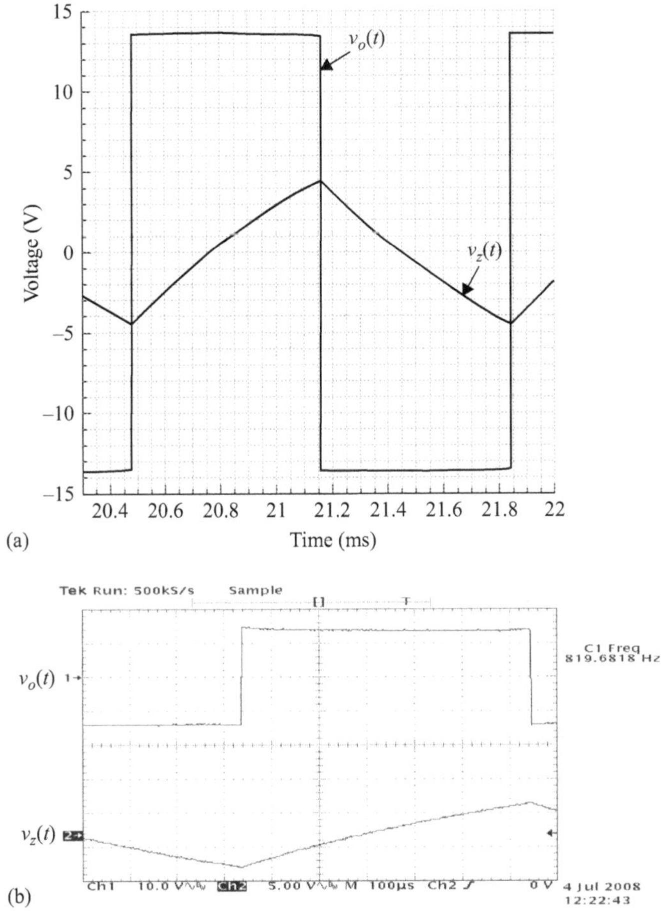

Figure 3.13 *(a) SPICE simulation of the multivibrator using a normal 1 μF capacitor, $R = 1$ kΩ and $\beta = \frac{1}{3}$. The observed oscillation frequency was 731.5 Hz. (b) Experimental results of the multivibrator using a normal 1 μF capacitor, $R = 1$ kΩ and $\beta = 0.3$. The measured oscillation frequency was 819.68 Hz. © 2009 Springer. Reprinted, with permission, from Reference 16*

capacitor, so long as the opamp is capable of handling the increased frequencies. Both values of the expected oscillation frequencies are therefore in agreement with the predicted results within limits of accuracy. Note that to generate the higher oscillation frequency using a normal integer order capacitor, a time constant $\tau = 0.514$ μs (rather than $\tau = 1$ ms) is needed. Finally, the results in Figures 3.12 and 3.13 for $v_z(t)$ also display the characteristic $1 - e^t \mathrm{erfc}\left(\sqrt{t}\right)$ response which can be derived from (3.32).

In conclusion, the usefulness of the fractional-order capacitor with multivibrators is demonstrated in this section. The frequency of oscillation is shown to be much higher for the same time constants in the simple multivibrator. This opens up the possibility of greater frequency control in multivibrators as physical fractional capacitors soon become available. It also shows that a capacitive behavior is needed to generate an oscillatory signal but this does not imply that an ideal capacitance is necessary.

3.2.3 Multivibrator with biological tissues

Biological tissues are known to contain fractional-order capacitive behavior [20]. Several electric models have been proposed to model the bio-impedance of biological materials but the most famous model is the single-dispersion Cole–Cole model shown in Figure 3.14. This model is composed of two resistors namely R_∞ and $R_0 - R_\infty$ in addition to a fractional-order capacitor. The total impedance is thus given by:

$$Z = R_\infty + \frac{R_0 - R_\infty}{1 + j(\omega\tau_c)^\alpha} \tag{3.33}$$

where R_0 is the DC resistance of the tissue, R_∞ is the very high frequency resistance of the tissue and τ_c is the characteristic time constant of the tissue given as $\tau_c = [(R_0 - R_\infty)C_\alpha]^{(1/\alpha)}$.

It is therefore possible to employ a biological tissue as a fractional-order capacitor taking into consideration its Cole–Cole impedance model. As an example, consider the classical relaxation oscillator shown in Figure 3.15. With a normal capacitor of value C, this oscillator is known to produce a square-wave signal at the output of the first opamp with a frequency $\omega = 1/RC$ where $R = R_1R_2/R_3$. Now consider replacing the normal capacitor with a biological tissue or a fruit sample. Using the Cole–Cole impedance model, we show that the oscillation frequency of this oscillator would be obtained by solving for x in the following equation:

$$\begin{aligned}&\left(1 + 2x^\alpha \cos\left(\frac{\alpha\pi}{2}\right) + x^{2\alpha}\right)^2 \\ &= \left(ax^\alpha\left(x^\alpha + \cos\left(\frac{\alpha\pi}{2}\right)\right) + b\left(1 + x^\alpha\cos\left(\frac{\alpha\pi}{2}\right)\right)\right)^2 + \left((b-a)x^\alpha \sin\left(\frac{\alpha\pi}{2}\right)\right)^2\end{aligned} \tag{3.34}$$

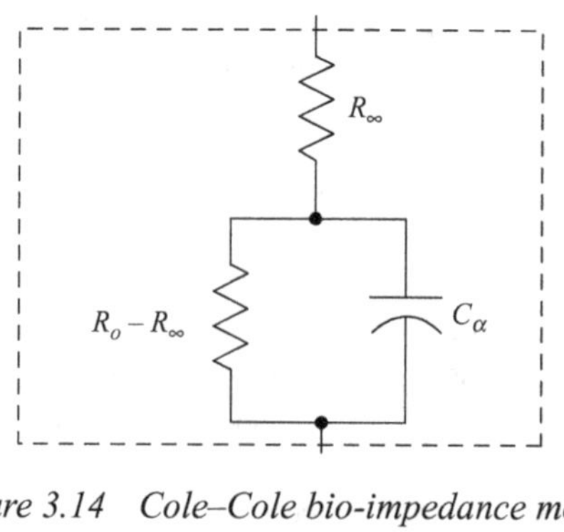

Figure 3.14 Cole–Cole bio-impedance model

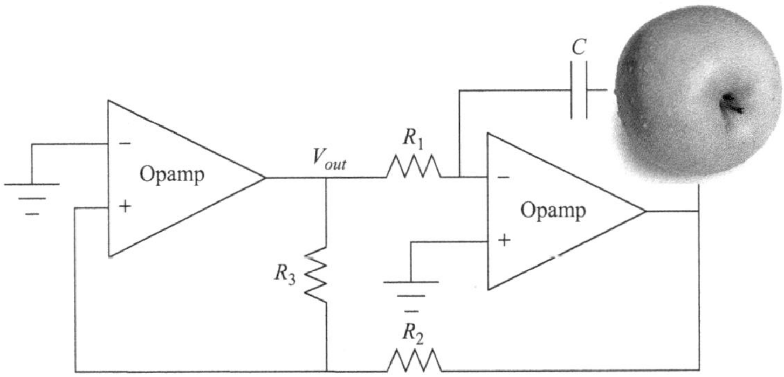

Figure 3.15 Relaxation oscillator employing a fruit sample

where x is the normalized frequency defined as $x = \omega\tau_c$ while the two constants (a, b) are resistor ratios defined as $a = R_\infty/R$ and $b = R_0/R$. Note that the constraints $x > 0$ and $b > a$ always hold. This equation is obtained as a solution to the equation $|Z| = R$ where Z is the apple impedance described by the Cole–Cole model. We can easily verify that with $\alpha = 1$ (ideal capacitor) and $R_\infty = 0\,(a = 0)$, the above equation simplifies to:

$$x = \sqrt{b^2 - 1} \tag{3.35}$$

and hence, the de-normalized oscillation frequency would be:

$$\omega = \frac{\sqrt{\left(\frac{R_0}{R}\right)^2 - 1}}{R_0 C} = \frac{1}{RC}|_{R_0 \to \infty} \tag{3.36}$$

as expected. Figure 3.16 shows the value profile of x versus α and b for two different values of a namely $a = 0$ and $a = 0.5$. Note that the smaller the value of a, the wider the range of possible solutions for x. Experimental results from the circuit with a fresh green apple are shown in Figure 3.17. Here, TL081 opamps were used after setting $R_2 = R_3 = 10$ kΩ and R_1 as 12 kΩ. The measured oscillation frequency is around 97 mHz. Impedance spectroscopy using a Biologic VSP-300 electrochemical station was used to estimate the Cole–Cole model parameters of the apple used in the experiments reporting $(R_\infty, R_0, \alpha, C_\alpha) = (8.2\text{ k}\Omega, 10.12\text{ k}\Omega, 0.46, 79.04\ \mu\text{F})$ based on a complex nonlinear least square fitting algorithm of the measured impedance data points to the model. With these values, the normalized oscillation frequency x is equal to 0.005 and the de-normalized frequency is 47 mHz. Figure 3.17 shows the observed waveform with a measured frequency of 97 mHz which is close enough to the predicted frequency given the fact that the Cole model is an approximation of the actual impedance and that the placement of the electrodes in different locations around the fruit as well as the insertion depth of the electrodes has an effect on the measured impedance. The applications of electronic circuits involving fruits and vegetables are a research topic which is still in its infancy yet has a significant application potential in the agricultural sector [21].

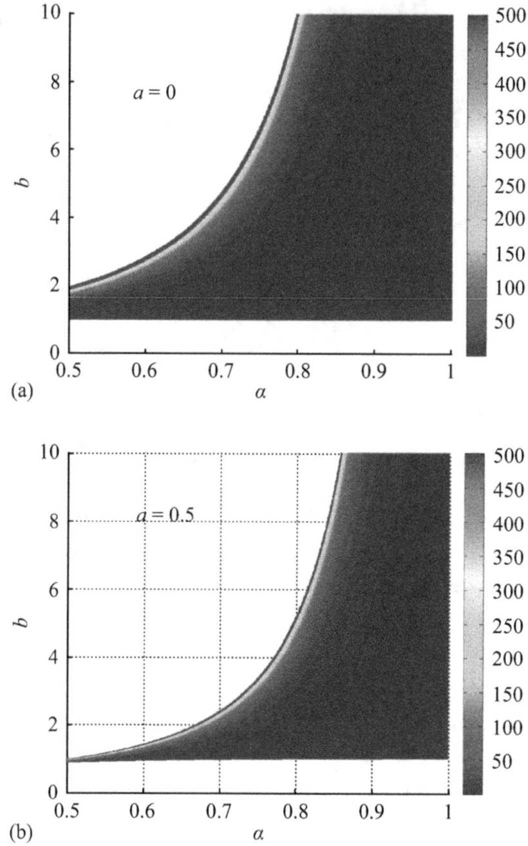

Figure 3.16 Numerical simulations showing the values of normalized oscillation frequency x versus α and b for two different values of a

From another perspective, biological tissues can be considered as distributed multi-port systems. If the tissue is considered homogenous and geometrically symmetrical, then the multi-port system is both symmetrical and reciprocal from an electrical point of view. Consider for example employing a lemon fruit as a two-port device. Let the input port be excited by a biploar 1 V_{pp} square-wave signal with variable frequency f_e. At the same time, the output port is connected as the fractional-order capacitor in Figure 3.15. Figure 3.18 shows the free running output of the oscillator when no signal is applied at the input port of the lemon fruit. A symmetrical waveform is observed in this case with measured frequency of approximately 2 kHz. However, upon applying the excitation signal at the input port, different periodic waveforms were observed at the oscillator output, as shown in Figure 3.19. In particular, signals periodic with periods 2, 3, 4, 5, 6, 8, 9 and even period 13 were observed for f_e = 300 Hz, 200 Hz, 120 Hz, 100 Hz, 70 Hz, 55 Hz, 40 Hz and 20 Hz, respectively. The effect of applying the excitation signal at the input port mimics a time-dependent

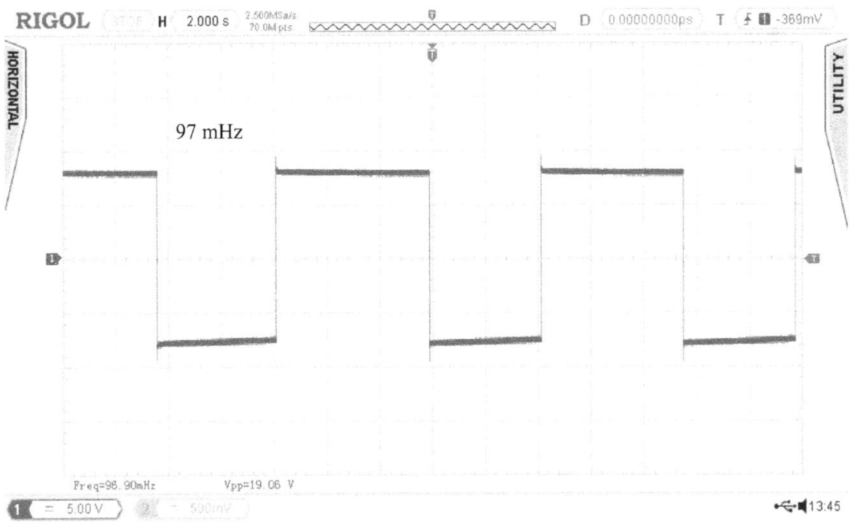

Figure 3.17 Experimental waveform with an apple fruit

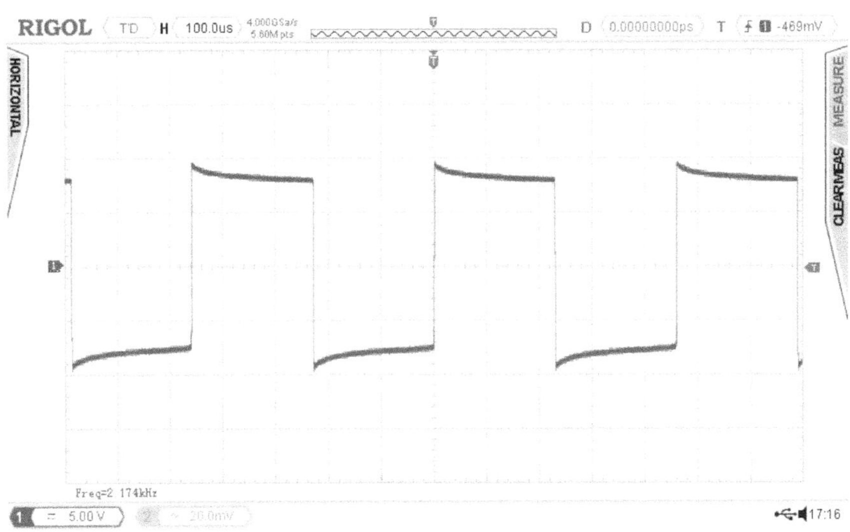

Figure 3.18 Experimental waveform with a lemon fruit

voltage-controlled capacitor. However, in this case, it is expected that all four variables in the Cole–Cole model are time and voltage dependent leading to a *dynamic* Cole–Cole model. This topic requires further investigation particularly with the distributed nature of the biological tissue and the practical ability to create many input ports with multiple excitations.

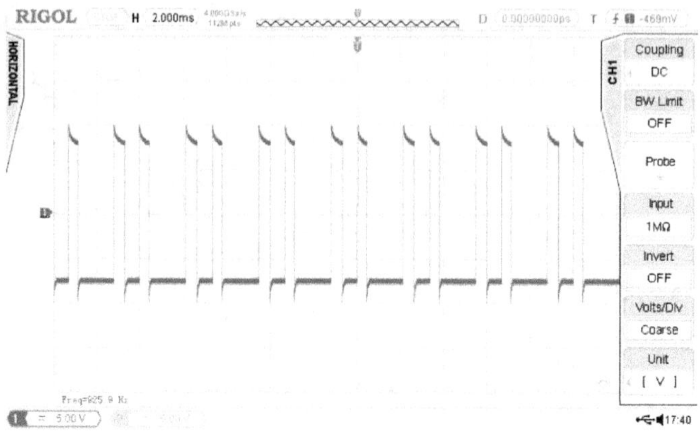

(a)

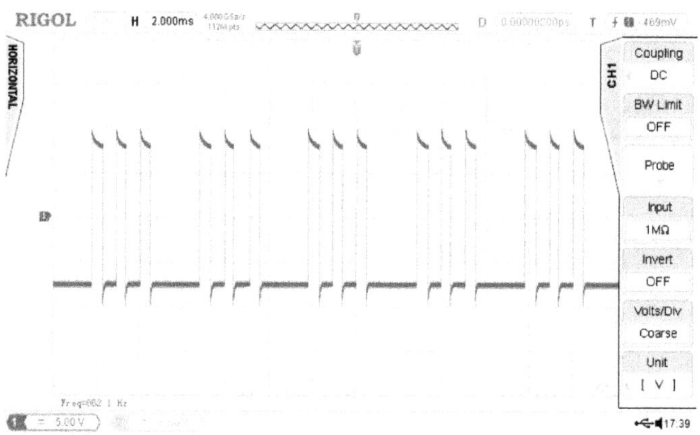

(b)

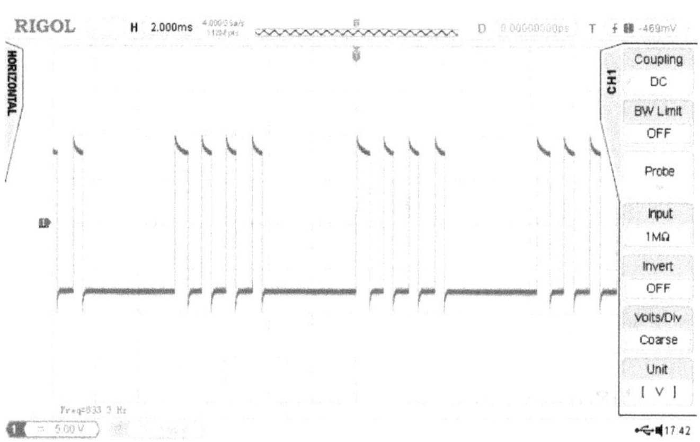

(c)

Figure 3.19 Experimental observations using a lemon fruit as a two-port network with $1V_{pp}$ square-wave excitation signal at input port with (a) $f_e = 300$ Hz, (b) $f_e = 200$ Hz, (c) $f_e = 120$ Hz, (d) $f_e = 100$ Hz, and (e) $f_e = 40$ Hz.

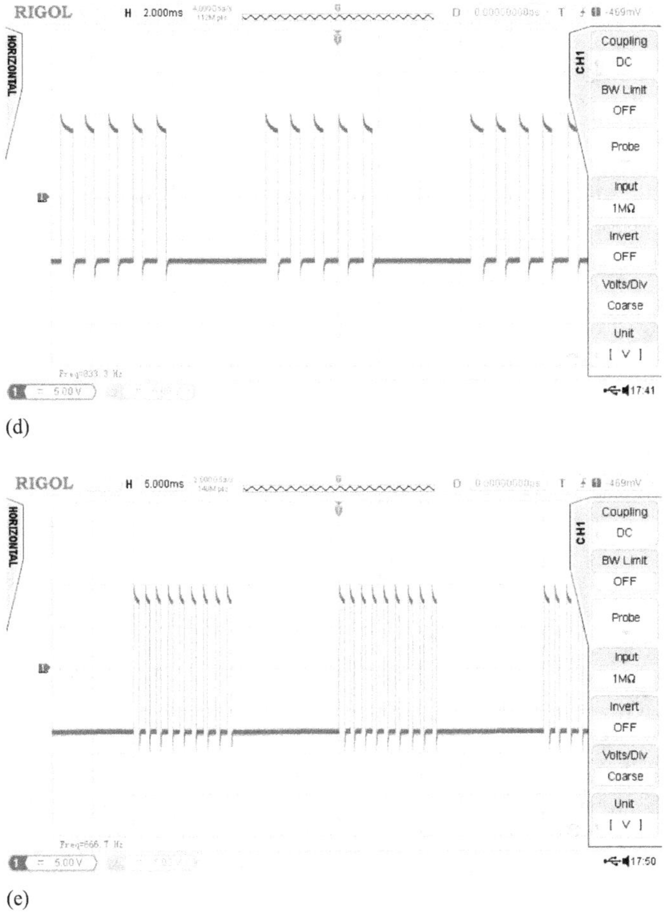

Figure 3.19 (Continued)

3.3 Conclusion

We have described the theory and practical implementations of fractional-order oscillators; both sinusoidal and relaxation-type oscillators. These oscillators offer the clear advantage of frequency tuning via extra degrees of freedom; namely the fractional order of the employed energy storage devices. Of specific interest are the applications in biomaterial characterization, a topic which is still in its infancy.

References

[1] A. G. Radwan, A. M. Soliman, A. S. Elwakil and A. El-Sedeek, "On the stability of linear systems with fractional-order elements," *Chaos Solitons & Fractals*, 40, 2317–2328, 2009.

[2] M. D. Ortigueira, "An introduction to the fractional continuous-time linear systems: The 21st century systems," *IEEE Circuits and Systems Magazine*, 8, 19–26, 2008.
[3] A. G. Radwan, A. M. Soliman and A. S. Elwakil, "First-order filters generalized to the fractional domain," *Journal of Circuits Systems and Computers*, 17, 55–66, 2008.
[4] A. G. Radwan, A. S. Elwakil and A. M. Soliman, "On the generalization of second-order filters to fractional-order domain," *Journal of Circuits Systems and Computers*, 18, 361–386, 2009.
[5] A. Soltan, A. G. Radwan and A. M. Soliman, "Fractional order filter with two fractional elements of dependent orders," *Microelectronics Journal*, 43, 818–827, 2012.
[6] A. Soltan, A. G. Radwan and A. M. Soliman, "Fractional order Sallen–Key and KHN filters stability and pole allocation," *Journal Circuits Systems and Signal Processing*, 34, 1461–1480, 2015.
[7] A. G. Radwan and M. E. Fouda, "Optimization of fractional-order RLC filters," *Journal of Circuits Systems and Signal Processing*, 32, 2097–2118, 2013.
[8] A. G. Radwan and K. N. Salama, "Fractional-order RC and RL circuits," *Journal of Circuits Systems and Signal Processing*, 31, 1901–1915, 2012.
[9] A. G. Radwan and K. N. Salama, "Passive and active elements using fractional $L_\beta C_\alpha$ circuit," *IEEE Transaction on Circuits and Systems-I*, 58, 10, 2388–2397, 2011.
[10] A. G. Radwan, "Resonance and quality factor of the fractional RLC circuit," *IEEE Journal on Emerging and Selected Topics in Circuits and Systems*, 3, 377–385, 2013.
[11] T. J. Freeborn, B. J. Maundy and A. S. Elwakil, "Fractional-step Tow-Thomas biquad filters," *IEICE Nonlinear Theory and Its Applications (NOLTA)*, 3, 357–374, 2012.
[12] T. J. Freeborn, B. J. Maundy and A. S. Elwakil, "Fractional step analog filter design," in *Analog/RF and Mixed Signal Circuit Systematic Design*, Chapter 11, 243–267, Springer, Berlin, 2013.
[13] W. Ahmed, R. Elkhazali and A. S. Elwakil, "Fractional-order Wien-bridge oscillator," *Electronics Letters*, 37, 1110–1112, 2001.
[14] A. G. Radwan, A. S. Elwakil and A. M. Soliman, "Fractional-order sinusoidal oscillators: Design procedure and practical examples," *IEEE Transactions on Circuits and Systems-I*, 55, 2051–2063, 2008.
[15] A. G. Radwan, A. M. Soliman and A. S. Elwakil, "Design equations for fractional-order oscillators: Four practical design examples," *International Journal of Circuit Theory and Applications*, 36, 473–492, 2008.
[16] B. J. Maundy, A. S. Elwakil and S. Gift, "On a multivibrator that employs a fractional capacitor," *Analog Integrated Circuits and Signal Processing*, 62, 99–103, 2010.

[17] J. Valsa and J. Vlach, "RC models of a constant phase element," *International Journal of Circuit Theory and Applications*, 41, 59–67, 2013.

[18] T. T. Hartley and C. F. Lorenzo, "A solution to the fundamental linear fractional order differential equation," Tech. Rep. 208693, National Aeronautics and Space Administration (NASA), Dec. 1998.

[19] G. Carlson and C. Halijak, "Approximation of fractional capacitors by regular newton process," *IEEE Transactions on Circuit Theory*, CT-11, 210–213, 1964.

[20] T. J. Freeborn, "A survey of fractional-order circuit models for biology and biomedicine," *IEEE Journal on Emerging and Selected Topics in Circuits and Systems*, 3, 416–424, 2013.

[21] B. Maundy, A. S. Elwakil and A. Allagui, "Extracting the parameters of the single-dispersion Cole bioimpedance model using a magnitude-only method," *Computers and Electronics in Agriculture*, 119, 153–157, 2015.

Chapter 4

Memristive and memcapacitive astable multivibrators

Dongsheng Yu[1]*, Herbert Ho-Ching Iu*[2]*, Tyrone Fernando*[2] *and Jason Eshraghian*[2]

In this chapter, two astable oscillators are introduced by making use of memristor and memcapacitor emulators, respectively. First, the oscillating characteristics of an astable multivibrator based on flux-controlled memristor emulator and 555 timer are explored. Then, another astable oscillator is newly implemented based on one floating memcapacitor emulator and two NOT gates. The inclusion of memristor or memcapacitor in the astable oscillators provides a new option allowing for large degree of flexible control over frequency as well as duty ratio.

4.1 Introduction

The memristor, which is defined as the fourth fundamental circuit element by Leon Chua in 1971, combines a persistent memory with electrical resistance constitutively related to the relationship between electric charge and flux [1]. Although memristive behaviors had been occasionally observed in the past decades, the first nanoscale device functioning as memristor was built in 2008 by Hewlett-Packard based on a bi-level titanium dioxide thin film containing dopants, which can be migrated by applied current [2]. Memcapacitor and meminductor are two newly generalized energy storage elements based on the conception of memristor in 2009 [3]. Together with memristor, their abilities of processing and storing information without the requirement of power source would open up new realms for future circuit investigations and applications in a range of areas including nonvolatile memories, low-power computation, neuromorphic and biological systems, signal processing, oscillators, filters, and so on [4–12].

[1]School of Information and Electrical Engineering, China University of Mining and Technology, University Road, Xuzhou, China

[2]School of Electrical, Electronic and Computer Engineering, The University of Western Australia, Crawley, WA, Australia

A number of emulating methods have been proposed to analyze the equivalent dynamical characteristics of these three mem-elements. Memristor has attracted relatively extensive research attentions. Most of the previous research related to modeling memristors is achieved on the basis of Personal Simulation Program with Integrated Circuit Emphasis (PSPICE) simulation [13–17]. Behavioral models for mimicking TiO_2 memristors are proposed in References 13,14, but can hardly be utilized for modeling memristive devices without precise physical equations. Hence, compact models, uniquely based on the constitutive relationship between charge and flux, are proposed in Reference 15, along with boundary assurance and global validation. By taking into account the very high electric fields in the thin film, a new model which can exactly mimic the boundary conditions of the HP memristor is discussed in Reference 16. A magnetic flux controlled memristor model is introduced in Reference 17 including a technical description and SPICE implementation.

Experimentally reported memristor emulators mainly take advantage of common electronic devices with variable resistance, such as programmable digital potentiometers [18], light-dependent diodes [19], light-dependent resistors [20], junction field effect transistors [21], etc. Analog multipliers are also frequently employed to create memristor emulators by experimentally realizing their constitutive relationships [7,22,23]. However, these practical emulators possess the limiting feature of one terminal connected to ground [7,18–23], which restricts their widespread use in circuit designs and applications. In view of grounded limitation, two methods of implementing floating memristor emulators have been proposed in Reference 24, of which the first method has a complex hardware implementation due to the requirement of a multiplier and the realization of a floating current source. Although the second method is more experimentally implementable, the utilization of a 6-bit Analog-to-Digital Converter and a floating 6-bit Resistive Digital-to-Analog Converter will obviously result in quantization errors and the corresponding memristance is incapable of being continuously adjusted.

Note that, in addition to information, memcapacitors and meminductors are capable of storing energy, and therefore could explore new orientations in the technologically important areas of energy storage and circuit operation [25]. Although nanoscale devices with memcapacitive and meminductive effect have been achieved [26,27], solid memcapacitors and meminductors are still commercially unavailable. Hence, research related to meminductor and memcapacitor emulators has been increasingly reported. The simulation analysis provided by Hu *et al.* [28,29] reflects that nonlinear memcapacitors are capable of being utilized to construct chaotic oscillators. In accordance with respective constitutive relations, behavioral models of meminductive and memcapacitive systems have been established on the basis of simulation [30,31]. Based on the requirement of active devices and memristor, several circuit mutators for emulating meminductor and memcapacitor have been proposed in References 32–38. By making use of the microprocessor-based memristive emulator proposed in Reference 18, meminductive and memcapacitive emulators inherently connected in series with a resistor are designed in Reference 32. In References 33,34, an improved memcapacitor emulator of high accuracy without parasitic resistance is proposed. Memcapacitor emulators in serial, parallel, hybrid and even Wye and Delta connections are proposed by making use of an expandable memristor emulator [35]. Meminductive and memcapacitive emulators with one terminal connected to ground are experimentally introduced in References 36,37. However, the memconductance can hardly be obtained with high precision due to the nonlinear response of the light-dependent resistor. Inspired by the mutators for transforming memristor into floating meminductor and memcapacitor, as conceptually discussed in Reference 38, a floating memcapacitor emulator is recently proposed and experimentally tested in Reference 39. Interestingly, by pointing out that most of the existing meminductor and memcapacitor emulators are designed necessarily based on memristor, a new set of memristor-less meminductor and memcapacitor emulators is proposed in References 40,41. Undoubtedly, these emulators have assisted researchers in discovering the dynamic behaviors of these memory elements.

The memristor-based oscillator is a novel topic recently emerging in nonlinear electronic systems. A thorough investigation of the nonlinear dynamics of memristor-based oscillators is a key step toward the applications of circuit systems based upon them [42]. A wide range of complex dynamic behavior, including chaos, has been observed even in simple circuit networks of memristor oscillators [7,42,43]. In Reference 43, a controller consisting of a Twin-T notch filter and a voltage-to-current converter is designed and employed to control the chaotic oscillation in a memristor-based nonlinear circuit. Memristors have been adopted to construct Wien and phase-shift oscillators in References 44,45, of which the results show that although the poles of the system oscillate, sustained oscillation is maintained owing to the properties of the memristors. By taking advantage of memristor characteristics, the first reactance-less oscillator without the requirement of any capacitor or inductor is proposed in Reference 46. In addition, generalized analysis of symmetric and asymmetric memristive two-gate relaxation oscillators is further presented in Reference 47. However, due to the lack of a practical floating memristor emulator, research related to memristive oscillators has been mainly carried out on the basis of theoretical and simulation analysis. A floating memristor emulator based relaxation oscillator has been analytically probed and experimentally tested in Reference 48.

The astable multivibrators (AMs) are primary components in a variety of electronic systems, due to its implementation convenience. In consideration of the memristor's unique properties, it is of high significance to display the realization possibilities when using memristors to structure multivibrators. In this chapter, the schematic of a floating memristor emulator is first theoretically analyzed and utilized to structure an AM based on a 555 timer. The floating memristor emulator is then modified into a memcapacitor emulator in order to achieve a memcapacitive AM on the basis of two NOT gates. The proposed multivibrators are validated by PSPICE simulation.

4.2 Circuit schematic of floating memristor emulator

Memristors can be classified as charge dependent or flux dependent, and also can be represented by two nonlinear functions $M(q)$ and $W(\varphi)$, called the *memristance* and *memductance*, respectively [1]. Here, an analog circuit, which can emulate the flux-dependent memristor emulator, is constructed by making use of four current conveyors, one operation amplifier, one multiplier, one capacitor, and several resistors. As shown in Figure 4.1, these four current conveyors are labeled U1, U2, U3, and U4. Based on the inherent input–output current performance of AD844, the terminal voltage v_m and current i_m can be expressed as:

$$v_m = i_1 R_2 = i_2 R_2, \quad i_m = \frac{v_{r9}}{R_9} = \frac{v_w}{R_9}, \tag{4.1}$$

where i_1 and i_2 are the current outputs from terminals x and z of U1, respectively, while v_{r9} is the voltage across resistor R_9 and v_w is the output voltage of multiplier.

Capacitor C_1 is the only energy-storage element adopted in this emulator design to provide the integral operation. The charge going through C_1 can be calculated by:

$$q_1 = \frac{\varphi_m}{R_2} = -C_1 v_{c1}, \tag{4.2}$$

where φ_m is the time integral of v_m, while v_{c1} denotes the voltage across capacitor C_1.

Operational amplifier U5 is introduced to realize an inverting adder circuit (IAC). Referring to the configuration of AD844, voltage of terminal z is transmitted to the

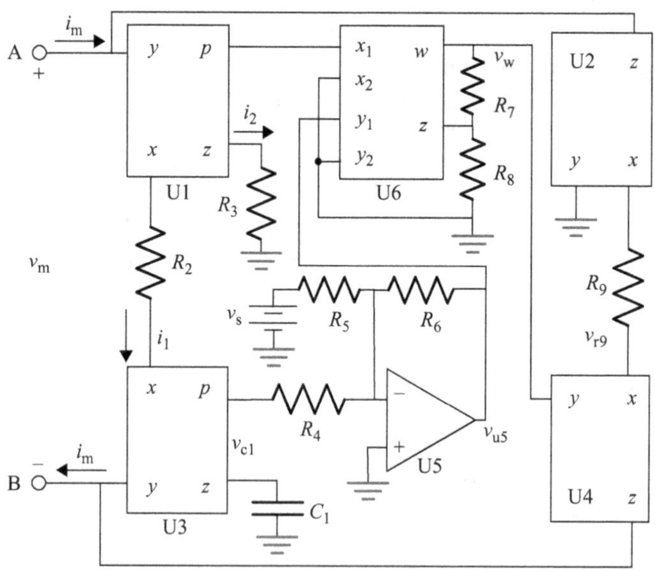

Figure 4.1 Circuit schematic of the floating memristor emulator

output terminal p via an internal voltage follower. Hence, the output voltage of U5 can be described as:

$$v_{u5} = -\left(\frac{R_6}{R_4}v_{c1} + \frac{R_6}{R_5}v_s\right), \tag{4.3}$$

where v_s is an adjustable direct-current voltage.

According to the input-to-output function of U6 (AD633), v_w can be calculated by

$$v_w = \frac{R_3(R_7 + R_8)}{10R_2R_7}v_m v_{u5}. \tag{4.4}$$

By combining (4.1)–(4.4), the current going through the emulator can be rewritten as:

$$i_m = v_m \cdot \frac{R_3(R_7 + R_8)}{10R_2R_7R_9}\left(\frac{R_6}{R_4}\frac{1}{R_2C_1}\varphi_{AB} - \frac{R_6}{R_5}v_s\right). \tag{4.5}$$

The flux-controlled *memductance* can therefore be mathematically derived from (4.5) as:

$$W(\varphi_m) = \alpha\varphi_m + \beta. \tag{4.6}$$

where parameters α and β are:

$$\alpha = \frac{R_3R_6(R_7 + R_8)}{10R_2^2R_4C_1R_7R_9}, \quad \beta = -\frac{R_3R_6(R_7 + R_8)}{10R_2R_5R_7R_9}v_s.$$

Assuming that U1, U2, U4, and U6 avoid saturation by properly configuring R_2, R_3, R_7, and R_8, the output voltage of U5 must be restricted to (0, 10) V to ensure positive memductance and effective operation of the AD633. Hence, the following condition must hold:

$$\frac{R_2R_4C_1}{R_5}v_s < \varphi_m < \frac{10R_2R_4C_1}{R_6} + \frac{R_2R_4C_1}{R_5}v_s. \tag{4.7}$$

Equation (4.7) infers that this memristor emulator is conditionally valid in relation to terminal voltage v_m and parameters C_1, R_2, R_4, R_5, R_6, and v_s. For the case of sinusoidal input voltage, $v_m = V_m\sin(2\pi ft)$, the *memductance* can be calculated as:

$$W(V_m, f) = -\alpha\frac{V_m}{2\pi f}\cos(2\pi ft) + \beta. \tag{4.8}$$

Assuming V_{u3} is the maximal output voltage of U3 and $\beta/\alpha \le V_{u3}$, V_m and f must satisfy the following condition to guarantee positive *memductance*:

$$\frac{V_m}{f} < \frac{2\pi\beta}{\alpha}. \tag{4.9}$$

Based on (4.8) and (4.9), the value interval of W can be deduced by:

$$0 < W(\varphi_m) < 2\beta. \tag{4.10}$$

For the case of DC input voltage, φ_m will eventually evolve into a constant due to the output saturation of U3, namely, $-V_{u3} \le \varphi_m \le V_{u3}$. Note that, V_{u3} can be adjusted by altering the voltage amplitude of the power supply or by adding an external voltage limiter (e.g. Zener diodes). Assuming that V_{u3} is suitably configured to ensure unsaturated φ_m, satisfying (4.7), all possible values of *memductance* can be presented by:

$$\beta - \alpha V_{u3} < W(\varphi_m) < \beta + \alpha V_{u3}. \tag{4.11}$$

On the basis of the preceding analysis, either of the two terminals can be arbitrarily connected to ground, which reflects that this newly proposed circuit can be utilized to emulate the floating memristor without grounded restriction. This feature of connection flexibility will definitely enable a wide range of practical applications using this emulator.

4.3 Theoretical analysis of memristive astable multivibrator

An AM is an oscillator circuit, which can continuously produce a rectangular wave without the aid of external triggering. Internal circuitry with external connections of the proposed memristive AM is shown in Figure 4.2.

The details of this memristive AM circuit are given below considering the pins of 555 timer. Pin 1 is connected to power supply V_{ee}; pins 4 and 8 are shorted and tied to power supply V_{cc}, rectangular wave v_o is taken from pin 3; pins 2 and 6 are shorted and connected to V_{ee} through capacitor C; pin 7 is connected to supply V_{cc} through a resistor R_1. Between pin 6 and pin 7 a memristor R_M is inserted. Inside the 555 timer,

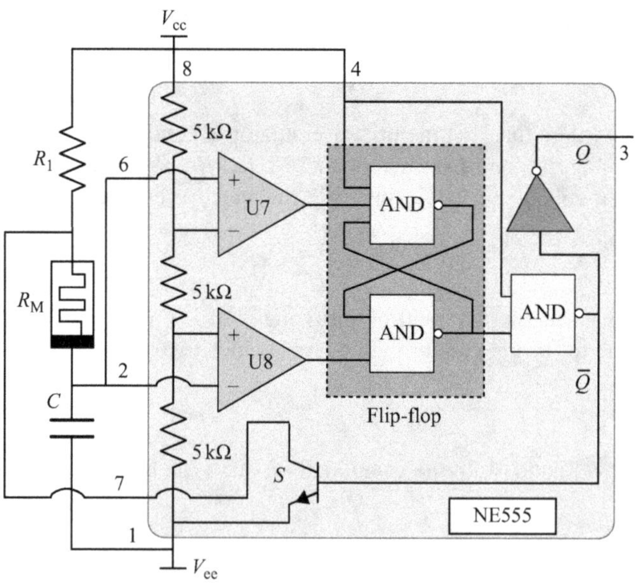

Figure 4.2 Memristive astable multivibrator

three 5 kΩ resistors are adopted to structure a voltage divider. Hence, the threshold voltages for flipping the output level are constants and can be calculated by:

$$v_{\mathrm{h}} = \frac{2}{3}V_{\mathrm{cc}} + \frac{1}{3}V_{\mathrm{ee}}, \quad v_{\mathrm{l}} = \frac{1}{3}V_{\mathrm{cc}} + \frac{2}{3}V_{\mathrm{ee}}. \tag{4.12}$$

Note that each oscillating period can be divided into two intervals with regards to the operating states of capacitor *C*.

4.3.1 Discharging interval

As shown in Figure 4.3(a), at the moment when output voltage level is flipped from high to low, the transistor *S* turns on and capacitor *C* begins discharging toward V_{ee} through R_{M}. In accordance with Kirchhoff's current law (KCL), the voltage across capacitor *C* can be derived as:

$$v_{\mathrm{c}} = \frac{1}{C}\int v_{\mathrm{m}} W(\varphi_{\mathrm{m}})\, dt = \frac{1}{2C}\alpha\varphi_{\mathrm{m}}^{2} + \frac{1}{C}\beta\varphi_{\mathrm{m}}. \tag{4.13}$$

Based on Kirchhoff's voltage law (KVL), the following equation holds:

$$v_{\mathrm{r}} + v_{\mathrm{m}} + v_{\mathrm{c}} = V_{\mathrm{cc}} - V_{\mathrm{ee}}. \tag{4.14}$$

By combining (4.13) with (4.14), we have:

$$\frac{d\varphi_{\mathrm{m}}}{dt} = V_{\mathrm{ee}} - \frac{1}{2C}\alpha\varphi_{\mathrm{m}}^{2} - \frac{1}{C}\beta\varphi_{\mathrm{m}}. \tag{4.15}$$

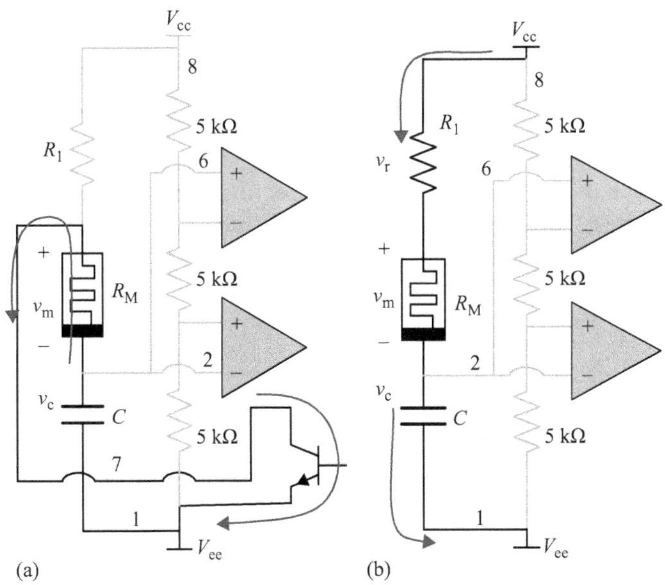

Figure 4.3 Operating states of AM

Here, φ_m can be analytically solved, namely:

$$\varphi_{\mathrm{m}} = \frac{C}{\alpha}\left(-\frac{\beta}{C} + \lambda_1 \tanh\left(\frac{\lambda_1}{2}t + \frac{1}{2}p_1\lambda_1\right)\right), \tag{4.16}$$

where tanh(·) is the hyperbolic tangent function and p_1 is a constant which depends on the initial value of φ_{m} or v_{m}, and:

$$\lambda_1 = \sqrt{\frac{2\alpha V_{\mathrm{ee}}}{C} + \frac{\beta^2}{C^2}}.$$

The voltage v_{m} across the memristor emulator can be easily obtained by carrying out differential operations on (4.16):

$$v_{\mathrm{m}} = \frac{C\lambda_1^2}{2\alpha}\left\{1 - \left[\tanh\left(\frac{\lambda_1}{2}t + \frac{1}{2}p_1\lambda_1\right)\right]^2\right\}. \tag{4.17}$$

It should be noted that, at the starting point of each discharging interval, v_{c} is equivalent to v_{h} and the initial value of v_{m} is $V_{\mathrm{ee}} - v_{\mathrm{h}}$. During this discharging process, v_{c} is decreased and hence, v_{m} should be monotonically increased with time, which infers that p_1 in (4.17) should be a positive constant, namely:

$$p_1 = \frac{2}{\lambda_1}\operatorname{arctan\,h}\sqrt{1 - \frac{4\alpha}{3C\lambda_1^2}(V_{\mathrm{ee}} - V_{\mathrm{cc}})}, \tag{4.18}$$

where arctanh(·) is the inverse hyperbolic tangent function. In discharging interval, v_{c} will eventually reach the lower threshold v_{l}. Hence, the time duration of the discharging interval can be obtained by:

$$t_{\mathrm{d}} = -p_1 + \frac{2}{\lambda_1}\operatorname{arctan\,h}\sqrt{1 - \frac{2\alpha}{3C\lambda_1^2}(V_{\mathrm{ee}} - V_{\mathrm{cc}})}. \tag{4.19}$$

4.3.2 Charging interval

When output v_{o} is varied from low to high level, the inside transistor S is switched off and the capacitor C is charged toward V_{cc} through R_1 and R_{M}, as shown in Figure 4.3(b). During this charging interval, voltage v_{c} can be derived as:

$$v_{\mathrm{c}} = \frac{1}{C}\int v_{\mathrm{m}} W(\varphi_{\mathrm{m}})\,dt = \frac{1}{2C}\alpha\varphi_{\mathrm{m}}^2 + \frac{1}{C}\beta\varphi_{\mathrm{m}}. \tag{4.20}$$

Likewise, referring to (4.14), the following equation holds:

$$\frac{d\varphi_{\mathrm{m}}}{dt} = \frac{V_{\mathrm{cc}}}{1 + R_1\alpha\varphi_{\mathrm{m}} + R_1\beta} - \frac{\alpha\varphi_{\mathrm{m}}^2 + 2\beta\varphi_{\mathrm{m}}}{2C(1 + R_1\alpha\varphi_{\mathrm{m}} + R_1\beta)}. \tag{4.21}$$

Unlike (4.15), (4.21) possesses much higher complication and can hardly be analytically solved. In consideration of voltage v_{c} increasing from initial value v_{l} to final value v_{h}, (4.21) can only be solved by taking advantage of numerical methods. The time duration of charging interval has no mathematical solution as well and hence numerical methods are required to obtain the oscillating period and duty cycle.

It should be stressed that in order to obtain the expected oscillating output, the circuit parameters must be properly configured to guarantee the results of square roots are in real numbers. In fact, improper parameter configuration leading to complex numbers indicates that the circuit is incapable of operating as AM.

4.4 Simulation validation for memristive astable multivibrator

In order to perform a comparative investigation with theoretical analysis, the PSPICE simulation data are collected and then transferred into OriginPro 8.0 software to draw the oscillating curves of state variables.

4.4.1 Testing the memristor emulator

The parameters for testing the floating memristor emulator are configured as: $R_2 = R_3 = 51\ \text{k}\Omega, R_4 = 10\ \text{k}\Omega, R_5 = 53\ \text{k}\Omega, R_6 = 10\ \text{k}\Omega, R_8 = 10\ \text{k}\Omega, R_9 = 90\ \text{k}\Omega, R_{10} = 20\ \text{k}\Omega, C_1 = 100$ nF, $v_s = -15$ V, and $V_{u3} = 15$ V. Adhikari *et al.* suggested tests to determine if a device can properly be categorized as a memristor [49]. The Lissajous curve in the voltage–current plane is a pinched hysteresis loop when driven by any bipolar and periodic voltage or current without respect to initial conditions. An AC voltage $v_{AB} = 1.5\sin(2\pi ft)$ is used to further test this hardware version of memristor emulator. Since i_m is proportional to v_{r9} with coefficient R_9, v_{r9} is used to characterize current i_m. Figure 4.4 displays three Lissajous curves of v_{r9} versus v_{AB} at $f = 20$ Hz, $f = 40$ Hz, and $f = 120$ Hz, respectively. Figure 4.4 clearly shows that all these three Lissajous curves in the voltage–current plane behave in pinched hysteresis loop as an inclined "8" passing through origin, of which the slopes at various points are

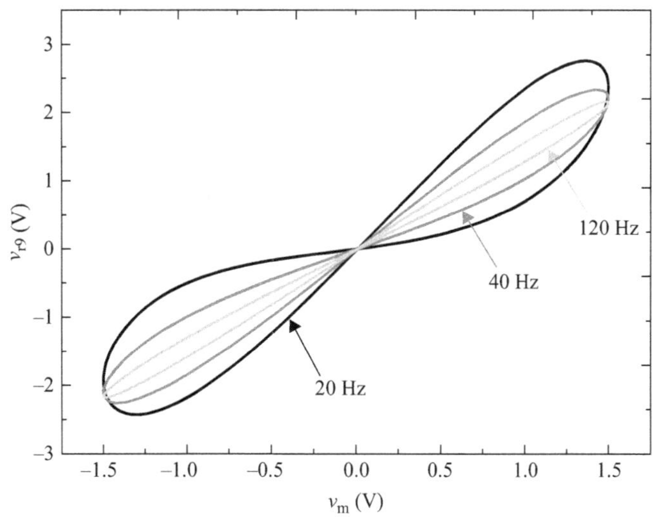

Figure 4.4 Pinched hysteresis loops

the memductances. This experimental result is also in good agreement with (4.6). Distinctly, the pinched hysteresis loop of $f = 120$ Hz is obviously shrunk as compared to the loop of $f = 20$ Hz. In fact, this hysteresis loop will be narrowed into a line segment as the imposed frequency increases toward infinity. This appearance reflects the unique characteristic of a memristor, namely, the frequency dependence of pinched hysteresis loops.

4.4.2 Observing the oscillation of memristive astable multivibrator

The parameters for implementing the floating memristor emulator are chosen as: $C_1 = 1$ μF, $R_2 = 100$ kΩ, $R_4 = 50$ kΩ, $R_3 = 50$ kΩ, $R_5 = 10$ kΩ, $R_6 = 10$ kΩ, $R_7 = 10$ kΩ, $R_8 = 40$ kΩ, $R_9 = 5$ kΩ, and $v_s = -15$ V. Under this parameter configuration, the floating emulator is supposed to possess the same dynamical characteristics as the memristor. To implement the memristive AM, the power supply voltages are configured as $V_{cc} = 5$ V and $V_{ee} = -5$ V, while $R_1 = 8$ kΩ and $C = 20$ μF. The oscillating voltages of v_r, v_m, and v_c are captured and presented in Figure 4.5 for discussion. It can be seen that all the voltage waveforms behave in stable periodicity.

With regards to each oscillating period, the waveforms can be divided into two intervals, corresponding to the charging and discharging states of capacitor C. During the charging interval, voltages v_r and v_m are both positive and decrease along with the increment of v_c. When v_c reaches the upper threshold voltage v_h, the transistor S inside the 555 timer is switched on and the capacitor starts to discharge through the memristor. Hence, the voltage across memristor evolves sharply into negative and then increases along with the decrement of v_c. Note that during the discharging interval, the voltage across resistor R_1 is a constant and equal to $V_{cc} - V_{ee}$. It can

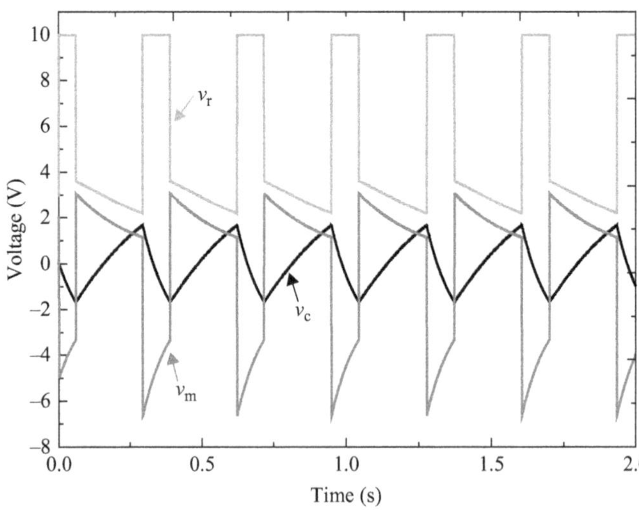

Figure 4.5 Simulated waveforms of voltages v_r, v_m, and v_c

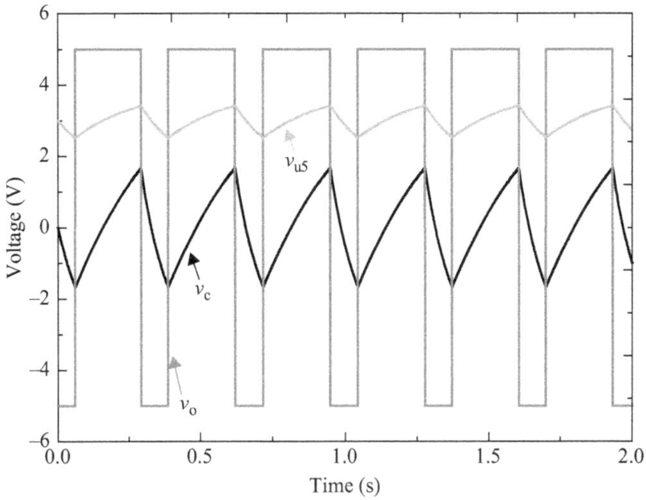

Figure 4.6 Simulated waveforms of voltages v_o, v_{u5}, and v_c

be clearly seen that the time interval of charging is much wider than the discharging interval. This is because resistor R_1 is no longer involved in the discharging operation and the discharging current only runs through the memristor and capacitor.

Equations (4.2) and (4.3) indicate that the memductance W_m is in proportion to the output voltage of amplifier U5. Hence, v_{u5} is captured here to depict the dynamic behaviors of W_m. Based on the above given memristor parameters, it can be calculated that the proportional coefficient is equal to 2×10^4 V·Ω. Simulated waveforms of voltages v_o and v_{u5} are shown in Figure 4.6, where v_c is sampled again for reference purpose. The rectangular waveform is the oscillating signal from pin 3, which maintains the high level of 5 V during the charging interval and steps into low level of −5 V when capacitor is discharged. The curve of v_{u5} periodically varies along with the output waveform, from 2.65 to 3.45 V. Hence, the variation range of memductance can be calculated as [0.1325, 0.1725] mS, which is in good agreement with (4.6).

Since the voltage across resistor R_9 is proportional to the current going through the memristor, v_{r9} can thus be adopted to represent i_m. The Lissajous curves of v_m versus v_{r9} are depicted in Figure 4.7, while capacitor C is configured as 5, 20, and 60 μF, respectively. Corresponding to these three capacitance values it can be numerically calculated that the output oscillating frequencies are 11.90, 3.05, and 0.93 Hz. Figure 4.7 clearly shows that all these three Lissajous curves in the voltage–current plane behave in pinched hysteresis loop passing through the origin, of which the slopes at various points are the memductances. The pinched hysteresis loop of 11.90 Hz ($C = 5$ μF) is obviously shrunk as compared to the loop of 0.93 Hz ($C = 60$ μF). This appearance reflects the unique characteristic of a memristor, namely, the frequency dependence of pinched hysteresis loops.

When symmetric voltage or current is employed to excite a memristor, the pinched hysteresis loop is symmetric in terms of origin, which has been disclosed in many

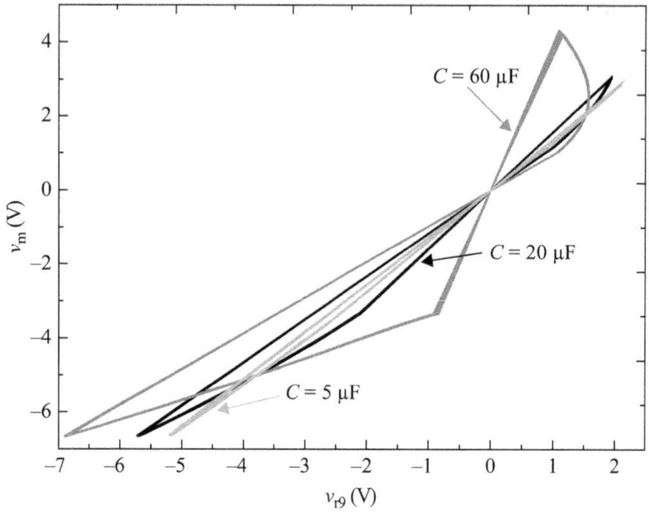

Figure 4.7 Lissajous curves of v_{r9} and v_m with different capacitance

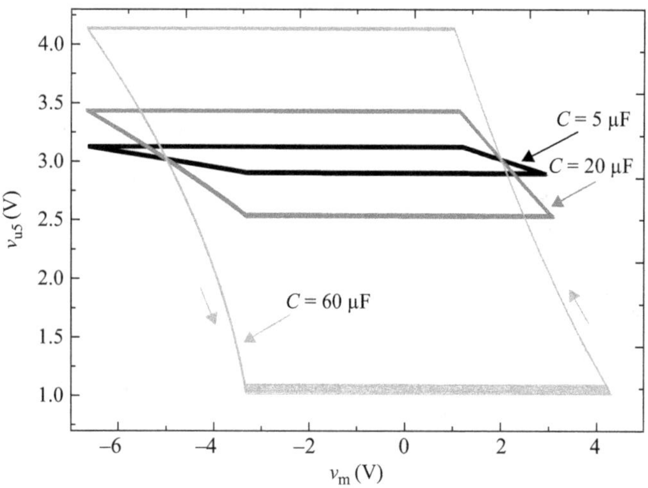

Figure 4.8 Lissajous curves of v_m and v_{u5} with different capacitance

preceding research papers. However, in the case of AM, corresponding to different output frequency, the area enclosed by the v_{r9} versus v_m curve in the first quadrant is less than that of the third quadrant, due to the voltage across the memristor is no longer symmetrical. The charging interval with smaller positive amplitude sustains clearly longer than the discharging interval with higher negative amplitude. This result is in good agreement with v_m curve as shown in Figure 4.5.

The simulated waveforms of v_m and W (v_{u5} instead) are exhibited in Figure 4.8, which indicate that W is positively altering with terminal voltage v_m, and W increases by positive v_m and decreases when v_m is negative. By taking the case of $C = 60\ \mu F$ for demonstration, it can be seen that although v_m can arbitrarily jump from −3.33 to 4.25 V or 1.02 to −6.66 V, the value of W cannot be changed abruptly. This is because W is dependent on the time integral of v_m, which shows good agreement with theoretical analysis of (4.6). The altering interval of v_{u5} can be measured as [1.10, 4.15] V, hence the variation interval of W can be calculated as [0.055, 0.207] mS. The measured Lissajous curve of v_m versus v_{u5} behaves as an irregular rectangle. Due to the dissymmetry of v_m, W develops an asymmetric shape in terms of $v_m = 0$ V.

4.5 Memcapacitor-based astable oscillator circuit

The memcapacitor links the flux $\varphi(t)$ and the time integral of the charge $\sigma(t)$, and has been shown to provide advances in nonvolatile memory-based devices. Memcapacitors open up a new area of possibilities and functionalities in electronics, and along with the other mem-elements, have been associated with neuromorphic devices to simulate adaptive behavior. Since there are very few works related to memcapacitive astable oscillator, this section discusses the feasibilities and characteristics of memcapacitor-based astable oscillator. The inclusion of the memcapacitor in the astable oscillator provides a new option allowing for large degree of control over frequency across a given range of circuit element values as well as an adjustable duty ratio.

The memcapacitor used in this analysis is based upon a compact mutator which is modified from the memristor emulator described in Figure 4.1 by replacing the resistor R_9 with a capacitor C_2, as shown in Figure 4.9. The equivalent flux-dependent memcapacitance can be computed by:

$$C_M(\varphi_m) = \frac{R_3R_6(R_7+R_8)C_2}{10R_2^2R_4C_1R_7}\varphi_m - \frac{R_3R_6(R_7+R_8)C_2}{10R_2R_5R_7}v_s. \tag{4.22}$$

Given there is a constant flux φ_m, C_M and C_2 are expected to retain a linear relationship.

The memcapacitive oscillator circuit consists of two NOT gates, a feedback system and one output terminal as shown in Figure 4.10 by connecting switch S to node a. A resistor with high resistance, R_{a1}, is introduced in order to restrict the current going into the NOT gate. The output voltages of the NOT gates are either 0 V or at the high voltage level of V_c, where V_c is the power supply voltage. This circuit possesses two repeatable states according to the charging or discharging operation of memcapacitor C_M. Assume that V_{Hth} and V_{Lth} are the threshold voltages for NOT gates to flip between high and low levels. (a) During the charging state, the voltage v_f is the same as the voltage v_c across the memcapacitor. Moreover, v_f is increased from $-V_{Lth}$ to V_{Lth} and the output voltage v_o is retained at low level until v_f reaches V_{Lth}. At the moment of $v_f = V_{Lth}$, the output voltage v_o flips from low to high level and the circuit steps into discharging state. (b) During the discharging state, v_f is decreased from $V_c + V_{Lth}$ toward V_{Hth} and the output voltage v_o is retained at high level until v_f reaches V_{Hth}. When $v_f = V_{Hth}$, the circuit evolves back into charging operation and

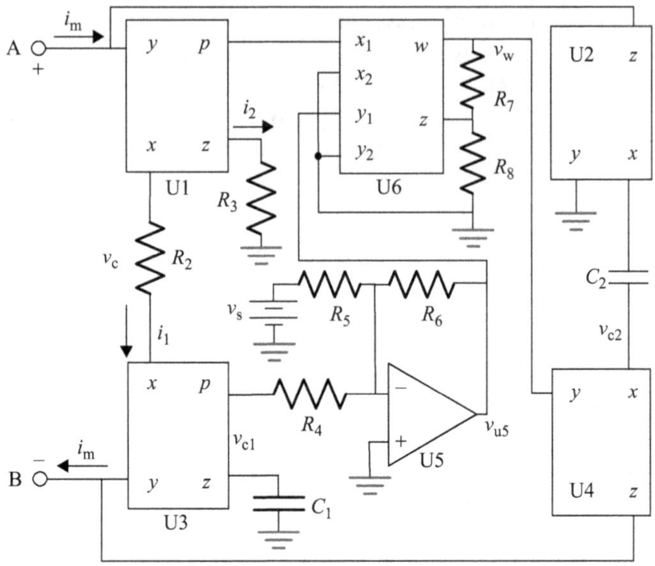

Figure 4.9 Floating memcapacitor emulator used in the oscillator circuit

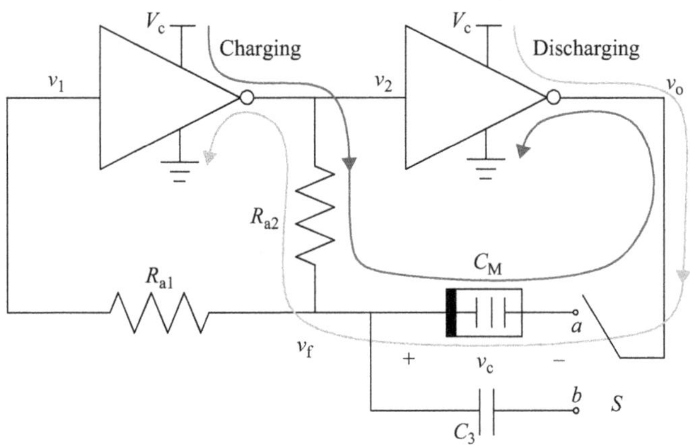

Figure 4.10 The proposed memcapacitor-based astable oscillator

the output voltage v_o varies from high to low level. Therefore, the frequency of the proposed circuit is highly related to the voltage v_c across the memcapacitor.

The output frequency of the astable oscillator circuit can be calculated using the following equation:

$$f_{output} = \frac{k}{R_{a2}C_M}, \quad (4.23)$$

where k is a constant ascertainable via experimental deduction, and C_M is the memcapacitance. Assuming k does not change between the memcapacitative and non-memcapacitative circuits, we expect to find a relationship linking C_2 and output frequency f which will highlight both the inverse nature of it, as well as the ability to fine-tune the desired frequency with greater control.

4.6 Simulation validation for memcapacitive astable multivibrator

The constant k is calculated for the circuit where the memcapacitor is replaced with a capacitor C_3 by connecting S to node b and is found to be $k = 1.18$ with an error of approximately 1% based on a sweep of C_3 values.

Note that, the voltage across capacitor C_2 of Figure 4.9 is in fact proportional to the charge going through the memcapacitor emulator, hence voltages $-v_c$ and v_{c2} can be used to display behavior characteristics of such a mem-element. The pinched hysteretic loop indicates the frequency-dependent feature of the memcapacitor, which also reveals the controllability of memcapacitance. The curve of variation of v_{c2} and v_c is plotted in Figure 4.11(a) under the condition of $f = 3.82$ Hz attained with the values $C_2 = 1\ \mu$F, $R_{a2} = 50\ \text{k}\Omega, R_{a1} = 1{,}000\ \text{k}\Omega, C_1 = 0.2\ \mu$F, $R_2 = 25\ \text{k}\Omega, R_3 = 51\ \text{k}\Omega, R_4 = 50\ \text{k}\Omega, R_5 = 29\ \text{k}\Omega, R_6 = 2\ \text{k}\Omega, R_7 = 10\ \text{k}\Omega$, and $R_8 = 90\ \text{k}\Omega$. It can be seen that the pinched hysteresis loop is irregular but symmetric and always passes through the origin.

The voltage waveforms of v_f, v_c, and v_o are exhibited in Figure 4.11(b), of which the discontinuous points of v_c and v_f are associated with the changes of the output, therefore confirming that the memcapacitative element must meet a threshold amount of stored potential before activating the effect of the NOT gate.

In order to further probe this AM circuit, voltage waveforms of v_{c2}, v_c, and v_{u5} are presented in Figure 4.12, where v_{u5} is the output voltage of amplifier U5. Note that v_{u5} is in proportion to the memcapacitance of C_M and hence is employed to depict the dynamic behavior of memcapacitance along with the variation of charge and terminal voltage v_c. It can be observed from Figure 4.12 that the memcapacitance is always positive, and can be increased by positive v_c and charge, and can be decreased when v_c and charge are negative.

As provided in the theoretical analysis, an increase in C_2 gives rise to an increase in C_M, which in turn produces a lower output frequency f. The oscillating frequencies are numerically calculated with various values of C_2 applied in order to affirm the expected relationship. The effect of C_2 on the output frequency f is shown in Figure 4.13, and where $R_{a1} = 1{,}000\ \text{k}\Omega, R_{a2} = 50\ \text{k}\Omega$, f is given by the following equation:

$$f_{\text{output}} = \frac{2.174 \times 10^{-6}}{C_2 + 1.264 \times 10^{-7}}. \tag{4.24}$$

By assuming the calculated constant $k = 1.18$ is consistent in the case where C_3 is replaced with the memcapacitor, the average value of C_M during the stable portion

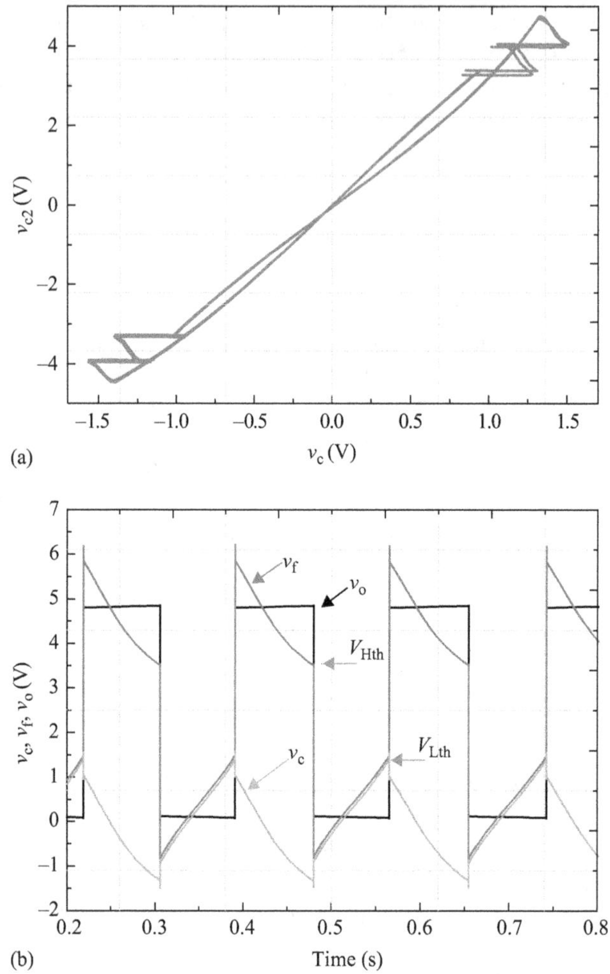

Figure 4.11 Behavior of the memcapacitor. Panel (a) shows measured pinched hysteretic loop and panel (b) shows voltage waveforms of v_f, v_c, *and* v_o

of oscillation can also be calculated using (4.23). The slightly nonlinear potential across C_M indicates the capacitance changes throughout oscillation due to fundamental principles. A constant of proportionality can thus be derived in order to calculate an average value for $C_{M(average)}$ from any selected value of C_2.

$$C_{M(average)} = \frac{k}{R_{a2}} T_{output}. \tag{4.25}$$

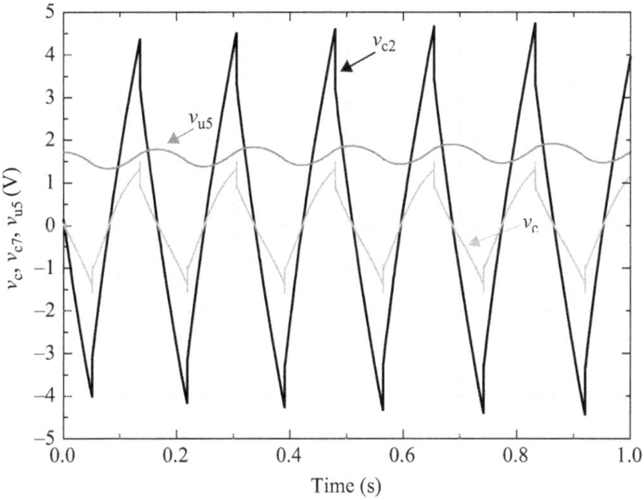

Figure 4.12 Voltage waveforms of v_{c2}, v_c, and v_{u5}

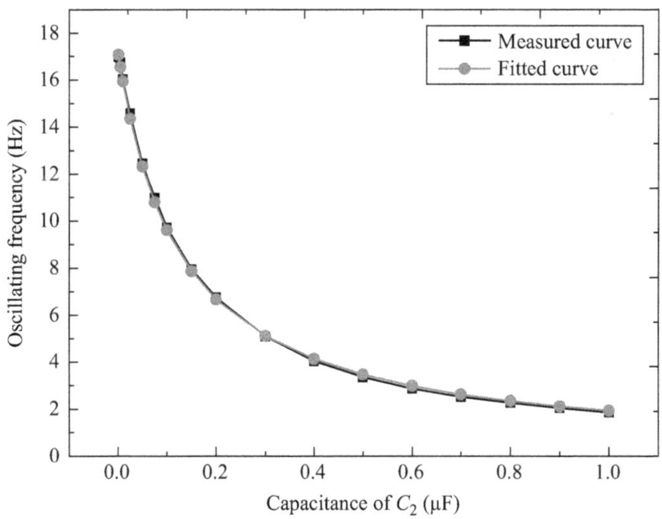

Figure 4.13 Altering the behavior of the astable oscillator by changing C_2 in the memcapacitor emulator

The error associated with this calculation arises due to the changing value of the memcapacitance which has been presumed to be constant for the purposes of (4.25), and the constant value of k presumed which actually fluctuates by 1% depending on the chosen value of C_2.

4.7 Conclusion

In this chapter, a floating flux controlled memristor emulator is presented and then utilized in AM based on 555 timer. The most attractive contribution of this new multivibrator circuit is its simplicity and potential practical application in electronic circuits. Also, this AM can be easily extended to memristive bistable and monostable multivibrators, which could be widely used in counting circuits, memory storage, frequency division, timers, delay circuits, and so on. Then, the memristor emulator is modified into a floating memcapacitor emulator by replacing a resistor with a regular capacitor. A novel memcapacitor-based oscillator is presented with a mathematical analysis and verification by circuit simulation. Low-frequency applications retain the memcapacitative properties which prevent a small range of parameters outputting an excessive range of frequencies, thus allowing for high degree of control over the output frequency, and the ability to fine-tune a desired frequency with simplicity. All the materials used in these designs are common off-the-shelf components and there is no requirement for any special or expensive components. The theoretical calculation and PSPICE simulation results manifest that these newly proposed AM circuits are capable of providing reliable output performance.

References

[1] L.O. Chua, "Memristor – the missing circuit element," *IEEE Trans. Circuit Theory*, vol. CT-18, no. 5, pp. 507–519, Sep. 1971.

[2] D.B. Strukov, G.S. Snider, D.R. Stewart, and R.S. Williams, "The missing memristor found," *Nature*, vol. 453, pp. 80–83, May 2008.

[3] M. Di Ventra, Y.V. Pershin, and L.O. Chua, "Circuit elements with memory: memristors, memcapacitors and meminductors," *Proc. IEEE*, vol. 97, no. 10, pp. 1717–1724, Oct. 2009.

[4] H. Kim, M.P. Sah, C. Yang, T. Roska, and L.O. Chua, "Memristor bridge synapses," *Proc. IEEE*, vol. 100, no. 6, pp. 2061–2070, Jun. 2012.

[5] Y.V. Pershin and M. Di Ventra, "Neuromorphic, digital, and quantum computation with memory circuit elements," *Proc. IEEE*, vol. 100, no. 6, pp. 2071–2080, Jun. 2012.

[6] F. Merrikh-Bayat, S. Bagheri-Shouraki, and A. Rohani, "Memristor crossbar-based hardware implementation of IDS method," *IEEE Trans. Fuzzy Syst.*, vol. 19, no. 6, pp. 1083–1096, Dec. 2011.

[7] B. Muthuswamy, "Implementing memristor based chaotic circuits," *Int. J. Bifurcat. Chaos*, vol. 20, no. 5, pp. 1335–1350, May 2010.

[8] S. Shin, K. Kim, and S.M. Kang, "Resistive computing: memristors-enabled signal multiplication," *IEEE Trans. Circuits Syst. I: Reg. Papers*, vol. 60, no. 5, pp. 1241–1249, May 2013.

[9] S.H. Jo, T. Chang, I. Ebong, B.B. Bhadviya, P. Mazumder, and W. Lu, "Nanoscale memristor device as synapse in neuromorphic systems," *Nano Lett.*, vol. 10, pp. 1297–1301, Mar. 2010.

[10] J. Borghetti, G.S. Snider, P.J. Kuekes, J.J. Yang, D.R. Stewart, and R.S. Williams, "'Memristive' switches enable 'stateful' logic operations via material implication," *Nature*, vol. 464, pp. 873–876, Apr. 2010.

[11] M.E. Fouda, M.A. Khatib, A.G. Mosad, and A.G. Radwan, "Generalized analysis of symmetric and asymmetric memristive two-gate relaxation oscillators," *IEEE Trans. Circuits Syst. I: Reg. Papers*, vol. 60, no. 10, pp. 2701–2708, Oct. 2013.

[12] T. Driscoll, J. Quinn, S. Klein, *et al.*, "Memristive adaptive filters," *Appl. Phys. Lett.*, vol. 97, no. 9, pp. 093502, 2010.

[13] S. Benderli and T.A. Wey, "On SPICE macromodelling of TiO_2 memristors," *Electron. Lett.*, vol. 45, no. 7, pp. 377–379, Mar. 2009.

[14] Z. Biolek, D. Biolek, and V. Biolková, "SPICE model of memristor with nonlinear dopant drift," *Radioengineering*, vol. 18, no. 2, pp. 210–214, Jun. 2009.

[15] S. Shin and S.M. Kang, "Compact models for memristors based on charge-flux constitutive relationships," *IEEE Trans. Comput.-Aided Design Integr. Circuits Syst.*, vol. 29, no. 4, pp. 590–598, Apr. 2010.

[16] A. Rák and G. Cserey, "Macromodeling of the memristor in SPICE," *IEEE Trans. Comput.-Aided Design Integr. Circuits Syst.*, vol. 29, no. 4, pp. 632–636, Apr. 2010.

[17] D. Batas and H. Fiedler, "A memristor SPICE implementation and a new approach for magnetic flux controlled memristor modeling," *IEEE Trans. Nanotechnol.*, vol. 10, no. 2, pp. 250–255, Mar. 2011.

[18] Y.V. Pershin and M. Di Ventra, "Practical approach to programmable analog circuits with memristors," *IEEE Trans. Circuits Syst. I: Reg. Papers*, vol. 57, no. 8, pp. 1857–1864, Aug. 2010.

[19] D. Biolek, J. Bajer, V. Biolkova, and Z. Kolka, "Mutators for transforming nonlinear resistor into memristor," in *20th European Conference on Circuit Theory and Design*, August 2011, Linkoping, Sweden, pp. 488–491.

[20] A.L. Fitch, H.H.C. Iu, X.Y. Wang, V. Sreeram, and W.G. Qi, "Realization of an analog model of memristor based on light dependent resistor," in *IEEE International Symposium on Circuits and Systems*, Seoul, Korea, May 2012, pp. 1139–1142.

[21] J. Valsa, D. Biolek, and Z. Biolek, "An analogue model of the memristor," *Int. J. Numer. Model.*, vol. 24, no. 3, pp. 400–408, Aug. 2011.

[22] H. Kim, M.P. Sah, C. Yang, S. Cho, and L.O. Chua, "Memristor emulator for memristor circuit applications," *IEEE Trans. Circuits Syst. I: Reg. Papers*, vol. 59, no. 10, pp. 2422–2431, Oct. 2012.

[23] A.L. Fitch, H.H.C. Iu, and C.K. Tse, "Hardware memristor emulators," in A. Adamatzky and G. Chen (Eds.), *Chaos, CNN, Memristors and Beyond*, World Scientific, Singapore, pp. 540–547, Dec. 2012.

[24] S. Shin, L. Zheng, G. Weickhardt, S. Cho, and S.M. Kang, "Compact circuit model and hardware emulation for floating memristor devices," *IEEE Circuits Syst. Mag.*, vol. 13, no. 2, pp. 42–55, May 2013.

[25] G.Z. Gohen, Y.V. Pershin, and M. Di Ventra, "Lagrange formalism of memory circuit elements: classical and quantum formulations," *Phys. Rev. B*, vol. 85, 165428, 2012.

[26] M. Krems, Y.V. Pershin, and M. Di Ventra, "Ionic memcapacitive effects in nanopores," *Nano Lett.*, vol. 10, no. 7, pp. 2674–2678, Jun. 2010.

[27] J. Han, C. Song, and S. Gao, "Realization of the meminductor," *ACS Nano*, vol. 8, no. 10, pp. 10043–10047, 2014.

[28] Z. Hu, Y. Li, L. Jia, and J. Yu, "Chaos in a charge-controlled memcapacitor circuit," in *International Conference on Communications, Circuits and Systems*, July 2010, Chengdu, China, pp. 828–831.

[29] Z. Hu, Y. Li, L. Jia, and J. Yu, "Chaotic oscillator based on voltage-controlled memcapacitor," in *International Conference on Communications, Circuits and Systems*, July 2010, Chengdu, China, pp. 824–827.

[30] D. Biolek, Z. Biolek, and V. Biolkova, "PSPICE modeling of meminductor," *Analog Integr. Circ. Sig. Process*, vol. 66, no. 1, pp. 129–137, Jul. 2011.

[31] D. Biolek, Z. Biolek, and V. Biolkova, "SPICE modeling of memcapacitor," *Electron. Lett.*, vol. 46, no. 7, pp. 520–522, Apr. 2010.

[32] Y.V. Pershin and M. Di Ventra, "Memristive circuits simulate memcapacitors and meminductors," *Electron. Lett.*, vol. 46, no. 7, pp. 517–518, Apr. 2010.

[33] D. Biolek, V. Biolkova, and Z. Kolka, "Mutators simulating memcapacitors and meminductors," in *IEEE Asia-Pacific Conference on Circuits and Systems*, December 2010, Kuala Lumpur, Malaysia, pp. 800–803.

[34] D. Biolek and V. Biolkova, "Mutator for transforming memristor into memcapacitor," *Electron. Lett.*, vol. 46, no. 21, pp. 1428–1429, Oct. 2010.

[35] M.P. Sah, R.K. Budhathoki, C.J. Yang, and H. Kim, "Expandable circuits of mutator-based memcapacitor emulator," *Int. J. Bifurcation Chaos*, vol. 23, no. 5, 1330017, May 2013.

[36] X.Y. Wang, A.L. Fitch, H.H.C. Iu, and W.G. Qi, "Design of a memcapacitor emulator based on a memristor," *Phys. Lett. A*, vol. 376, no. 4, pp. 394–399, Nov. 2012.

[37] Y. Liang, D.S. Yu, and H. Chen, "A novel meminductor emulator based on analog circuits," *Acta Phys. Sin.*, vol. 62, no. 15, 158501, Aug. 2013.

[38] Y.V. Pershin and M. Di Ventra, "Emulation of floating memcapacitors and meminductors using current conveyors," *Electron. Lett.*, vol. 47, no. 4, pp. 243–244, Jan. 2011.

[39] D. Yu, Y. Liang, H. Chen, and H.H.C. Iu, "Design of a practical floating memcapacitor emulator without grounded restriction," *IEEE Trans. Circuits Syst. II: Express Briefs*, vol. 60, no. 4, pp. 207–211, Apr. 2013.

[40] M.E. Fouda and A.G. Radwan, "Charge controlled memristor-less memcapacitor emulator," *Electron. Lett.*, vol. 48, no. 23, pp. 1454–1455, Nov. 2012.

[41] Y. Liang, H. Chen, and D.S. Yu, "A practical implementation of a floating memristor-less meminductor emulator," *IEEE Trans. Circuits Syst. II: Express Briefs*, vol. 61, no. 5, pp. 209–303, May 2014.

[42] F. Corinto, A. Ascoli, and M. Gilli, "Nonlinear dynamics of memristor oscillators," *IEEE Trans. Circuits Syst. I: Reg. Papers*, vol. 58, no. 6, pp. 1323–1336, 2011.

[43] H.H.C. Iu, D. Yu, A. Fitch, V. Sreeram, and H. Chen, "Controlling chaos in a memristor based circuit using a twin-T notch filter," *IEEE Trans. Circuits Syst. I: Reg. Papers*, vol. 58, no. 6, pp. 1337–1344, Jun. 2011.

[44] A. Talukdar, A.G. Radwan, and K.N. Salama, "A memristor-based third-order oscillator: beyond oscillation," *Appl. Nanosci.*, vol. 1, no. 3, pp. 143–145, 2011.

[45] A. Talukdar, A.G. Radwan, and K.N. Salama, "Non-linear dynamics of memristor based third-order oscillatory system," *Microelectron. J*, vol. 43, no. 3, pp. 169–175, Mar. 2012.

[46] M.A. Zidan, H. Omran, A.G. Radwan, and K.N. Salama, "Memristor-based reactance-less oscillators," *Electron. Lett.*, vol. 47, no. 22, pp. 1220–1221, 2011.

[47] M.E. Fouda, M.A. Khatib, A.G. Mosad, and A.G. Radwan, "Generalized analysis of symmetric and asymmetric memristive two-gate relaxation oscillators," *IEEE Trans. Circuits Syst. I: Reg. Papers*, vol. 60, no. 10, pp. 2701–2708, Oct. 2013.

[48] D.S. Yu, H.H.C. Iu, A.L Fitch, and Y. Liang, "A floating memristor emulator based relaxation oscillator," *IEEE Trans. Circuits Syst. I: Reg. Papers*, vol. 61, no. 10, pp. 2888–2896, Oct. 2014.

[49] S.P. Adhikari, M.Pd. Sah, H. Kim, and L.O. Chua, "Three fingerprints of memristor," *IEEE Trans. Circuits Syst. I: Reg. Papers*, vol. 60, no. 11, pp. 3008–3021, Nov. 2013.

Chapter 5

Piecewise-constant oscillators and their applications

Tadashi Tsubone[1], *Keisuke Suzuki*[1] *and Takahiro Aoki*[1]

The analysis of nonlinear phenomena in continuous-time dynamics is one of important topics in the field of engineering, and it has attracted many researchers in recent years. The researchers have tried to solve this problem by simplifying the nonlinear dynamics. We introduce extremely simple oscillators whose dynamics are represented by piecewise-constant equations, and show two examples. One of them is a chaotic spiking oscillator with piecewise-constant vector field. We analytically prove the generation of chaos by using Poincaré map which is derived through a simple systematic procedure. Another is a coupled system of piecewise-constant oscillators. The parameter regions of in-phase and anti-phase synchronization are clarified by using a fast calculation algorithm. Some theoretical results are verified in the experimental circuits.

5.1 Basic concept of piecewise-constant oscillations

In this section, we introduce the mechanism that causes the oscillation of the piecewise-constant (PWC) system. First of all, to emphasize the some resemblance and different characters between linear systems and PWC systems, we consider the following two-dimensional linear dynamics:

$$\begin{aligned} \dot{x} &= y \\ \dot{y} &= -x + \tfrac{1}{a}y \end{aligned} \tag{5.1}$$

where "$\cdot$" represents the derivative of a normalized time τ. If $|\frac{1}{2a}| < 1$, (5.1) has complex-conjugate eigenvalues, and the eigenvalues are denoted by $\frac{1}{2a} \pm j\sqrt{1 - (\frac{1}{2a})^2}$. The solution of (5.1) is oscillatory as shown in Figure 5.1. The positive a value ($a > \frac{1}{2}$) realizes expanding oscillations, namely unstable focus, as shown in Figure 5.1(a), and the negative a ($a < -\frac{1}{2}$) value realizes converging oscillations, namely stable focus, as shown in Figure 5.1(b).

[1]Department of Electrical, Electronics and Information Engineering, Nagaoka University of Technology, Niigata, Japan

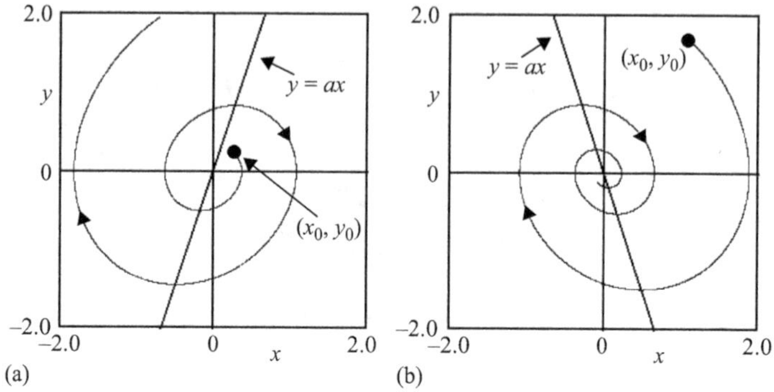

Figure 5.1 Oscillations of a linear system (5.1): (a) expanding oscillation in the case of $a > \frac{1}{2}$ ($a = 3$, $x_0 = 0.3$, $y_0 = 0.2$) and (b) converging oscillation in the case of $a < -\frac{1}{2}$ ($a = -3$, $x_0 = 1.1$, $y_0 = 1.7$)

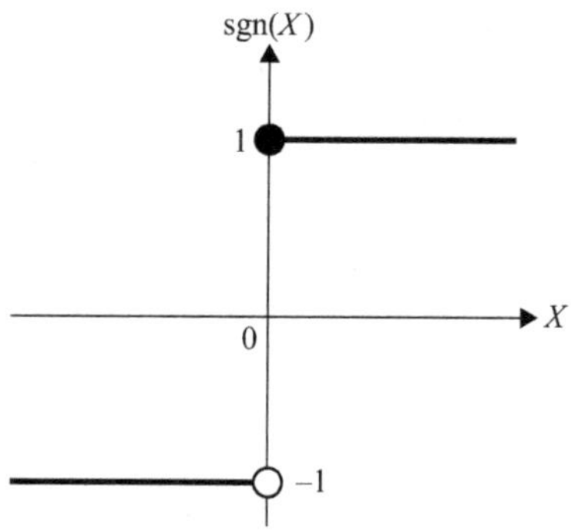

Figure 5.2 The characteristic of the signum function

Here, in order to explain the basic concept of piecewise-constant oscillations, we introduce the following nonlinear equations:

$$\begin{aligned} \dot{x} &= \mathrm{sgn}(y) \\ \dot{y} &= \mathrm{sgn}\left(-x + \frac{1}{a}y\right) \end{aligned} \tag{5.2}$$

where the function sgn(X) is a signum function defined as follows:

$$\mathrm{sgn}(X) = \begin{cases} 1 & \text{for } X \geq 0 \\ -1 & \text{for } X < 0 \end{cases} \tag{5.3}$$

The characteristic shape of the signum function is depicted as shown in Figure 5.2. It is noted that (5.2) is a significantly simplified form of the dynamics of (5.1). Because $\dot{x}$ and $\dot{y}$ take only constant values, 1 or -1, such dynamics is called piecewise-constant.

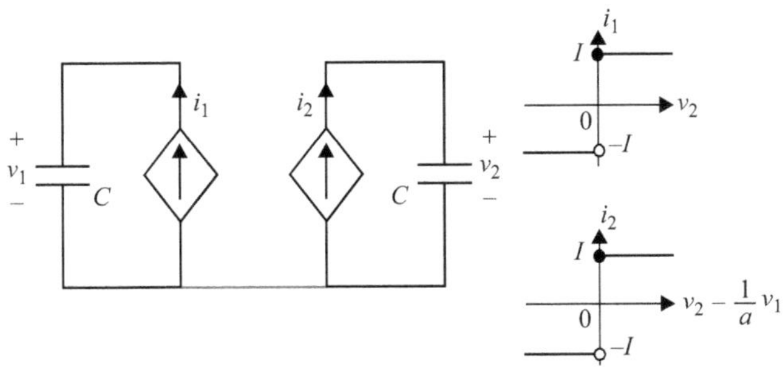

Figure 5.3 An example of circuit diagram for generating piecewise-constant oscillations

The piecewise-constant dynamics of (5.2) is realized by simple electrical circuits, such as an example shown in Figure 5.3. The circuit consists of only two capacitors and two voltage-controlled current sources (VCCSs) with signum characteristics. Applying Kirchhoff's voltage law, we obtain the circuit dynamics represented by:

$$\begin{aligned} C\frac{\mathrm{d}v_1}{\mathrm{d}t} &= I \cdot \mathrm{sgn}(v_2) \\ C\frac{\mathrm{d}v_2}{\mathrm{d}t} &= I \cdot \mathrm{sgn}\left(v_2 - \frac{1}{a}v_1\right) \end{aligned} \tag{5.4}$$

By using the following dimensionless and normalizing transformation:

$$\tau = \frac{I}{CE}t, \quad x = \frac{1}{E}v_1, \quad y = \frac{1}{E}v_2 \tag{5.5}$$

the circuit dynamics of (5.4) is equivalent to (5.2), where E is a dummy variable for rescaling the state variables. The solution of (5.2) represents piecewise-constant oscillation, which is similar to linear oscillation in Figure 5.1, as shown in Figure 5.4. Here, we assume the following parameter condition:

$$|a| > 1 \tag{5.6}$$

If $a > 1$, the system exhibits expanding oscillations with piecewise-linear trajectory as shown in Figure 5.4(a), and if $a < -1$, the system exhibits converging oscillations as shown in Figure 5.4(b).

Next, we introduce an another example of piecewise-constant oscillations. The dynamics is given by the following equations:

$$\begin{aligned} \dot{x} &= \mathrm{sgn}(y) \\ \dot{y} &= -\mathrm{sgn}(x) + \frac{1}{a}\mathrm{sgn}(y) \end{aligned} \tag{5.7}$$

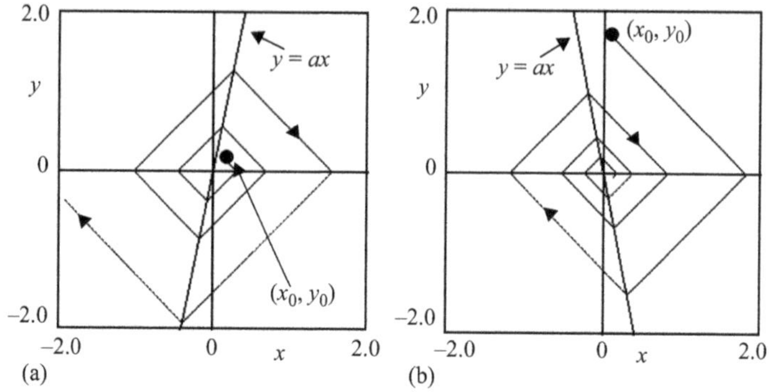

Figure 5.4 Oscillations of a PWC system (5.2): (a) expanding oscillation in the case of $a > 1$ ($a = 5$, $x_0 = 0.1$, $y_0 = 0.2$) and (b) converging oscillation in the case of $a < -1$ ($a = -5$, $x_0 = 0.1$, $y_0 = 1.7$)

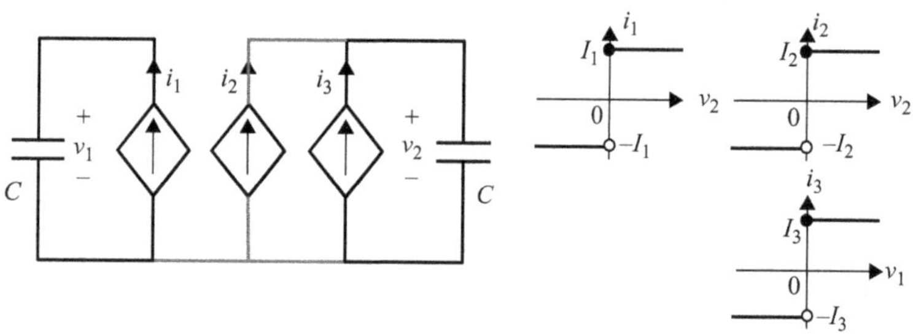

Figure 5.5 Another example of circuit diagram for generating piecewise-constant oscillations

The dynamics of (5.7) is also realized by a simple electrical circuit as shown in Figure 5.5. Likewise, from Kirchhoff's voltage law, we obtain the circuit dynamics represented by:

$$
\begin{aligned}
C\frac{\mathrm{d}v_1}{\mathrm{d}t} &= I_1 \cdot \mathrm{sgn}(v_2) \\
C\frac{\mathrm{d}v_2}{\mathrm{d}t} &= I_3 \cdot \mathrm{sgn}(v_1) + I_2 \cdot \mathrm{sgn}(v_2)
\end{aligned}
\tag{5.8}
$$

By using the following dimensionless and normalizing variables:

$$
\tau = \frac{I_1}{CE}t, \quad x = \frac{1}{E}v_1, \quad y = \frac{1}{E}v_2 \tag{5.9}
$$

and supposing $I_1 = -I_3$, the circuit dynamics of (5.8) is equivalent to (5.7), where $a = \frac{I_1}{I_2}$. Note that the piecewise-constant vector field of (5.7) is divided into four constant vector fields depending on the values of sgn(x) and sgn(y). In order to

Table 5.1 Local vector fields (values of **a**(*i*)*)*

i	**sgn**(x)	**sgn**(y)	**a**(i)
0	1	1	${}^t\left(1, -1 + \frac{1}{a}\right)$
1	1	−1	${}^t\left(-1, -1 - \frac{1}{a}\right)$
2	−1	1	${}^t\left(1, 1 + \frac{1}{a}\right)$
3	−1	−1	${}^t\left(-1, 1 - \frac{1}{a}\right)$

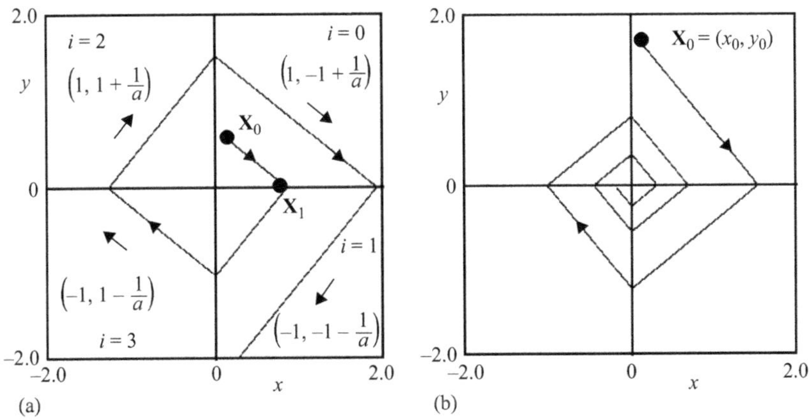

Figure 5.6 Oscillations of a PWC system (5.7): (a) expanding oscillation in the case of $a > 1$ ($a = 5$, $x_0 = 0.1$, $y_0 = 0.2$) and (b) converging oscillation in the case of $a < -1$ ($a = -5$, $x_0 = 0.1$, $y_0 = 1.7$)

describe the four constant vector fields, we introduce an integer-dependent variable i which is defined as follows:

$$i = \frac{-\mathrm{sgn}(x) + 1}{2} \cdot 2^1 + \frac{-\mathrm{sgn}(y) + 1}{2} \cdot 2^0, \quad i \in \{0, 1, 2, 3\} \tag{5.10}$$

Letting $\mathbf{a}(i)$ be a local vector field for i, the dynamics (5.7) can be represented as:

$$\dot{\mathbf{x}} = \mathbf{a}(i) \tag{5.11}$$

where $\mathbf{x} = (x, y)^\top$ and $\top$ denotes the transpose of vectors, and $\mathbf{a}(i)$ is described in Table 5.1. For example, if $i = 0$, that is, $\mathrm{sgn}(x) = 1$ and $\mathrm{sgn}(y) = 1$, then $\mathbf{a}(0) = (1, -1 + \frac{1}{a})^\top$. Let us consider the condition of $|a| > 1$, same as (5.6). If $a > 1$, the system exhibits expanding oscillations as shown in Figure 5.6(a), and if $a < -1$, the system exhibits converging oscillations as shown in Figure 5.6(b). The local solution of (5.11) with an initial condition $\mathbf{x}(0) = \mathbf{x}_0 = (x_0, y_0)^\top$ is explicitly given as follows:

$$\mathbf{x} = \mathbf{x}_0 + \mathbf{a}_i\tau \tag{5.12}$$

Let us consider the situation where the solution started from the initial point $\mathbf{x}_0$ hits the boundary at $\tau = \tau_1$, and let the hitting point be $\mathbf{x}_1$ as shown in Figure 5.6(a). The boundary is expressed by the following linear plane (line, in this case):

$$\mathbf{n}\mathbf{x} = D \tag{5.13}$$

where $\mathbf{n}^\top$ is a normal vector of the boundary, and D is a scalar corresponding to an intercept. For example, see Figure 5.6(a), if the trajectory hits $y = 0$, $\mathbf{n} = (0,\ 1)$ and $D = 0$ represent the boundary. By substituting (5.12) into (5.13), we obtain:

$$\mathbf{n}(\mathbf{x}_0 + \mathbf{a}_i\tau_1) = D \tag{5.14}$$

and thus, we obtain the time of the arrival, τ_1:

$$\tau_1 = \frac{D - \mathbf{n}\mathbf{x}_0}{\mathbf{n}\mathbf{a}_i} \tag{5.15}$$

Hence, the solution on the boundary, which is denoted by $\mathbf{x}_1$, is represented by:

$$\mathbf{x}_1 = \left(I - \frac{\mathbf{a}_i\mathbf{n}}{\mathbf{n}\mathbf{a}_i}\right)\mathbf{x}_0 + \frac{D}{\mathbf{n}\mathbf{a}_i}\mathbf{a}_i \tag{5.16}$$

where I is an identity matrix. Furthermore, note that the solution of the variational equation can be also explicitly derived in piecewise-constant dynamics. For example, it is clear from (5.16) that Jacobian matrix is represented as follows:

$$\frac{d\mathbf{x}_1}{d\mathbf{x}_0} = I - \frac{\mathbf{a}_i\mathbf{n}}{\mathbf{n}\mathbf{a}_i} \tag{5.17}$$

5.2 Example 1: a piecewise-constant chaotic spiking oscillator

In this section, we introduce a chaotic spiking oscillator (CSO) with piecewise-constant vector field (PWCCSO). Some researches on CSOs have been reported in the field of nonlinear problem and neurocomputing [1–6]. These CSOs have rotate-and-fire dynamics. The dynamics is considerably related to some spiking neuron models as pointed out by the literatures [7,8]. Motivation of study on PWCCSO is its simplicity. Although some implementations of analog neuron models [9–12] have been developed, it is sometime difficult to synthesize small-scale circuit and to set up parameters. PWCCSO has an advantage to overcome these problems [13–15] and theoretical analysis of bifurcation behavior and stability can be possible by discontinuous procedure [3,16].

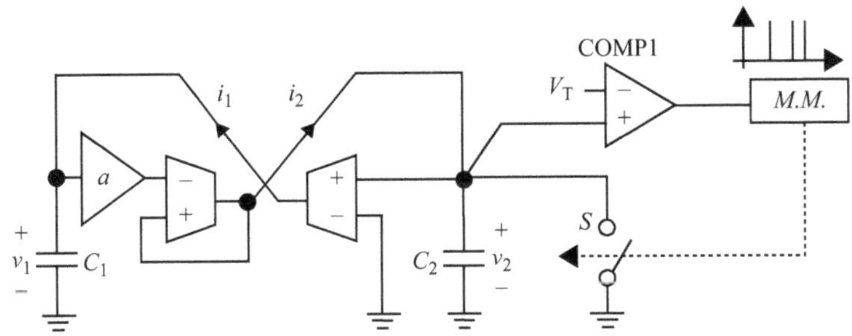

Figure 5.7 Schematic diagram of piecewise-constant chaotic spiking oscillator

5.2.1 Circuit and dynamics

Figure 5.7 shows the circuit model of PWCCSO. Triangle labeled a is linear voltage amplifier with gain a, which is realized by operational amplifier, TL084. The triangle and trapezoids labeled $+$ and $-$ are nonlinear voltage-controlled comparators. The trapezoids are differential amplification current sources and their output currents are i_1 and i_2, respectively. They are realized by operational transconductance amplifiers (OTAs), NJM13600, and they are characterized by:

$$
\begin{aligned}
i_1 &= I_o \cdot \mathrm{sgn}(v_2) \\
i_2 &= I_o \cdot \mathrm{sgn}(v_2 - av_1) = I_o \cdot \mathrm{sgn}\left(-v_1 + \frac{1}{a}v_2\right)
\end{aligned}
\tag{5.18}
$$

where I_o is the absolute value of output current of OTAs. Therefore, the circuit block which consists of the pair of two OTAs and two capacitors in Figure 5.7 is a realization example of the VCCSs circuit in Figure 5.3. When S is opened, the circuit dynamics is described by (5.2) by using the following dimensionless variables:

$$
\tau = \frac{I_o}{CV_T}t, \quad x = \frac{v_1}{V_T}, \quad y = \frac{v_2}{V_T} \tag{5.19}
$$

where V_T is a spiking threshold voltage as will be shown later. In a parameter range $a > 1$, system (5.2) has unstable spiral trajectories as explained in the previous section. For the following discussions in this subsection, we consider the parameter range of $a > 1$. The trajectory on the phase space moves around the origin divergently and it must reach to $y = 1$ as shown in Figure 5.8. In the circuit in Figure 5.7, $M.M.$ is a monostable multivibrator which outputs pulse signals, "spike," to close the switch S instantaneously. A comparator COMP1 detects the impulsive switching condition. If $v_2 < V_T$, the switch S is opened. If $v_2 = V_T$, then $M.M.$ is triggered by the comparator, and the switch S is closed instantaneously. At that time, v_2 is reset instantaneously to 0. On the phase space as shown in Figure 5.8, trajectories started from any initial condition must reach the threshold $y = 1$, then the switching occurs. The normalized

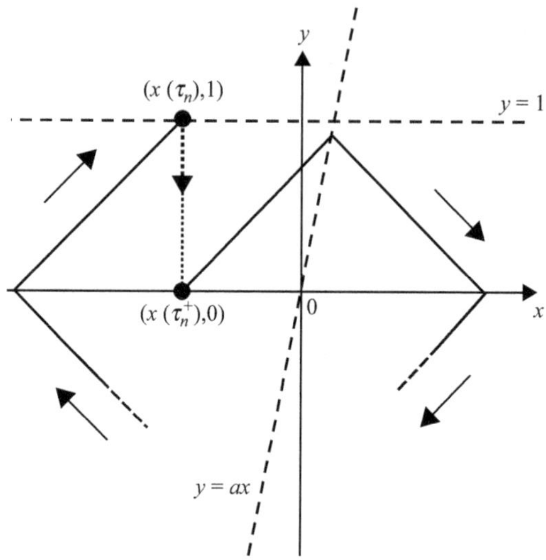

Figure 5.8 Behavior of trajectories on the phase space

trajectory must jump from $(x(\tau_n), 1)$ to $(x(\tau_n^+), 0)$ as shown in Figure 5.8, where τ_n is the nth switching moment. Consequently, the dynamics is expressed by:

$$\begin{cases} \dot{x} = \mathrm{sgn}(y) \\ \dot{y} = \mathrm{sgn}\left(-x + \dfrac{1}{a}y\right) \end{cases} \quad \text{for } y < 1 \tag{5.20}$$

$$(x(\tau_n^+), y(\tau_n^+)) = (x(\tau_n), 0) \text{ for } y(\tau_n) = 1 \tag{5.21}$$

Now the system is characterized by only one parameter a. Figure 5.9 shows a typical chaos attractor.

5.2.2 Embedded return map

In order to consider embedded return map, we define a line $L \equiv \{(x, y) | y = 0\}$ as shown in Figure 5.10. In the parameter condition $a > 1$, the trajectory starting from L moves divergently around the origin and reaches the switching threshold, and returns to L. If we choose L as Poincaré section, we can define one-dimensional return map f from L to itself. Let $(x_m, 0)$ be the starting point, and let $(x_{m+1}, 0)$ be the returning point. Letting any points on L be represented by its x-coordinate, f is defined by:

$$f : L \mapsto L, \quad x_{m+1} = f(x_m) \tag{5.22}$$

$$f(x_m) = \begin{cases} f_1(x_m) = x_m + 1 & \text{for } x_m \leq X_{\text{th}} \\ f_2(x_m) = Ax_m & \text{for } x_m > X_{\text{th}} \end{cases} \tag{5.23}$$

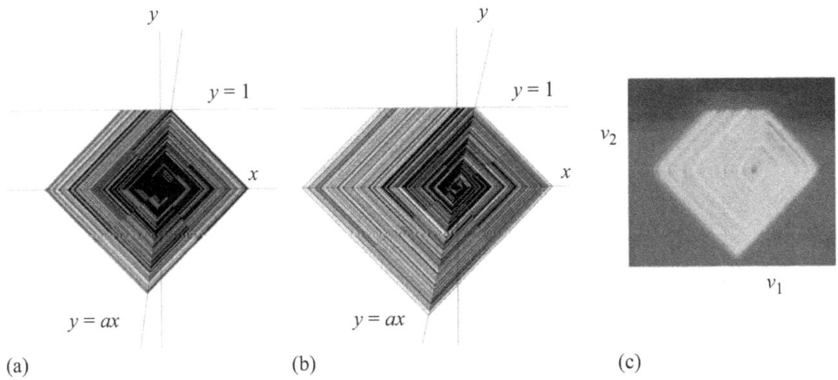

Figure 5.9 Typical chaos attractors: (a) a = 7, (b) a = 5, and (c) laboratory measurement (vertical axis: v_1 [0.5 V/div], horizontal axis: v_2 [0.5 V/div], a = 5)

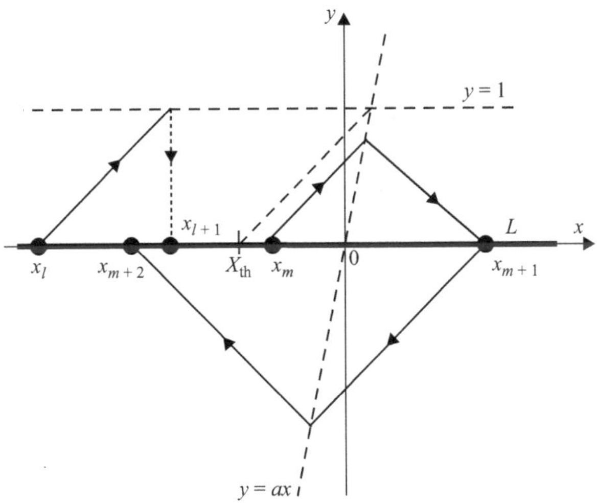

Figure 5.10 Definition of return map

where $A = \frac{1+a}{1-a}$ and $X_{\text{th}} = \frac{1}{a} - 1$. Typical examples of the shape of return map f are shown in Figure 5.11. To give the proof for chaos generation, we define two border values as shown in Figure 5.11:

$$X_{\text{R}} = f_2(X_{\text{th}}), \quad X_{\text{L}} = f_2(X_{\text{R}}) \tag{5.24}$$

Note that $|\frac{\partial f}{\partial x_m}| \geq 1$ is satisfied without discontinuous points and $f(I) \subseteq I$ is obvious, where $I = [X_{\text{L}}, X_{\text{R}}]$. Hence, f exhibits chaos under the condition $a > 1$.

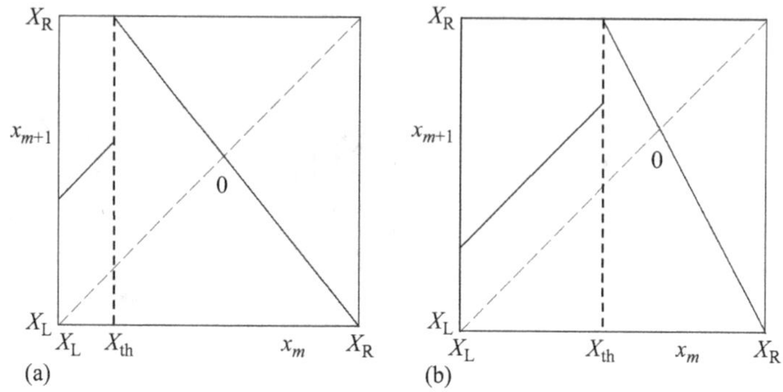

Figure 5.11 Typical examples of shape of return map: (a) a = 10 and (b) a = 3.3

5.3 Example 2: coupled systems of piecewise-constant oscillators

In this section, we consider synchronization phenomena in the coupled systems of piecewise-constant oscillators. Both in-phase and anti-phase synchronization phenomena are observed in the oscillators coupled by VCCS with signum characteristic. On the other hand, their coexistence is observed in the oscillators coupled by VCCS with hysteresis characteristic. We introduce the stability analysis of the synchronization phenomena in the coupled systems by using a fast calculation algorithm for the rigorous solutions and the parameter regions of in-phase and anti-phase synchronization. The result suggests that the synchronization phenomena of the coupled PWC oscillators qualitatively correspond to one of van der Pol oscillators coupled by passive elements.

5.3.1 A piecewise-constant oscillator exhibiting limit cycle

In this section, we introduce a piecewise-constant oscillator as shown in Figure 5.12. The circuit consists of two capacitors whose capacitance is C, and OTAs that are represented by two binary characteristics. One is signum function sgn as shown in Figure 5.2, and the other is hysteresis characteristic $H(V)$ as shown in Figure 5.13. H is switched from 1 to -1 if V reaches to the threshold $-v_{th}$ and H switched from -1 to 1 if V reaches to v_{th}. We can describe the output currents of the OTAs as follows:

$$i_1 = I_1 \cdot H(v_2), \quad i_2 = I_2 \cdot H(v_2), \quad i_3 = I_3 \cdot \mathrm{sgn}(v_1), \quad i_h = -I_h \cdot H(v_2) \tag{5.25}$$

where I_1, I_2, I_3, and I_h are the absolute values of output currents of OTAs, and v_{th} is determined by I_h and R_{th}; $v_{th} = I_h R_{th}$. The circuit dynamics is described by a two-dimensional piecewise-constant equation:

$$\begin{cases} C\dfrac{\mathrm{d}v_1}{\mathrm{d}t} = I_1 \cdot H(v_2) \\ C\dfrac{\mathrm{d}v_2}{\mathrm{d}t} = I_3 \cdot \mathrm{sgn}(v_1) + I_2 \cdot H(v_2) \end{cases} \tag{5.26}$$

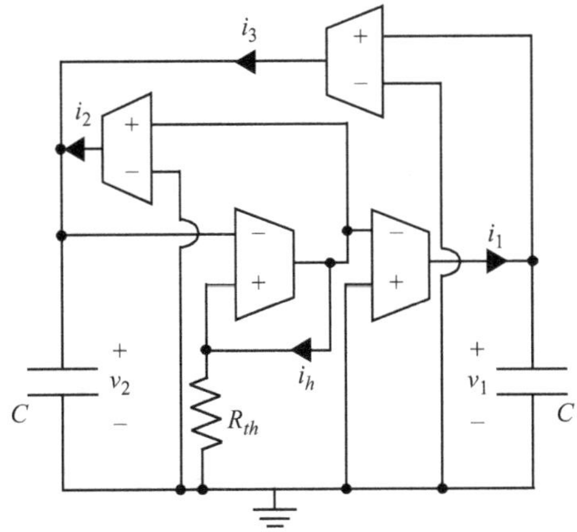

Figure 5.12 Schematic diagram of a piecewise-constant oscillator exhibiting limit cycle

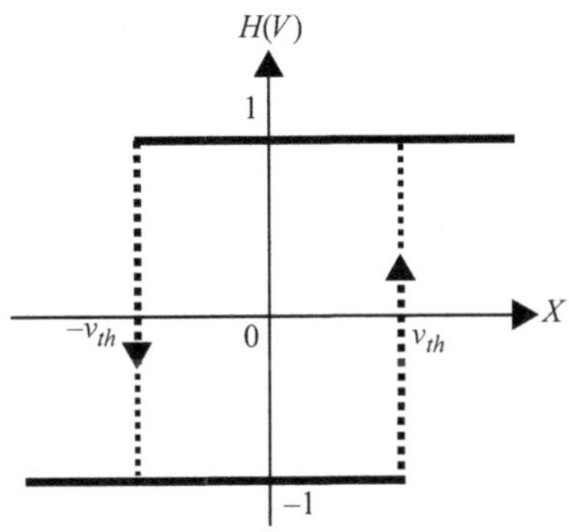

Figure 5.13 The characteristic of the hysteresis

If $v_{th} \to 0$ which corresponds to shorting the resistor R_{th}, then (5.26) is equal to (5.8). Therefore, the circuit in Figure 5.12 is a realization example of the VCCSs circuit in Figure 5.5. In order to realize oscillatory behavior, we consider the following conditions:

$$I_1 = -I_3, \quad I_1 \cdot I_2 < 0 \tag{5.27}$$

By using the normalized variables of (5.9), we can rewrite the circuit dynamics by normalized equation with only one parameter a,

$$\begin{cases} \dot{x} = h(y) \\ \dot{y} = -\mathrm{sgn}(x) + \dfrac{1}{a}h(y) \end{cases} \tag{5.28}$$

where $h(y)$ is a normalized hysteresis; h is switched from 1 to -1 if y reaches to the threshold -1 and is switched from -1 to 1 if y reaches to 1. In order to describe the four constant vector fields, we define an integer-dependent variable i in the same way as (5.10):

$$i = \frac{-\mathrm{sgn}(x)+1}{2}\cdot 2^1 + \frac{-h(y)+1}{2}\cdot 2^0 \in \{0,1,2,3\} \tag{5.29}$$

The velocity vector $\mathbf{a}(i)$, a local vector field for i, is given likewise Table 5.1, and the dynamics (5.28) can be represented by (5.11). On the condition $a < -1$, a trajectory started from an initial state $\mathbf{x}_0$ and $i = 0$ as shown in Figure 5.14 moves in a straight line in accordance with the local vector field $\mathbf{a}(0) = (1, -1 + 1/a)^\top$, x is increasing and y is decreasing. If the trajectory reaches the switching border at $\mathbf{x}_1$, i changes from 0 to 1 because $h(y)$ changes from 1 to -1. The trajectory started from $\mathbf{x}_1$ subsequently moves by $\mathbf{a}(1) = (-1, -1 - 1/a)^\top$, x is decreasing and y is also decreasing. Here, we define the threshold of x as $E_y(0) = -1$ when $h(y)$ switches from 1 to -1 for $i = 0$. $E_x(0) = 0$ is also defined with respect to sgn(x), where there is no situation in which a trajectory hits $y = E_y(0)$ for $i = 0$ because x is increasing for $i = 0$. Consequently, if a trajectory reaches the threshold $E_y(i)$, then the output of $h(y)$ inverts from ±1 to ∓1. If a trajectory reaches the threshold $E_x(i)$, then the output of sgn(x) inverts from ±1 to ∓1. The system repeats the manner of linear uniform motions and switching. As shown in Figure 5.14(a), all trajectories started from any initial states converge to a square limit cycle, as long as the condition $a < -1$ is satisfied. Figure 5.14(b) shows a laboratory measurement of an experimental circuit. VCCSs in Figure 5.12 are implemented by OTAs, NJM13700.

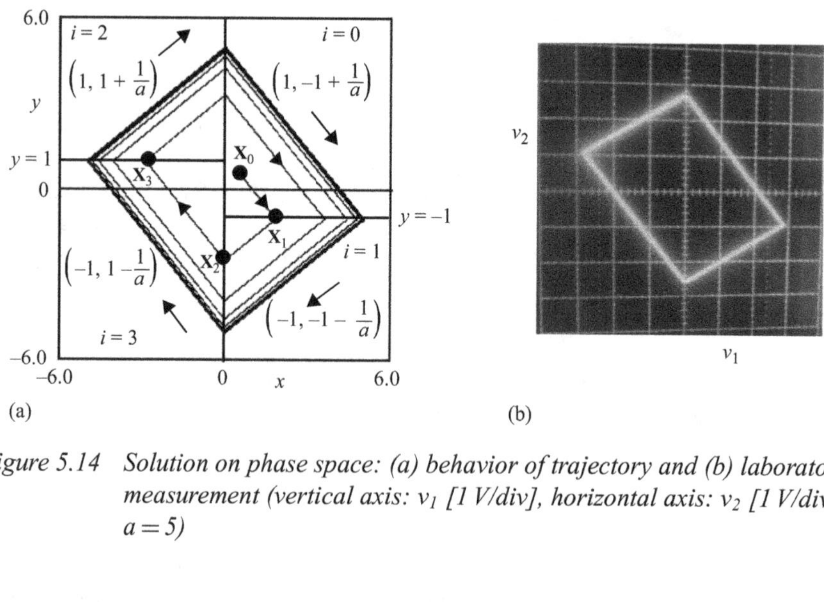

Figure 5.14 Solution on phase space: (a) behavior of trajectory and (b) laboratory measurement (vertical axis: v_1 [1 V/div], horizontal axis: v_2 [1 V/div], $a = 5$)

5.3.2 Coupled system of piecewise-constant oscillators

In this subsection, we make a comparison between a PWC oscillator with limit cycle and a well-known oscillator, van der Pol oscillator, and we explain the motivation of study of Coupled system of PWC oscillators. Coupled oscillatory systems are good models to explain complex nonlinear phenomena of high dimension in the field of natural sciences. These systems have been extensively studied in the various fields such as physical [17], biological [18], and electric systems [19]. Many studies on coupled oscillators modeled by electrical circuits have been reported [20–29]. Especially, studies on coupled van der Pol oscillators have been often investigated by many researchers [21–28]. The circuit equation of the uncoupled oscillator is described as follows:

$$\begin{cases} L\dfrac{\mathrm{d}i}{\mathrm{d}t} = v \\[2ex] C\dfrac{\mathrm{d}v}{\mathrm{d}t} = -i - i_r \end{cases} \tag{5.30}$$

where i_r with third-order nonlinear characteristic is given by $i_r = -g_1 v + g_3 v^3$ and $g_1, g_3 > 0$. The circuit model is shown in Figure 5.15. Suezaki and Mori have investigated mutual synchronization phenomena in coupled two van der Pol oscillators in Reference 21. Typical synchronization phenomena can be observed in two oscillators coupled by inductor as shown in Figure 5.16. By using the following dimensionless variables and parameters:

$$\begin{aligned} &\tau = \frac{1}{\sqrt{LC}}t, \quad x_1 = \sqrt{\frac{Lg_3}{Cg_1}}i_1, \quad x_2 = \sqrt{\frac{Lg_3}{Cg_1}}i_2, \quad y_1 = \sqrt{\frac{g_3}{g_1}}v_1, \\ &y_2 = \sqrt{\frac{g_3}{g_1}}v_2, \quad \varepsilon = g_1\sqrt{\frac{L}{C}}, \quad \gamma_L = \frac{L}{L_c} \end{aligned} \tag{5.31}$$

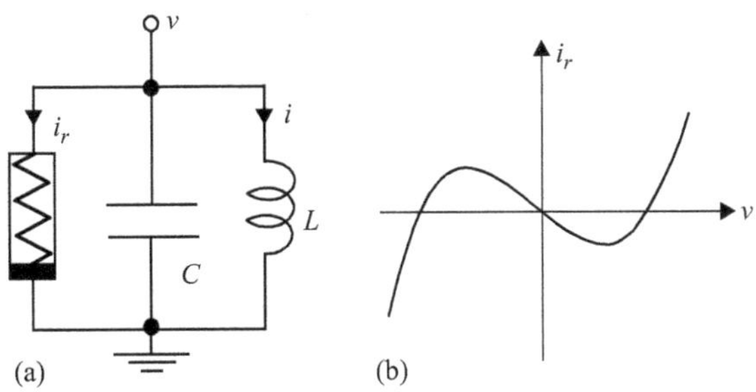

Figure 5.15 van der Pol oscillator: (a) circuit model and (b) v–i characteristic of nonlinear resistor

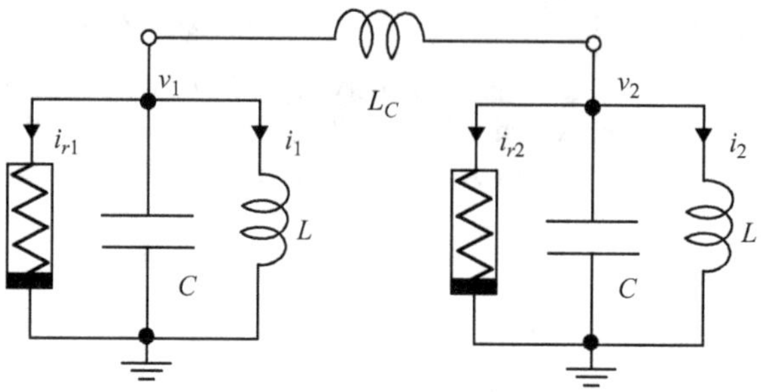

Figure 5.16 Coupled van der Pol oscillators by an inductor

the dynamic of the van der Pol oscillators by a coupled inductor is described by coupled Rayleigh equations:

$$\begin{cases} \dfrac{\mathrm{d}x_1}{\mathrm{d}\tau} = y_1 \\ \dfrac{\mathrm{d}y_1}{\mathrm{d}\tau} = -x_1 + \varepsilon(1 - y_1^2)y_1 - \gamma_L \displaystyle\int_{\tau_0}^{\tau_0+\tau} (y_1 - y_2)\mathrm{d}\tau \\ \dfrac{\mathrm{d}x_2}{\mathrm{d}\tau} = y_2 \\ \dfrac{\mathrm{d}y_2}{\mathrm{d}\tau} = -x_2 + \varepsilon(1 - y_2^2)y_2 + \gamma_L \displaystyle\int_{\tau_0}^{\tau_0+\tau} (y_1 - y_2)\mathrm{d}\tau \end{cases} \tag{5.32}$$

where γ_L is a coupling intensity corresponding to the coupling inductance L_c. Both in-phase and anti-phase synchronization are observed from the oscillators coupled by an inductor, and coexistence of in-phase and anti-phase synchronization is confirmed. It is noted that, in the case of the oscillators coupled by a conductor, although both in-phase and anti-phase synchronization are observed [30], the coexistence phenomenon does not appear. Some pioneering approaches to analyze synchronization phenomena on many oscillators connected in a ladder, a two-dimensional array, and a ring topologies, are found in References 22–24. In these large-scale systems, phase states between the coupled two adjacent oscillators are very interesting. For example, on the oscillators connected in a two-dimensional array with inductors, phase states between two adjacent oscillators can become both in-phase and anti-phase synchronization. These phenomena have been confirmed by laboratory measurements [23]. Another example is the case of the oscillators connected in a ladder with inductors. In the system, phase-inversion wave phenomena are observed. Phase differences between two adjacent oscillators are propagated and their phase states change from in-phase to anti-phase and vice versa. The phenomena have been investigated in the literature [25]. The interesting behavior of in-phase and anti-phase synchronization in the oscillatory network is largely due to coupling by inductors. On the other hand, the study in

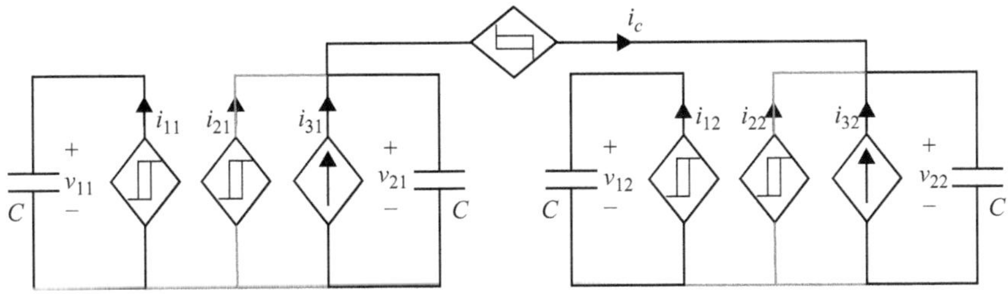

Figure 5.17 Piecewise-constant oscillators coupled by a hysteresis VCCS

Reference 27 has considered the oscillators coupled by a time-varying resistor which switches positive and negative resistance periodically. In this system, coexistence of in-phase and anti-phase synchronization has been confirmed by laboratory measurements. On the oscillators coupled by the time-varying resistors, phase-inversion wave phenomena can be confirmed in a ring topology network [28]. Therefore, coexistence of in-phase and anti-phase synchronization is essentially important.

We now consider a coupled system of piecewise-constant oscillators. The circuit model is shown in Figure 5.17. The each oscillator corresponds to a limit cycle generator in Figure 5.12, and i_h is a coupling current and is given by the hysteresis characteristic in Figure 5.13. The output currents of VCCSs are represented as follows:

$$\begin{aligned} &i_{11} = I_{11}\cdot H(v_{21}), \quad i_{21} = I_{21}\cdot H(v_{21}), \quad i_{31} = I_{31}\cdot \mathrm{sgn}(v_{11}),\\ &i_{12} = I_{12}\cdot H(v_{22}), \quad i_{22} = I_{22}\cdot H(v_{22}), \quad i_{32} = I_{32}\cdot \mathrm{sgn}(v_{11}),\\ &i_h = I_h\cdot H(v_{21}-v_{22}) \end{aligned} \tag{5.33}$$

The circuit equation is described by:

$$\begin{aligned} \dot{x_1} &= H(y_1)\\ \dot{y_1} &= -\mathrm{sgn}(x_1) + \frac{1}{a_1}H(y_1) - \gamma H(y_1 - y_2)\\ \dot{x_2} &= H(y_2)\\ \dot{y_2} &= -\mathrm{sgn}(x_2) + \frac{1}{a_2}H(y_2) + \gamma H(y_1 - y_2) \end{aligned} \tag{5.34}$$

where $I_{11} = -I_{31}$, $I_{12} = -I_{32}$, and the following dimensionless parameters and normalizing variables are used:

$$\begin{aligned} &\tau = \frac{I_{11}}{Cv_{th}}t, \quad x_1 = \frac{1}{v_{th}}v_{11}, \quad y_1 = \frac{1}{v_{th}}v_{21}, \quad x_2 = \frac{1}{v_{th}}v_{12},\\ &y_1 = \frac{1}{v_{th}}v_{22}, \quad a_1 = \frac{I_{11}}{I_{21}}, \quad a_2 = \frac{I_{12}}{I_{21}}, \quad \gamma = \frac{I_c}{I_{11}} \end{aligned} \tag{5.35}$$

Here, we can find some similarities between (5.32) and (5.34). Their whole structures are similar and the coupling terms are influenced by past histories, an integration term in the case of von del Pol oscillators and a hysteresis in the case of

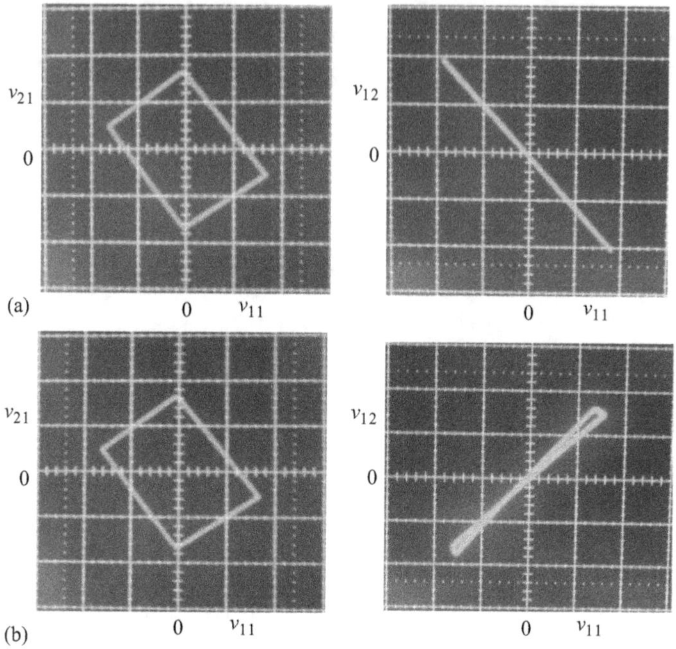

Figure 5.18 Laboratory measurements of PWC oscillators coupled by a hysteresis VCCS: (a) anti-phase synchronization and (b) in-phase synchronization (vertical axis: 2 V/div], horizontal axis: [2 V/div], $a_1 = a_2 = 3.3$, $\gamma = -0.03$)

PWC oscillators. Especially, from the coupled PWC oscillators, in-phase and anti-phase synchronization phenomena and their coexistence have also been observed in laboratory as shown in Figure 5.18. We can observe in-phase and anti-phase synchronization phenomena from the system coupled by a VCCS with signum characteristic where the case corresponds to $v_{th} \to 0$. If the coupling coefficient γ is positive, in-phase synchronization is observed, and if the γ is negative, anti-phase synchronization is observed. As compared with coupled van der Pol oscillators, these show that mutual synchronization phenomena in coupled systems of PWC oscillators qualitatively correspond to one of van der Pol oscillators.

5.3.3 Analysis of PWC oscillators

In this section, we introduce a summary of the calculation algorithm for rigorous solutions in the coupled PWC oscillators. The detail can be found in Reference 30. Let us focus on a state $\mathbf{x}_k$ at the kth moment when the integer state i switches, where the integer state i is defined in the same way to (5.10). A trajectory started from $\mathbf{x}_k$ goes straightforwardly to $\mathbf{x}_{k+1}$ at the $(k+1)$th switching moment. Namely, if we calculate the state $\mathbf{x}_{k+1}$ when the switching of i occurs, we obtain rigorous solutions by connecting $\mathbf{x}_k$ and $\mathbf{x}_{k+1}$. We show the algorithm to calculate $\mathbf{x}_{k+1}$ from $\mathbf{x}_k$ as shown in the following four steps.

Step 1.

We set the initial state $\mathbf{x}_0$ and get the dependent variable i.

Step 2.

We calculate a time t_k until the i switches to another value. Assuming a trajectory started from $\mathbf{x}_k$ arrives one of thresholds $E_{y_1-y_2}(i)$, $E_{x_1}(i)$, $E_{y_1}(i)$, $E_{x_2}(i)$, and $E_{y_2}(i)$, each virtual arrival times $t_{y_1-y_2}$, t_{x_1}, t_{y_1}, t_{x_2}, and t_{y_2} can be calculated by linear algebra approaches. Then, the actual arrival time t_k that means the switching time of i is given by the minimum of them omitting zero and negative:

$$t_k = \min\{\{t_{x_1-x_2}, t_{x_1}, t_{y_1}, t_{x_2}, t_{y_2}\} \cap \{\xi \in \mathbf{R} | \xi > 0\}\} \tag{5.36}$$

If all of $t_{x_1-x_2}$, t_{x_1}, t_{y_1}, t_{x_2}, and t_{y_2} are not positive, it means that the switching of i_k does not occur. In such case, the trajectory diverges. However, the situation never happen on parameter conditions $a < -1$ and satisfyingly small γ.

Step 3.

We calculate $\mathbf{x}_{k+1}$ by:

$$x_{k+1} = x_k + a(i) \cdot t_k \tag{5.37}$$

We get the integer variable i after switching. For example, if $t_k = t_{y_1-y_2}$, the sign of output of $H(y_1 - y_2)$ inverts. Namely, by substituting $-H(y_1 - y_2)$ instead of $H(y_1 - y_2)$ into (5.10), we can update the integer variable i.

Step 4.

Replace $\mathbf{x}_{k+1}$ with $\mathbf{x}_k$. Subsequently, return to Step 2.

By using the calculation algorithm, we obtain discrete time series $(\mathbf{x}_0, \mathbf{x}_1, \ldots, \mathbf{x}_N)$ by N iterations. The time series corresponds to the rigorous solutions.

Next, we consider to derive Lyapunov exponents from the time series based on the above calculation algorithm. Noting (5.37), $\mathbf{x}_{k+1}$ is given by a map F_k of $\mathbf{x}_k$

$$\mathbf{x}_{k+1} = F_k(\mathbf{x}_k) \tag{5.38}$$

because i and t_k are depending on $\mathbf{x}_k$. Thus, we can obtain the Jacobi matrix:

$$J_k = \left. \frac{\partial F_k(\mathbf{x})}{\partial \mathbf{x}} \right|_{\mathbf{x}=\mathbf{x}_k} \tag{5.39}$$

It is described explicitly. Consequently, we obtain the J_k systematically by referring a value table like Table 5.1. Letting λ denote the maximum one-dimensional Lyapunov exponent, it is given by:

$$\begin{aligned} \lambda &= \lim_{N\to\infty} \frac{1}{N} \sum_{k=0}^{N} \ln \|J_k \mathbf{e}_k\| \\ \mathbf{e}_{k+1} &= \frac{J_k \mathbf{e}_k}{\|J_k \mathbf{e}_k\|} \end{aligned} \tag{5.40}$$

where $\mathbf{e}_k$ is an orthonormal base [31]. In practical calculations, we set $N = 10{,}000$ which is assumed as sufficiently large. The initial state $\mathbf{x}_0$ is given adequately to

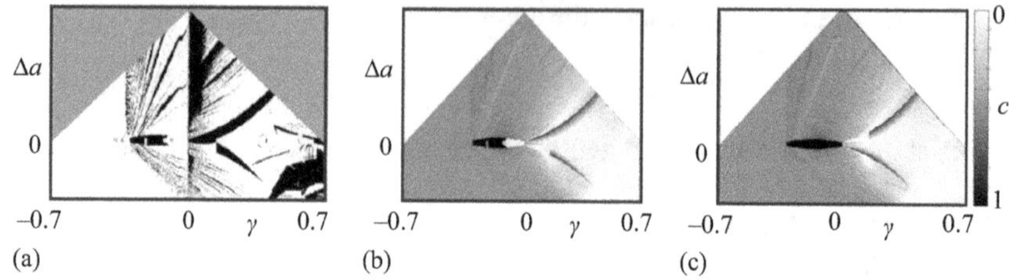

Figure 5.19 Lyapunov exponent and correlation coefficient on parameter space: (a) Lyapunov exponent ($\lambda > 0$ for white region and $\lambda \leq 0$ for black region), (b) correlation coefficient (initial conditions: $x_1 = x_2$ and $y_1 = y_2$), and (c) correlation coefficient (initial conditions: $x_1 = -x_2$ and $y_1 = -y_2$)

omit transient behavior. Figure 5.19(a) shows the sign of λ on the parameter space of γ versus $\Delta a = a_1 - a_2$ with $a = 3.3$. Gray region does not satisfy the parameter condition of attractor existence, and the gray region does not be considered here. The Lyapunov exponent λ has non-negative value in white region. In this case, the system exhibits non-periodic behavior. The λ has negative value in black region. In this case, the system exhibits periodic behavior. In-phase synchronization as shown in Figure 5.18(b) is observed with $\gamma = -0.03$ and $\Delta\alpha = 0$. Since the parameters are in white region, the value of λ is positive, and the in-phase synchronization has a chaotic motion.

In order to classify parameter regions of in-phase and anti-phase synchronization, we introduce correlation coefficients c with regard to y_1 and y_2. It is given by:

$$c = \frac{\frac{1}{N}\sum_{k=0}^{N}(y_{1k} - \bar{y_1})(y_{2k} - \bar{y_2})}{\sqrt{\frac{1}{N}\sum_{k=0}^{N}(y_{1k} - \bar{y_1})^2}\sqrt{\frac{1}{N}\sum_{k=0}^{N}(y_{2k} - \bar{y_2})^2}} \tag{5.41}$$

where $\bar{y_1}$ and $\bar{y_2}$ are the mean of N time series data of y_1 and y_2, respectively. We use $N = 2,000$ which is assumed as sufficiently large. Figure 5.19(b) and (c) shows the value of the c for different initial states by using a gray scale on the parameter spaces. In the white region which c is close to 1, in-phase synchronization is observed. As compared with Figure 5.19(a) and noting the area around the origin with a negative value of γ, we confirm that the region of the positive Lyapunov exponent λ corresponds to one of in-phase synchronization and the coexistence region of in-phase and anti-phase synchronization.

5.4 Conclusions

In this chapter, we investigated extremely simple oscillators whose dynamics are represented by piecewise-constant equations. More specifically, two examples of

PWC systems, a CSO with piecewise-constant dynamics and a coupled system of piecewise-constant oscillators, are considered. We introduced the methods used to obtain the rigorous solutions and to analyze the stability of some phenomena in these systems.

References

[1] K. Mitsubori and T. Saito, "Dependent switched capacitor chaos generator and its synchronization," *IEEE Trans. Circuits Syst. I*, vol. 44, no. 12, pp. 1122–1128, 1997.

[2] H. Nakano and T. Saito, "Basic dynamics from an integrated-and-fire chaotic circuits with a periodic input," *IEICE Trans. Fundamentals*, vol. E84-A, no. 5, pp. 1293–1300, 2001.

[3] T. Tsubone and Y. Wada, "Stabilizing unknown periodic orbits of a chaotic spiking oscillator," *IEICE Trans. Fundamentals*, vol. E82-A, no. 5, pp. 1316–1321, 2009.

[4] T. Matsubara, H. Torikai and T. Hishiki, "A generalized rotate-and-fire digital spiking neuron model and its on-FPGA learning," *IEEE Trans. Circuits Syst. II: Express Briefs*, vol. 58, no. 10, pp. 677–681, 2011.

[5] T. Hishiki and H. Torikai, "A novel rotate-and-fire digital spiking neuron and its neuron-like bifurcations and responses," *IEEE Trans. Neural Networks*, vol. 22, no. 5, pp. 752–767, 2011.

[6] Y. Matsuoka, "Master–slave coupled piecewise constant spiking oscillators," *IEICE Trans. Fundamentals*, vol. E94-A, no. 9, pp. 1860–1863, 2011.

[7] E. M. Izhikevich, "Which model to use for cortical spiking neurons?," *IEEE Trans. Neural Networks*, vol. 15, no. 5, pp. 1063–1070, 2004.

[8] E. M. Izhikevich, *Dynamical Systems in Neuroscience*, MIT Press, Cambridge, MA, 2006.

[9] M. F. Simoni and S. P. DeWeerth, "Adaptation in an a VLSI model of a neuron," *IEEE Trans. Circuits Syst. II*, vol. 46, no. 7, pp. 967–970, 1999.

[10] E. Farquhar and P. Hasler, "A bio-physically inspired silicon neuron," *IEEE Trans. Circuits Syst. I*, vol. 52, no. 3, pp. 477–488, 2005.

[11] T. Kohno and K. Aihara, "A MOSFET-based model of a class 2 nerve membrane," *IEEE Trans. Neural Networks*, vol. 16, no. 3, pp. 754–773, 2005.

[12] A. Basu and P. E. Hasler, "Nullcline-based design of a silicon neuron," *IEEE Trans. Circuits Syst. I*, vol. 57, no. 11, pp. 2938–2947, 2010.

[13] Y. Yamashita and H. Torikai, "A generalized PWC spiking neuron model and its neuron-like activities and burst-related bifurcations," *IEICE Trans. Fundamentals*, vol. E95-A, no. 7, pp. 1125–1135, 2012.

[14] Y. Yamashita and H. Torikai, "Bifurcation analysis of inhibitory responses of a PWC spiking neuron model," *NOLTA IEICE*, vol. 3, no. 4, pp. 557–572, 2012.

[15] Y. Yamashita and H. Torikai, "A novel PWC spiking neuron model: Neuron-like bifurcation scenarios and responses," *IEEE Trans. Circuits Syst. I*, vol. 59, no. 11, pp. 2678–2691, 2012.

[16] T. Tsubone and T. Saito, "Manifold piecewise constant system and chaos," *IEICE Trans. Fundamentals*, vol. E82-A, no. 8, pp. 1619–1626, 1999.

[17] I. Belykh, M. Hasler, M. Lauret and H. Nijmeijer, "Synchronization and graph topology," *Int. J. Bifurcat. Chaos*, vol. 15, no. 11, pp. 3423–3433, Nov. 2005.

[18] J. Cosp, J. Madrenas, E. Alarcon, E. Vidal and G. Villar, "Synchronization of nonlinear electronic oscillators for neural computation," *IEEE Trans. Neural Networks*, vol. 15, no. 5, pp. 1315–1327, Sep. 2004.

[19] T. Yang and L. O. Chua, "Impulsive stabilization for control and synchronization of chaotic systems theory and application to secure communication," *IEEE Trans. Circuit and Syst. I*, vol. 44, no. 10, pp. 976–988, Oct. 1997.

[20] H.B. Fontsina and J. Daafouza, "Adaptive synchronization of uncertain chaotic Colpitts oscillators based on parameter identification," *Phys. Lett. A*, vol. 339, pp. 304–315, May 2005.

[21] T. Suezaki and S. Mori, "Mutual synchronization of two oscillators," *Trans. IECE*, vol. 48, no. 9, pp. 1551–1557, Sep. 1965.

[22] T. Endo and S. Mori, "Mode analysis of a multimode ladder oscillator," *IEEE Trans. Circuits Syst.*, vol. 23, pp. 100–113, Feb. 1976.

[23] T. Endo and S. Mori, "Mode analysis of two-dimensional low-pass multimode oscillator," *IEEE Trans. Circuits Syst.*, vol. 23, pp. 517–530, Sep. 1976.

[24] T. Endo and S. Mori, "Mode analysis of a ring of a large number of mutually coupled van der Pol oscillators," *IEEE Trans. Circuits Syst.*, vol. 25, no. 1, pp. 7–18, Jan. 1978.

[25] M. Yamauchi, Y. Nishio and A. Ushida, "Phase-waves in a ladder of oscillators," *IEICE Trans. Fundamentals*, vol. E86-A, no. 4, pp. 891–899, Apr. 2003.

[26] Y. Uwate and Y. Nishio, "Synchronization in several types of coupled polygonal oscillatory networks," *IEEE Trans. Circuits Syst. I*, vol. 59, no. 5, pp. 1042–1050, May 2012.

[27] Y. Uwate and Y. Nishio, "Synchronization phenomena in van der Pol oscillators coupled by a time-varying resistor," *Int. J. Bifurcat. Chaos*, vol. 17, no. 10, pp. 3565–3569, 2007.

[28] Y. Uwate, Y. Nishio and R. Stoop, "Complex pattern in a ring of van der Pol oscillators coupled by time-varying resistors," *J. Circuit Syst. Comput.*, vol. 19, no. 4, pp. 819–834, Jun. 2010.

[29] T. Honda, T. Tsubone and N. Kambayashi, "A simple piecewise-constant oscillator coupled by a capacitor," in *Proceedings of International symposium on nonlinear theory and its applications*, pp. 463–466, Xi'an, China, 2002.

[30] K. Suzuki and T. Tsubone, "In-phase and anti-phase synchronization phenomena in coupled systems of piecewise constant oscillators," *IEICE Trans. Fundamentals*, vol. E98-A, no. 1, pp. 340–353, 2015.

[31] I. Shimada and T. Nagashima, "A numerical approach to ergodic problem of dissipative dynamical systems," *Prog. Theor. Phys.*, vol. 61, no. 6, pp. 1605–1616, 1979.

Chapter 6

Master–slave synchronization of hysteresis neural-type oscillators

Kenya Jin'no[1], *Takuya Kurihara*[2] *and Toshimichi Saito*[3]

In this chapter, we focus on the case where the common external force of some relaxation oscillators is a non-periodic signal. The objective relaxation oscillator is regarded as an electronic firefly (EFF) circuit. By using the implementation EFF circuits, we observe the synchronization phenomena induced by the common external force. First, we consider the case where the external force is a periodic pulse signal. Next, we investigate the case where the external force is a non-periodic pulse signal. We clarify that the EFF circuits are achieved the complete synchronization and the stochastic synchronization depending on the external force.

6.1 Introduction

Synchronization phenomena can be universally observed in nature. For example, we can observe a phase synchronization behavior in Southeast Asian fireflies [1]. A circadian rhythm is one example of such synchronization phenomena [2]. The circadian rhythm is that plants and animals are driven by the solar cycle.

Synchronization phenomena are analyzed by oscillator model. We focus on a relaxation oscillator that the threshold is driven by binary external force. The rigorous analysis of this model is relatively easy. For such relaxation oscillator systems, the synchronization phenomena of the coupled relaxation oscillators have analyzed [3,4]. Kohari *et al.* have analyzed the characteristic of the attractors when the system is driven by a periodic external force [5–7]. They have clarified that the coupled relaxation oscillators exhibited a phase-lock by a periodic external force. Also, when the periodic external force is injected the relaxation oscillator, we derived Arnold tongue in the two-parameter bifurcation diagram [6,7].

[1]Department of Electrical and Electronics Engineering, Nippon Institute of Technology, Saitama, Japan
[2]Department of Electrical and Electronics Engineering, Nippon Institute of Technology, Saitama, Japan
[3]Department of Electrical and Electronics Engineering, Hosei University, Tokyo, Japan

In this chapter, we focus on the case where the common external force of some relaxation oscillators is non-periodic signal. The objective relaxation oscillator is regarded as an electronic firefly (EFF) circuit because an optical signal is applied as the input and output of the circuit. By using the implementation EFF circuits, we observe the synchronization phenomena induced by the common non-periodic external force. First, we confirm that EFFs achieve the synchronization by a common periodic pulse signal. Next, we investigate synchronization phenomena of that EFFs are injected a common non-periodic pulse signal whose pulse interval is a uniform random. We examine the relationship of periodic signal's synchronization and non-periodic signal's synchronization.

6.2 Relaxation oscillator with a time-variant threshold

In this chapter, we pay attention to a relaxation oscillator whose dynamics is described by the following equation:

$$\frac{\mathrm{d}}{\mathrm{d}t}x(t) = -x(t) + y(t), \tag{6.1}$$

$$y(t) = h(t, x(t)) = \begin{cases} 1 & x(t) < S(t), \\ -1 & x(t) > W(t), \end{cases} \tag{6.2}$$

$$S(t) = (a_1 - a_2)u(t) + a_2 > W(t), \tag{6.3}$$

$$W(t) = b < S(t), \tag{6.4}$$

where $x(t)$ denotes a state variable of the relaxation oscillator, $y(t)$ denotes a bipolar hysteresis output. $h(t, x(t))$ represents a bipolar hysteresis. $S(t)$ and $W(t)$ are threshold of the binary hysteresis. $S(t)$ is time-variant that is driven by a binary external force of $u(t)$.

Figure 6.1 shows an example of the state variable $x(t)$ and the corresponding bipolar hysteresis output $y(t)$. The lower dashed line in the upper of Figure 6.1 is threshold $W(t)$, and the upper dotted rectangle waveform in Figure 6.1 is the threshold $S(t)$ that is driven by the periodic external force $u(t)$ of period T. $x(t)$ is a state variable that varies between the upper and the lower threshold. Since this system is described by a piecewise linear first-order differential equations, the theoretical analysis of the dynamics of the system is relatively easy.

6.2.1 *Periodic fluctuation threshold*

When the periodic external force injects to the relaxation oscillator, the system exhibits injection locking. In this case, the observed synchronization phenomena can be classified into various switching ratios depending on the intensity and the period of the external force. Figure 6.2 shows two-parameter bifurcation diagrams. The horizontal axis denotes the period T of the injecting external force, and the vertical axis denotes the intensity of the external force that is represented as $(a_1 - a_2)$. In the case of

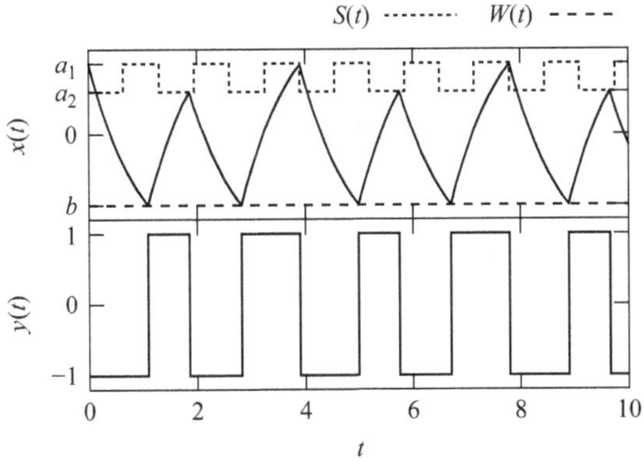

Figure 6.1 Examples of relaxation oscillation waveform ($a_1 = 0.5, a_2 = 0.3, b = -0.5, T = 1.3$)
$x(t)$: state, $y(t)$: bipolar output

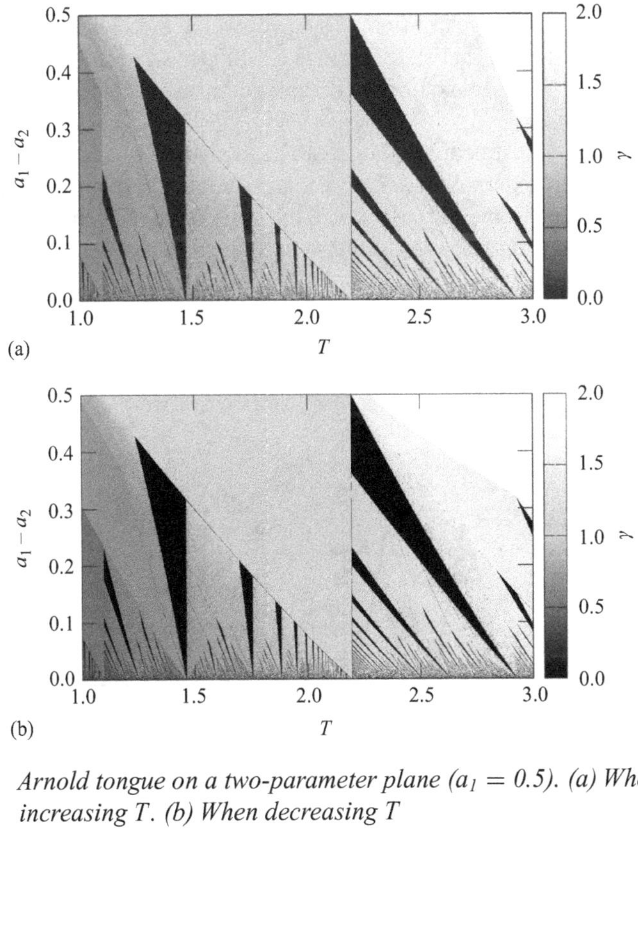

Figure 6.2 Arnold tongue on a two-parameter plane ($a_1 = 0.5$). (a) When increasing T. (b) When decreasing T

Figure 6.2, a_1 is set as 0.5. The shade represents the switching ratio of the external force and relaxation oscillator. The switching ratio is defined by the following equation:

$$\gamma = \frac{\text{The number of switchings of } y(t) \text{ in synchronization period}}{\text{The number of switchings of } u(t) \text{ in synchronization period}}. \tag{6.5}$$

Black region in Figure 6.2 represents a quasi-periodic state, and the other parts are synchronized in some switching rates to the external force. Thus, the relaxation oscillator when the periodic external force is injected has a synchronization area called the Arnold tongue.

6.2.2 Period adjustment capability

The reason why the Arnold tongue is observed is that the relaxation oscillator has a period adjustment capability. Figure 6.3 shows an example of two oscillation waveforms in the case of $\gamma = 1$. Figure 6.3(a) shows the case where the period of the external force is long, and Figure 6.3(b) shows the case where the period of the external force is short. The interval of the frequencies between Figure 6.3(a) and Figure 6.3(b) means the synchronization range of which the switching ratio is 1. The switching point of the internal state depends on the period of the external force. As a result, the period of the relaxation oscillator is adjusted to match the period of the external force.

The frequency adjustment capability can be explained by a formula. We assume that the period of the external force is T, and the switching point in the upper of $x(t)$ is z. In addition, we assume the case where the system archives the synchronization of $\gamma = 1$. The following relationship between the external force period T and the switching point z is derived:

$$z = -\frac{(b-1) + (b+1)\exp(T)}{(b-1) - (b+1)\exp(T)}, \tag{6.6}$$

where b denotes the threshold.

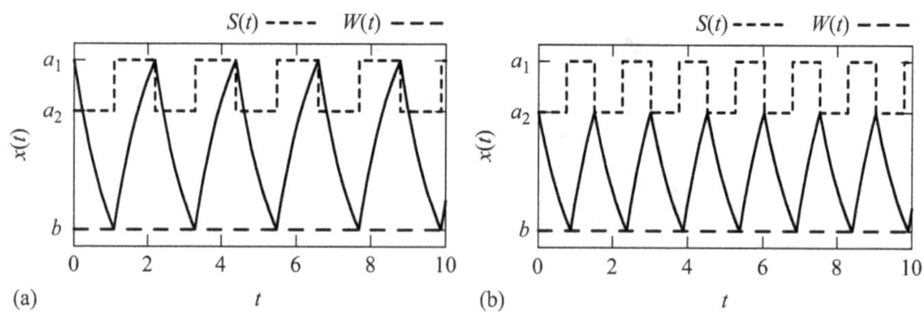

Figure 6.3 The oscillation example of $\gamma = 1$ ($a_1 = 0.5$, $a_2 = 0.2$, $b = -0.5$). (a) The period of external force is long ($T = 2.19$). (b) The period of external force is short ($T = 1.51$)

Equation (6.6) indicates that the switch point z is changed when the external force period T is changed. This is why the synchronization is maintained if external force period is changed. Namely, the changing of the switching point causes the period adjustment capability. z must be located into the interval between a_1 and a_2 to avoid changing the switching ratio. Therefore, the following constraint condition of x is derived:

$$a_2 \leq z \leq a_1. \tag{6.7}$$

The constraint condition as shown in (6.7) gives the synchronization range. The synchronization range is represented as follows:

$$T_s \leq T \leq T_l, \tag{6.8}$$

$$T_l = \ln \frac{a_1 + 1}{b + 1} \frac{b - 1}{a_1 - 1}, \tag{6.9}$$

$$T_s = \ln \frac{a_2 + 1}{b + 1} \frac{b - 1}{a_2 - 1}, \tag{6.10}$$

where T_l and T_s are the upper limit and the lower limit, respectively. T_l is equivalent to the oscillation period when the upper threshold is a_1. T_s is equivalent to the oscillation period when the upper threshold is a_2.

This synchronization range is the case where the switching ratio is 1. By using the analogous consideration for all switch ratio, we derive the two-parameter bifurcation diagram as shown in Figure 6.2.

6.3 The response to non-periodic external force

In this section, we investigate the response of the relaxation oscillators when a common non-periodic external force is injected to the relaxation oscillators. Figure 6.4 shows an example of non-periodic external force. The non-periodic external force is a pulse waveform that has random period, and the duty of the pulse is fixed to 50%. The distribution of the random periods corresponds to the frequency distribution in the external force.

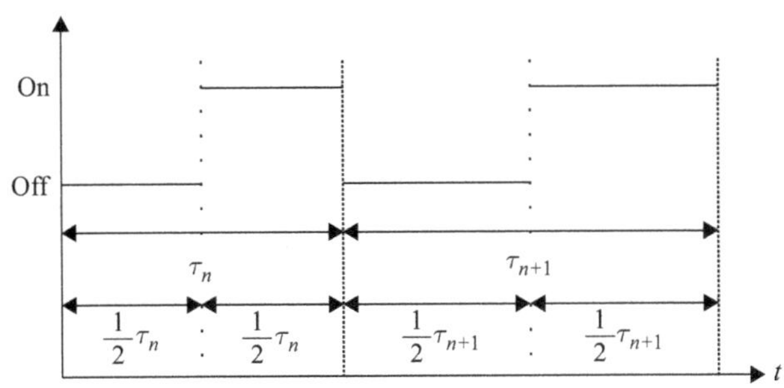

Figure 6.4 Overview of a non-periodic external force

In order to clarify the state of the oscillator when the non-periodic external force is induced, we measure the synchronization state by using an implementation circuit of the relaxation oscillator. Figure 6.5 shows a circuit diagram of the relaxation oscillator model. The threshold of the relaxation oscillator is driven by the binary optical signal. The reason why the optical signal is applied as the injection external force is each oscillator circuit must be insulated. Thus the input and output of each implementation relaxation oscillator are operated optical signals. This circuit can be said to be a type of EFFs [4]. Therefore, this relaxation oscillator circuit is called an EFF in this chapter. Figure 6.6 shows the experimental system. We investigate the phenomena of

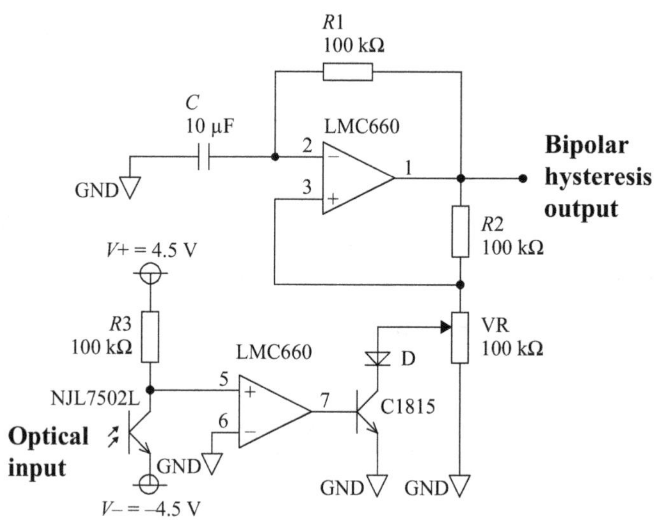

Figure 6.5 Experimental EFF circuit diagram

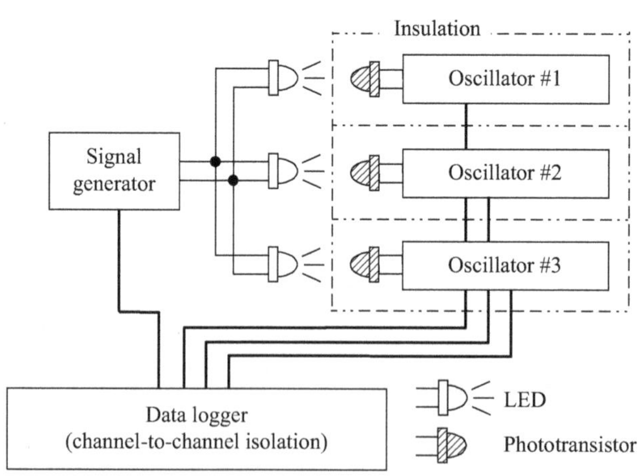

Figure 6.6 Our measurement system

oscillators that a common external force are injected. Our developed signal generation circuit generates the common external force. The data logger (GRAPHTECH GL900-4) records the oscillation states of EFFs and the external force signal. For accurate measurements, each recording channel of the data logger must be insulated each other. The used data logger satisfies this requirement.

6.3.1 Without external force

First, we measure the characteristics of the EFFs without the external force. The lower threshold of each EFF is set to $b = -0.5$. The higher thresholds are set to $a_1 = 0.5$ or $a_2 = 0.4$. Note that a_1 corresponds to the threshold in the case where the binary external force is high state. On the other hand, a_2 corresponds to the threshold in the case where the binary external force is low state. Table 6.1 shows the measured actual oscillation period of each EFF. In Table 6.1, "on" represents the oscillation period in the case where the binary external force is always high; "off" denotes the oscillation period in the case where the binary external force is always low. The oscillation period of each EFF is different since the quality of the actual electronic elements is uneven. These oscillation periods correspond to the synchronization range of $\gamma = 1$. Figure 6.7 shows the synchronization range of each EFF. The each synchronization range is called as the fundamental synchronization range of each EFF in this chapter.

Figure 6.8 shows the oscillation waveforms of EFFs. Figure 6.8(a) shows the temporal binary output waveform. In Figure 6.8, "oscX" indicates the output of the Xth EFF. Figure 6.8(b) shows the pulse train that the pulse corresponds to the falling timing of the temporal binary output waveform. Figure 6.9 shows the enlargement

Table 6.1 Steady-state oscillation period of each EFFs

	Oscillation period (ms)	
	On	**Off**
osc1	230.9	200.5
osc2	234.6	208.2
osc3	224.3	196.5

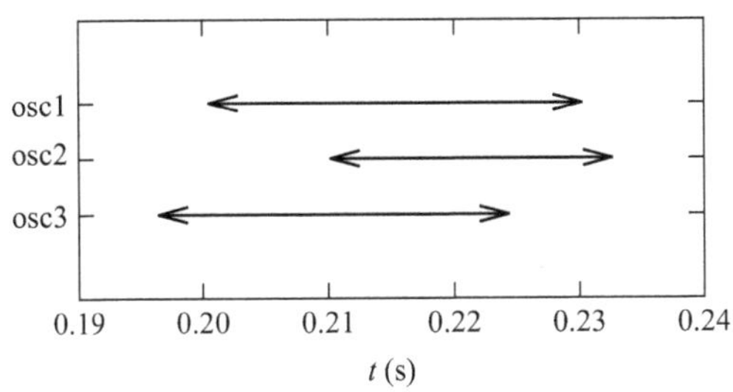

Figure 6.7 Fundamental synchronization range (γ = 1)

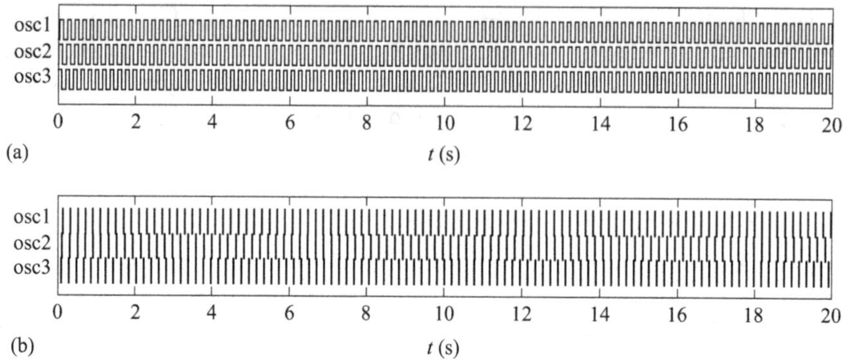

Figure 6.8 The oscillation waveforms without the light external force. (a) The temporal binary output waveform of the EFFs. (b) The pulse train corresponding to the falling timing of the waveform of panel (a)

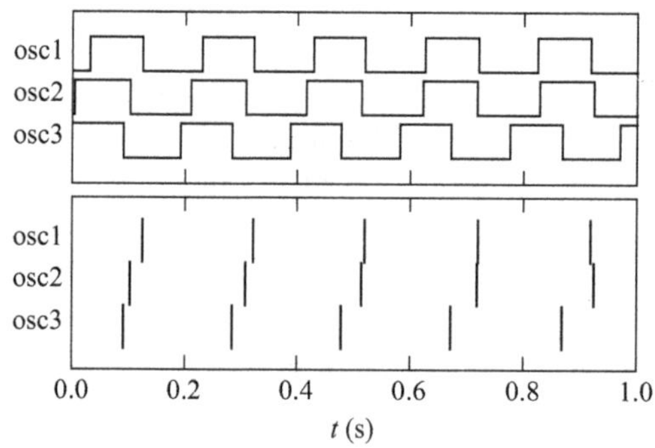

Figure 6.9 Enlargement figure of a part of the temporal binary output waveform and the corresponding pulse train in Figure 6.8

figure of a part of the temporal wave and the corresponding pulse train in Figure 6.8. By using the falling pulse train, we confirm the synchronization states, because the pulse train is more suitable than the temporal binary output waveform. Note that synchronization is not achieved in the case of Figure 6.8.

In order to confirm the synchronization state numerically, we calculate the correlation coefficient by using the temporal binary output waveform. The sampling interval of the measured waveform is 2 ms, and the measurement time is 3 min. Table 6.2 shows the results of the case where the oscillation of each EFF is independent.

The correlation coefficient quantifies the long-term synchronization state. Therefore, the correlation coefficient cannot be detected the short-time synchronization state. So, we introduce a coincidence rate as follows:

$$CR_{ij}(n) = \frac{1}{M}\sum_{\tau=0}^{M-1} s_i(n+\tau)s_j(n+\tau), \tag{6.11}$$

Table 6.2 The correlation coefficient between the binary output of the EFFs

	Correlation coefficient
osc1–osc2	−0.00019
osc1–osc3	0.00015
osc2–osc3	−0.00321

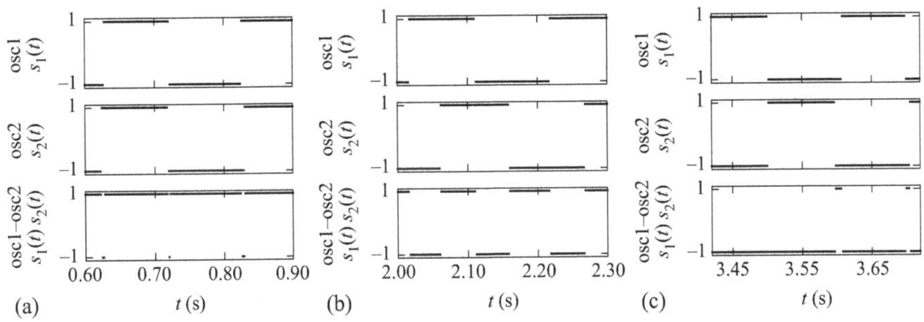

Figure 6.10 Sampled binary outputs of first and second EFF, and the coincidence between these EFFs. (a) $CR_{12} = 0.94$; (b) $CR_{12} = 0.06$; and (c) $CR_{12} = -0.90$

where $s_i(n)$ represents the nth sampled binary output of the ith EFF by the data logger. The sampling interval of the data logger is 2 ms. $CR_{ij}(n)$ denotes the nth coincidence rate between the ith EFF and the jth EFF. M denotes the sample window size to calculate the coincidence rate.

The coincidence rate is equivalent to the moving average of the product of two sampled binary outputs. Figure 6.10 shows the relationship between the coincidence and the sampled binary output. In Figure 6.10, CR_{12} denotes the coincidence rate of these sampled binary output. In this case, the sample window size M is 150 that is equivalent to 300 ms. If the coincidence ratio is close to 1, these waveforms exhibit nearly in-phase synchronization as shown in Figure 6.10(a). On the other hand, if the coincidence ratio is close to −1, these waveforms exhibit nearly the anti-phase synchronization as shown in Figure 6.10(c). Also, if the coincidence rate is a constant, these waveforms exhibit complete synchronization. The coincidence rate of Figure 6.11 is observed in the case where all EFFs do not receive the external force. This result indicates that the coincidence rate continues to change. Therefore, we confirm that the synchronization is not achieved in this situation.

6.3.2 Periodic external force

Next, we measure the case where the EFF received a periodic square wave as an external force. In the experiment, we set parameters as follows: the period of the external

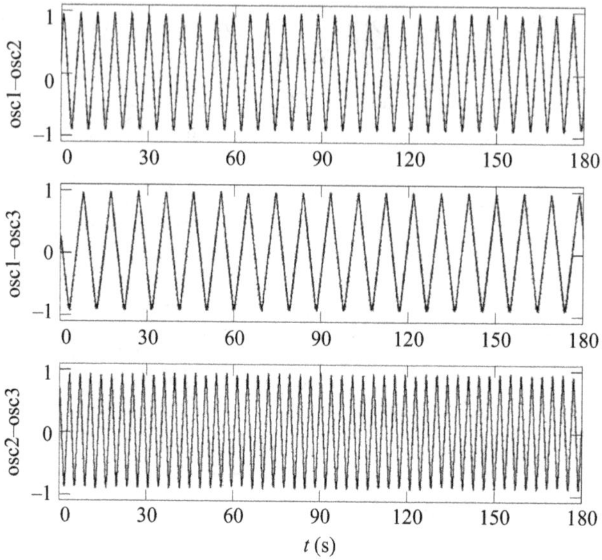

Figure 6.11 Coincidence rates without the common optical external force

force is varied from 50 ms to 350 ms at the interval of 2 ms. The data logger records the binary outputs of three EFFs and the external force. The sampling interval and the measurement time are 2 ms and 3 min, respectively. When a periodic external force is injected into the EFF, EFFs exhibit a phase lock synchronization phenomena which is called an injection locking. Therefore, EFFs achieve synchronization. However, since the quality of the actual electronic elements is uneven, the shape of Arnold tongue of each EFF slightly vary. The reason is that each EFF cannot achieve the synchronization at the edge of Arnold tongue.

Figure 6.12(a) shows the temporal correlation coefficient against the periodic external force. EFFs are synchronized to an external force around $T = 210$ ms. In addition, as expected, the fundamental synchronization range is different in each EFF. Figure 6.12(b) shows the mutual correlation coefficient between the EFFs. These two figures indicate that the synchronization between each EFF with overlap range of the fundamental synchronization range. The overlap range is called as an overlap synchronization range in this chapter.

Figure 6.13 represents the falling pulse train of the measured waveform. Figure 6.13(b) is an example pulse train which is observed at the center of the overlap synchronization range; $T = 210$ ms. In this case, because the pulse is aligned, three EFFs achieve complete synchronization. Figure 6.13(a) and (c) is example trains which are observed at the edge of the overlap synchronization range. The period of the external force of Figure 6.13(a) is 200 ms, and the period of Figure 6.13(c) is 222 ms. In these cases, a part of the pulse train does not exhibit synchronization. This fact indicates that the complete synchronization is unlikely to occur at the edge of the overlap synchronization range.

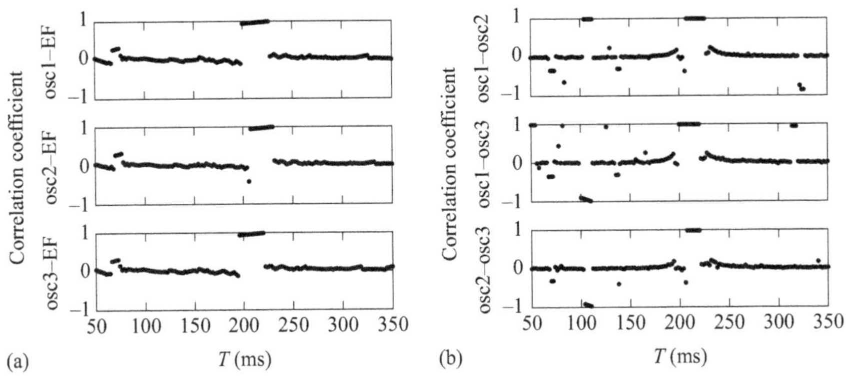

Figure 6.12 Relationship between the period of the external force and each correlation coefficient. (a) Correlation coefficient between each EFF and external force. (b) Correlation coefficient between EFFs

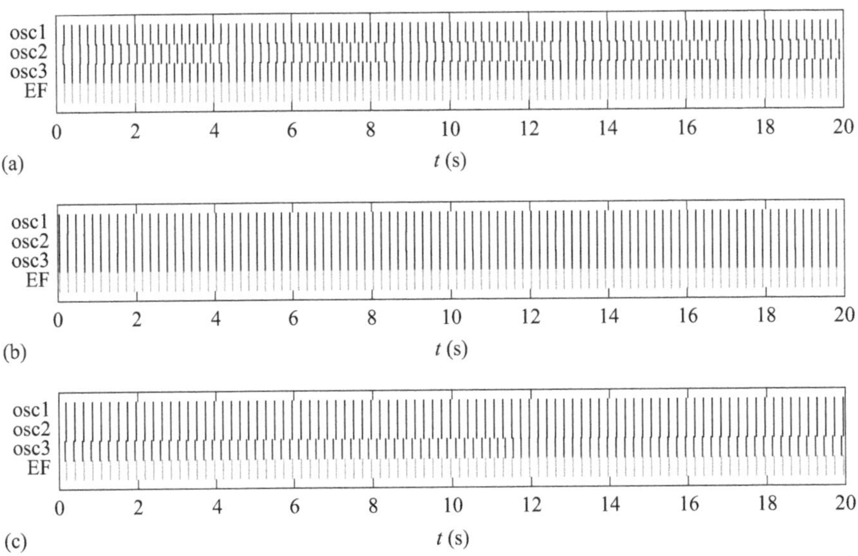

Figure 6.13 Pulse trains of EFFs and the periodic external force. (a) T = 200 ms; (b) T = 210 ms; and (c) T = 222 ms

Figure 6.14(a)–(c) illustrates the temporal coincidence rates of period 200, 210, and 222 ms, respectively. In the case of $T = 210$ ms, all temporal coincidence rate becomes almost 1. Namely, these EFFs achieve the synchronization. On the other hand, in the cases of $T = 200$ ms and $T = 222$ ms, the temporal coincidence rate continue to change periodically excepting a certain EFF. Therefore, the synchronization is not achieved in these cases. These results indicate that to achieve the synchronization is quite difficult at the edge of the overlap synchronization range.

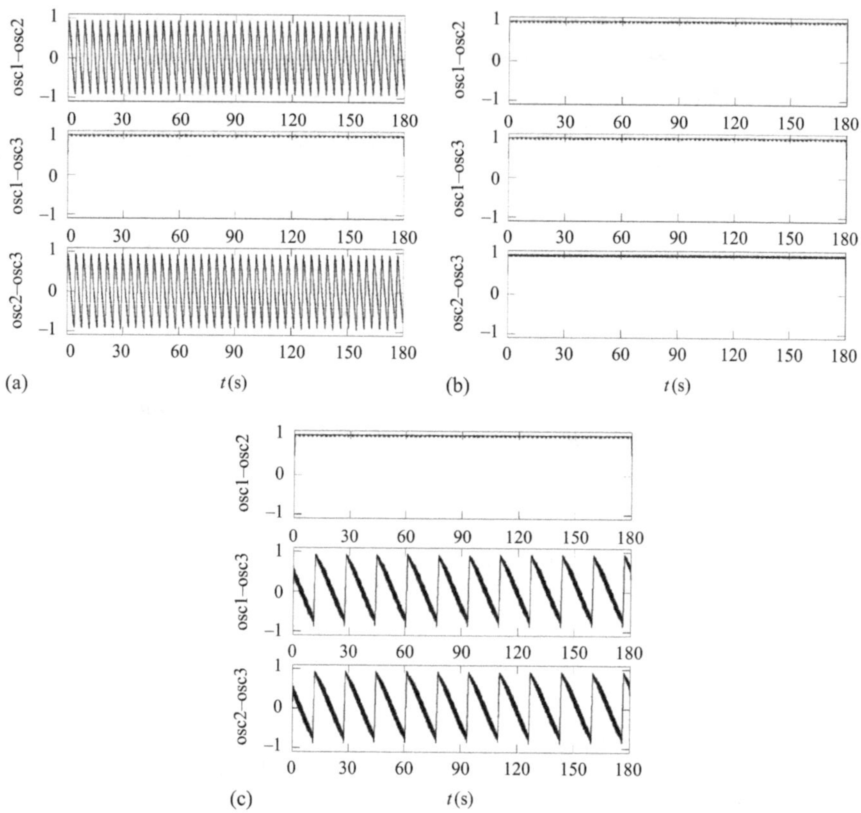

Figure 6.14 The temporal coincidence rate of each EFF when periodic external force is injected. (a) T = 200 ms; (b) T = 210 ms; and (c) T = 222 ms

6.3.3 Uniform random period external force

We measure synchronization phenomena when a non-periodic pulse train is injected into EFFs. The period of the non-periodic external force is driven by the uniform distribution random noise. The period τ_n is fluctuated as follows:

$$\tau_n = \text{Uniform}(T - \Delta T, T + \Delta T), \tag{6.12}$$

where τ_n is a uniform distribution random number in the range from $T - \Delta T$ to $T + \Delta T$, and the expected value is T. ΔT is set to 10 ms. The excepted period T is varied from 50 to 350 ms in interval at 2 ms. The data logger records the measurement data. The sampling interval is 2 ms, and the measurement time 3 min.

Figure 6.15 shows the temporal correlation coefficient with respect to the non-periodic external force. In this case, the synchronization range becomes narrow comparing with the case of periodic external force.

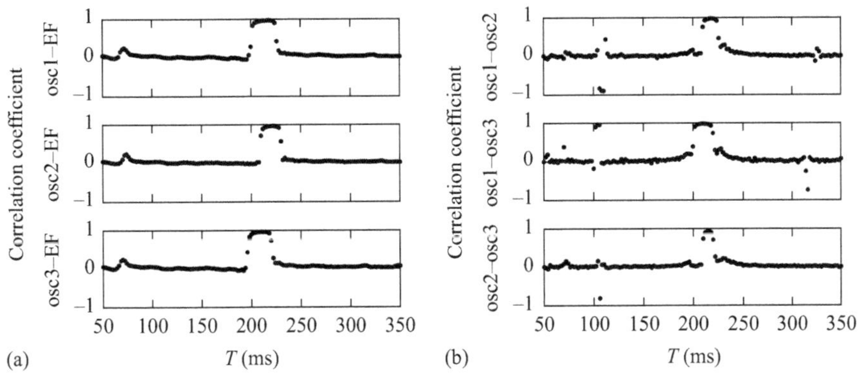

Figure 6.15 Relationship between the expected period value of external force and each correlation coefficient. (a) Correlation coefficient between each EFF and external force. (b) Correlation coefficient between each EFF

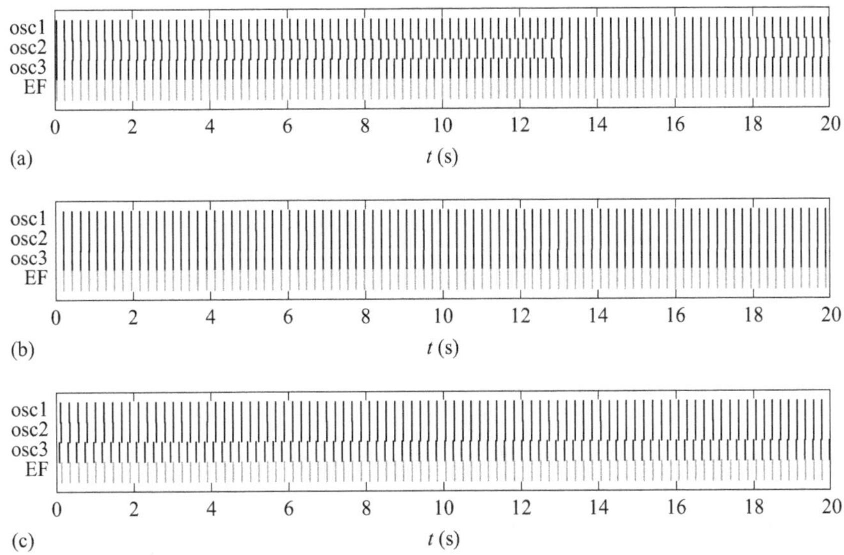

Figure 6.16 Falling pulse train of each EFF with the non-periodic external force. T means the expected period value of the non-periodic external force. (a) T = 208 ms; (b) T = 216 ms; and (c) T = 222 ms

Figure 6.16 shows an example temporal pulse train of each EFF. Figure 6.16(b) shows the pulse train of $T = 216$ ms. The falling pulse train is aligned despite the external force is non-periodic. In this case, the EFFs achieve complete synchronization. Figure 6.16(a) and (c) illustrates the falling pulse trains of $T = 208$ ms and $T = 222$ ms, respectively. These pulse trains are observed at the edge of the

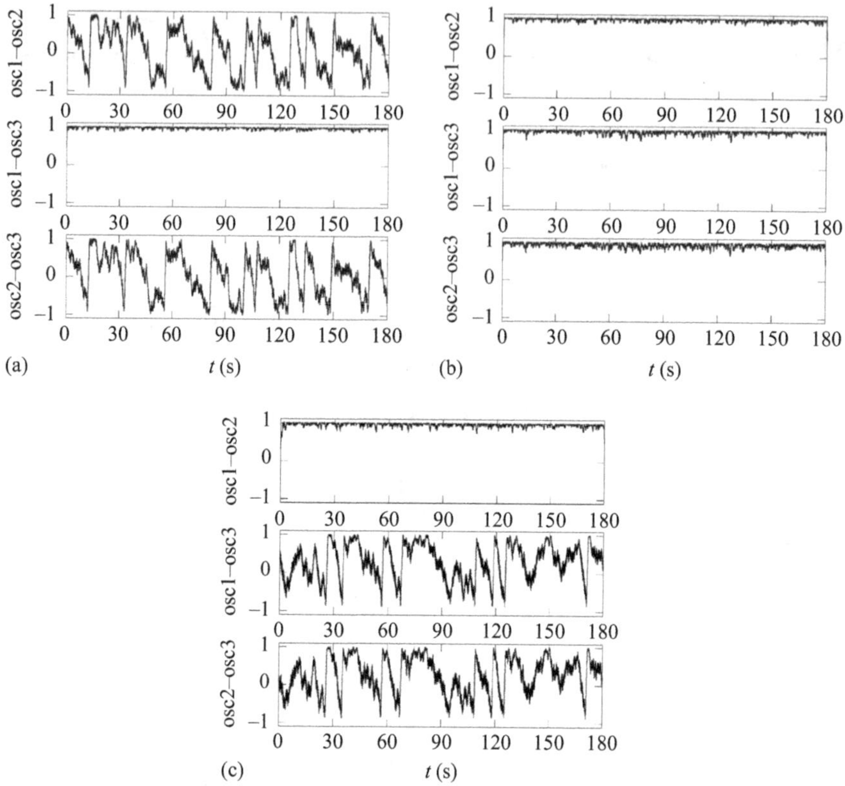

Figure 6.17 The period distribution in the non-periodic external force. T denotes the expected value of the non-periodic external force. (a) T = 208 ms; (b) T = 216 ms; and (c) T = 222 ms

overlap synchronization range. In these cases, the synchronization state is broken occasionally; however, the broken synchronization state is recovered quickly. Such synchronization phenomena are classified into a stochastic synchronization.

Figure 6.17 represents the temporal coincidence rate. Figure 6.17(b) is the temporal coincidence rate in the case of $T = 216$ ms. In this case, the coincidence rate is almost 1. Therefore, we can say that the EFFs achieve the complete synchronization robustly. Figure 6.17(a) and (c) shows the temporal coincidence rate in $T = 208$ ms and $T = 222$ ms, respectively. In these cases, the temporal coincidence rates continue to change irregularly excepting a certain EFF. Therefore, the synchronization is not achieved in these cases. However, the temporal pulse trains as shown in Figure 6.17(a) and (c) indicate that the EFFs achieve the stochastic synchronization.

We pay attention to the period distribution in the injecting non-periodic external force. Figure 6.18 represents the actual period histogram that corresponds to the period distribution in the non-periodic external force. T denotes the excepted value of the non-periodic external force. The width of each arrow in Figure 6.18 indicates

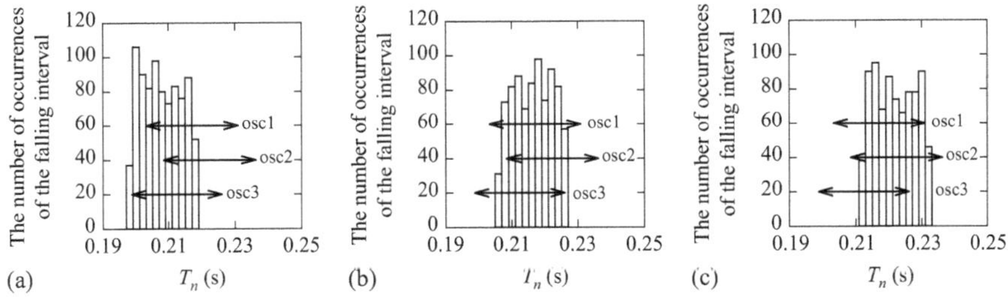

Figure 6.18 Histogram that represents the period distribution of external force. Each arrow denotes the fundamental synchronization range. (a) T = 208 ms; (b) T = 216 ms; and (c) T = 222 ms

the fundamental synchronization range when the periodic external force is injected to each EFF. The derived switching rate γ is 1. In the case of $T = 216$ ms, the period distribution fits into the fundamental synchronization range of each EFF. We consider this is the cause of generation of the complete synchronization. Namely, if the period distribution of the non-periodic external force fits into the overlap synchronization range, EFFs exhibit the complete synchronization phenomena in spite of the external forces are non-periodic. In the cases of $T = 208$ ms and 222 ms, the period distributions are out of the overlap synchronization range. We think that it is difficult to synchronize in the case where the period distribution of the injected non-periodic external force is out of the overlap synchronization range of EFFs.

6.4 Conclusions

In this chapter, we observed the synchronization phenomena by EFFs that these thresholds are driven by common non-periodic external signals. As a result, we confirmed that EFFs exhibit the synchronization phenomena induced by common non-periodic signals. We clarified the relationship between the individual fundamental synchronization range and the noise synchronization range by the implementation measurements. In other words, the synchronization range by the periodic signal injection affects the achievement of the noise synchronization. In the synchronization range, we confirmed the complete synchronization and the stochastic synchronization. The rigorous theoretical analysis is our future problem.

References

[1] H. Tanaka, A. Hasegawa, and N. Ohba, “Percolation effect on population synchrony,” *IEICE Trans. Fundamentals (Japanese Edition)* vol. J84-A, no. 6, pp. 870–874, 2001 (in Japanese).

[2] S. Kai, "Onset of circadian rhythms and noise-induced phenomena in plants," *J. Plasma Fusion Res.*, vol. 88, no. 7, pp. 368–373, 2012 (in Japanese).

[3] K. Jin'no, "Synchronization phenomena for graph coloring problem," In: *Visarath In, Patrick Longhini, Antonio Palacios (eds.) Applications of Nonlinear Dynamics: Model and Design of Complex Systems*, Springer, pp. 331–340, Berlin, 2009.

[4] M. Sekikawa, K. Kimoto, T. Kohno, H. Kawakami, and K. Aihara, "Synchronization phenomena in square-wave oscillators with optical coupling," The Institute of Electronics, Information and Communication Engineers (IEICE) Technical Report, CAS2011-45, 2010–11 (in Japanese).

[5] K. Kohari, T. Saito, and H. Kawakami, "On a hysteresis oscillator including periodic thresholds," *IEICE Trans. Fundamentals*, vol. E76-A, no. 12, pp. 2102–2107, 1993.

[6] T. Kurihara and K. Jin'no, "A study of hysteresis relaxation oscillator driven by a periodic square wave," The Institute of Electronics, Information and Communication Engineers (IEICE) Technical Report, NLP2013-167, 2013–14 (in Japanese).

[7] T. Kurihara and K. Jin'no, "Analysis of a relaxation oscillator driven by a square wave external force," in *Proceedings of 2014 International Conference on Nonlinear Theory and its Applications (NOLTA2014)*, pp. 727–730, Luzern, Switzerland, September, 2014.

Chapter 7
Multimode oscillations in coupled hard oscillators

Kuniyasu Shimizu[1] *and Tetsuro Endo*[2]

This chapter investigates multimode oscillations and localized oscillation modes in inductor-coupled hard oscillators, which are considered as a form of a van der Pol oscillators with relatively higher-order nonlinearity. First, the method of averaging for weakly nonlinear oscillators is surveyed. Then, the averaged equation of the two inductor-coupled oscillators is derived. In particular, we show that a double-mode oscillation is stably excited in the coupled system for weakly nonlinear oscillators. When the nonlinearity of the coupled system becomes stronger, the double-mode oscillation disappears, whereas a localized oscillation mode emerges. Furthermore, we investigate a propagating wave phenomenon and a localized mode observed in the six inductor-coupled hard-oscillator rings. We show that the propagating wave corresponds to a multimode oscillation with several dominant peaks in a spectral distribution, and that the solution originates in the neighbourhood of a local bifurcation point of the localized mode.

7.1 Introduction

Recently, localized oscillation modes in nonlinear lattices have attracted considerable attention [1,2]. In particular, studies of moving localized modes in nonlinear lattices have attracted constant interest for many decades, examples include propagating waves in reaction–diffusion systems [3], FitzHugh–Nagumo dynamics [4,5], and micro-electro-mechanical systems [6,7]. In general, a waveform of a propagating wave at a nonlinear site undergoes a multimode oscillation because a large-amplitude excitation in one oscillator propagates to several adjacent oscillators. This chapter focuses on a multimode oscillation observed in coupled 'hard oscillators', which are considered as a form of a van der Pol oscillator with relatively higher-order nonlinearity.

[1]Department of Information and Communication Systems Engineering, Chiba Institute of Technology, Narashino, Chiba, Japan
[2]Department of Electronics and Bioinformatics, Meiji University, Kawasaki, Kanagawa, Japan

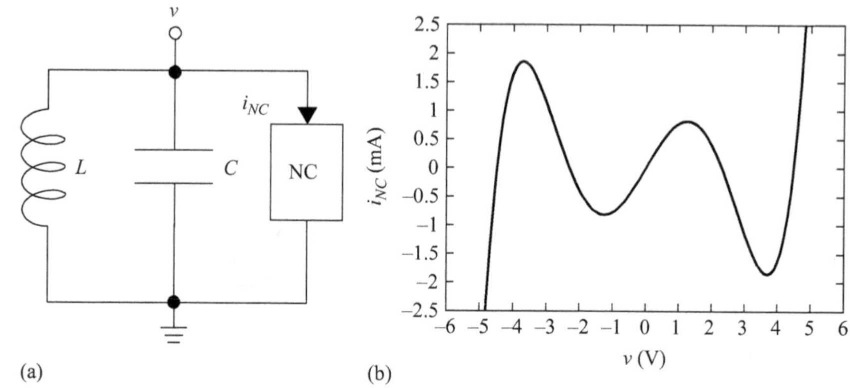

Figure 7.1 Hard oscillator. (a) Schematic circuit diagram and (b) voltage–current (v–i_{NC}) characteristics of NC for $g_1 = 1.000 \times 10^{-3}$ A/V, $g_3 = 2.216 \times 10^{-4}$ A/V^3, and $g_5 = 8.645 \times 10^{-6}$ A/V^5

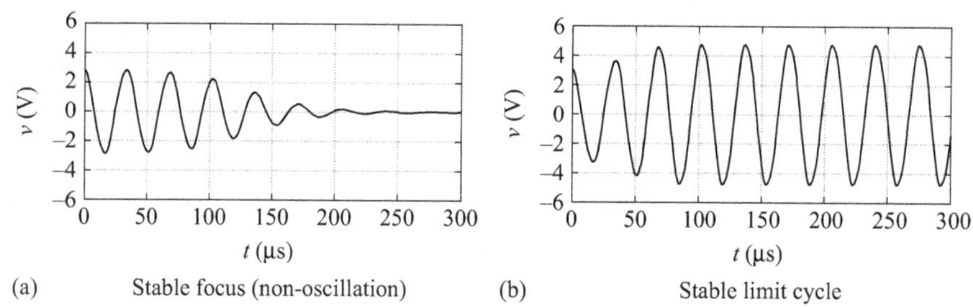

Figure 7.2 Two numerically obtained steady states in the hard oscillator for L = 20 mH, C = 15 nF with the NC in Figure 7.1(b). The initial voltages of v(0) for the panels (a) and (b) are 2.85 V and 3.07 V, respectively

A hard oscillator consists of an inductor (L), capacitor (C), and nonlinear conductance (NC) as shown in Figure 7.1(a). The voltage–current (v–i_{NC}) characteristic of the NC is assumed to be given by a fifth-order polynomial:

$$i_{NC}(v) = g_1 v - g_3 v^3 + g_5 v^5, \quad g_1, g_3, g_5 > 0. \tag{7.1}$$

Figure 7.1(b) shows the characteristic curve of the NC for $g_1 = 1.000 \times 10^{-3}$ A/V, $g_3 = 2.216 \times 10^{-4}$ A/V^3, and $g_5 = 8.645 \times 10^{-6}$ A/V^5. The NC operates as a passive resistor when a low voltage is applied to the oscillator, and the LC resonance oscillation gradually decays to a non-oscillating state (Figure 7.2(a)). On the other hand, a limit cycle oscillation exists when a high voltage is injected into the capacitor because the voltage across the nonlinear element passes through both end points of the extrema of the characteristic curve for the NC as shown in Figure 7.2(b). Therefore, the oscillator has two steady states, i.e. a stable focus (non-oscillation) and a limit cycle oscillation, and is called a hard oscillator (also known as a bistable oscillator) [8].

In this chapter, to introduce the multimode oscillation and localized mode in coupled hard oscillators, two inductor-coupled hard oscillators are investigated for weakly and strongly nonlinear oscillators. First, the method of averaging for weakly nonlinear oscillators, known as the Krylov–Bogoliubov technique, is explained. Then, the averaged equation of the two inductor-coupled hard oscillators is derived. Specifically, we show that a double-mode oscillation is stably excited in a coupled system. When the nonlinearity of the coupled system becomes stronger, the double-mode oscillation disappears, whereas a localized oscillation mode, which cannot be obtained by the averaged equation, emerges. We investigate a quasi-periodic oscillation that is observed immediately after the localized mode disappears by comparing the waveform and frequency distribution in a power spectrum with the result in the usual double-mode oscillation for weakly nonlinear oscillators.

Moreover, we investigate a propagating wave observed in six inductor-coupled hard-oscillator rings. Because of the rotational symmetry of this system, there exist diverse localized phenomena. The propagating wave is a travelling localized mode where a localized large-amplitude excitation propagates in an oscillator array and emerges near a stationary localized mode for strongly nonlinear oscillators. This solution corresponds to a multimode oscillation with several dominant peaks in a power spectrum.

7.2 Two inductor-coupled hard oscillators

7.2.1 *Weakly nonlinear oscillators*

A perturbation method provides important insight into the solutions for this system in the weakly nonlinear case. The method of averaging uses the idea of slow and fast variables to study slow variations of the amplitudes and phases and is a powerful tool to investigate the oscillation modes of weakly coupled oscillators [9–12]. In general, the weakly N-coupled oscillators are described by:

$$\ddot{y}_k + \lambda_k y_k = \varepsilon f_k(\boldsymbol{y}, \dot{\boldsymbol{y}}) \quad (k = 1, 2, \ldots, N) \quad \left(\cdot = \frac{\mathrm{d}}{\mathrm{d}\tau}, \cdot\cdot = \frac{\mathrm{d}^2}{\mathrm{d}\tau^2} \right), \tag{7.2}$$

where $\boldsymbol{y} = [y_1, y_2, \ldots, y_k]^{\mathrm{T}}$ $(k = 1, 2, \ldots, N)$ and τ correspond to the state variables and a (normalized) time, respectively. We assume that the value of the parameter ε is sufficiently small. The $\sqrt{\lambda_k}$ corresponds to the natural frequency of the kth individual oscillator when $\varepsilon = 0$.

7.2.1.1 Averaged equation

The idea of two time scales, $(T_n = \varepsilon^n \tau,\ n = 0, 1)$, is applied to derive the averaged equation for the weakly nonlinear oscillators [9]. That is, we assume that the variables

y_k, $k = 1, 2, \ldots, N$, are a function of the two independent first and slow time scales, $y_k(T_0, T_1)$. Then, the second-order time derivative is written as:

$$\begin{aligned}\frac{\mathrm{d}^2 y_k}{\mathrm{d}\tau^2} &= \frac{\partial^2 y_k}{\partial T_0^2}\left(\frac{\mathrm{d}T_0}{\mathrm{d}\tau}\right)^2 + 2\frac{\partial^2 y_k}{\partial T_0 \partial T_1}\frac{\mathrm{d}T_0 \mathrm{d}T_1}{\mathrm{d}\tau^2} + \frac{\partial^2 y_k}{\partial T_1^2}\left(\frac{\mathrm{d}T_1}{\mathrm{d}\tau}\right)^2 \\ &= \delta_{T_0T_0} y_k + 2\varepsilon \delta_{T_0T_1} y_k + \varepsilon^2 \delta_{T_1T_1} y_k, \\ &\left(\delta_{T_nT_n} \equiv \frac{\partial^2}{\partial T_n^2},\ n = 1, 2 \quad \delta_{T_0T_1} \equiv \frac{\partial^2}{\partial T_0 \partial T_1}\right).\end{aligned} \tag{7.3}$$

In addition, based on perturbation theory, the solutions of (7.2) are approximated in the form of a power series of ε as follows:

$$y_k = y_k^0 + \varepsilon y_k^1. \tag{7.4}$$

Substituting (7.4) into (7.3), the second-order derivative is given by:

$$\ddot{y}_k = \delta_{T_0T_0} y_k^0 + \varepsilon(\delta_{T_0T_0} y_k^1 + 2\delta_{T_0T_1} y_k^0) + O(\varepsilon^2). \tag{7.5}$$

Substituting (7.4) and (7.5) into (7.2) (the magnitude of $O(\varepsilon^2)$ is ignored in the following because ε is considered to be sufficiently small), we obtain the following equation:

$$\delta_{T_0T_0} y_k^0 + \varepsilon(\delta_{T_0T_0} y_k^1 + 2\delta_{T_0T_1} y_k^0) + \lambda_k(y_k^0 + \varepsilon y_k^1) = \varepsilon f_k(\boldsymbol{y}, \dot{\boldsymbol{y}}). \tag{7.6}$$

Equating the coefficients of equal powers ($O(\varepsilon^n)$, $n = 0, 1$) of ε to zero results in:

$$O(1) : \delta_{T_0T_0} y_k^0 + \lambda_k y_k^0 = 0, \tag{7.7}$$

$$O(\varepsilon) : \delta_{T_0T_0} y_k^1 + \lambda_k y_k^1 = -2\delta_{T_0T_1} y_k^0 + f_k(\boldsymbol{y}, \dot{y}_k). \tag{7.8}$$

The solutions of (7.7) are written in the form of a simple harmonic oscillation: $y_k^0 = a_k \sin(\omega_k \tau + \phi_k)$, where $\omega_k \equiv \sqrt{\lambda_k}$.

One of the averaging methods, known as the Krylov–Bogoliubov technique, supposes that the amplitude and the phase of the solutions are a function of the slow time scale, $T_1 (= \varepsilon\tau)$ [9]. Then, the solutions of (7.7) can be represented by:

$$y_k^0 = a_k(T_1) \sin(\omega_k \tau + \phi_k(T_1)). \tag{7.9}$$

Note that the amplitude $a_k(T_1)$ and the phase $\phi_k(T_1)$ change very little during the time $T^k = 2\pi/\omega_k$ because they are slowly varying functions of time τ. Substituting (7.9) into (7.8), we find:

$$\begin{aligned}\delta_{T_0T_0} y_k^1 + \omega_k^2 y_k^1 &= -2\delta_{T_0T_1}\{a_k(T_1)\sin(\omega_k\tau + \phi_k(T_1))\} + f_k(\boldsymbol{y}, \dot{\boldsymbol{y}}) \\ &= -2\left\{\omega_k \frac{\partial a_k(T_1)}{\partial T_1}\cos(\omega_k\tau + \phi_k(T_1))\right. \\ &\qquad \left. - \omega_k a_k(T_1)\sin(\omega_k\tau + \phi_k(T_1))\frac{\partial \phi(T_1)}{\partial T_1}\right\} + f_k(\boldsymbol{y}, \dot{y}_k).\end{aligned} \tag{7.10}$$

Because the function $f_k(\boldsymbol{y}, \dot{\boldsymbol{y}})$ is a 2π periodic function of θ_k ($\equiv \omega_k\tau + \phi_k$), it can be approximated by the Fourier series. Therefore, (7.10) can be represented by:

$$\begin{aligned} \delta_{T_0T_0}y_k^1 + \omega_k^2 y_k^1 &= -2\left\{\omega_k \frac{\partial a_k(T_1)}{\partial T_1}\cos\theta_k - \omega_k a_k(T_1)\sin\theta_k \frac{\partial\phi(T_1)}{\partial T_1}\right\} \\ &\quad + \sum_{n=0}^{\infty} a_n^k \cos n\theta_k + \sum_{m=1}^{\infty} b_m^k \sin m\theta_k, \end{aligned} \tag{7.11}$$

where the Fourier coefficients are given by:

$$\begin{aligned} a_0^k &= \frac{1}{T^k}\int_0^{T^k} f_k(\boldsymbol{y}, \dot{\boldsymbol{y}})\mathrm{d}\tau = \frac{\omega_k}{2\pi}\int_0^{2\pi} f_k(\boldsymbol{y}, \dot{\boldsymbol{y}})\mathrm{d}\theta, \\ a_n^k &= \frac{2}{T^k}\int_0^{T^k} f_k(\boldsymbol{y}, \dot{\boldsymbol{y}})\cos n\theta_k \,\mathrm{d}\tau = \frac{\omega_k}{\pi}\int_0^{2\pi} f_k(\boldsymbol{y}, \dot{\boldsymbol{y}})\cos n\theta_k \,\mathrm{d}\theta \quad (n \geqq 1), \\ b_n^k &= \frac{2}{T^k}\int_0^{T^k} f_k(\boldsymbol{y}, \dot{\boldsymbol{y}})\sin n\theta_k \,\mathrm{d}\tau = \frac{\omega_k}{\pi}\int_0^{2\pi} f_k(\boldsymbol{y}, \dot{\boldsymbol{y}})\sin n\theta_k \,\mathrm{d}\theta \quad (n \geqq 1). \end{aligned} \tag{7.12}$$

It should be noted that the resonant terms in the right-hand side of (7.11) are $\left(-2\omega_k \frac{\partial a_k(T_1)}{\partial T_1} + a_1^k\right)\cos\theta_k$ and $\left(2\omega_k a_k(T_1)\frac{\partial\phi(T_1)}{\partial T_1} + b_1^k\right)\sin\theta_k$. Therefore, to avoid the secular terms of the solutions, we need the following relationships:

$$\begin{aligned} \frac{\partial a_k}{\partial T_1} &= \frac{a_1^k}{2\omega_k}, \\ &= \frac{1}{2\pi}\int_0^{2\pi} f_k(\boldsymbol{y}, \dot{\boldsymbol{y}})\cos\theta_k \,\mathrm{d}\theta, \\ \frac{\partial\phi_k}{\partial T_1} &= \frac{-b_1^k}{2\omega_k a_k}, \\ &= \frac{-1}{2\pi a_k}\int_0^{2\pi} f_k(\boldsymbol{y}, \dot{\boldsymbol{y}})\sin\theta_k \,\mathrm{d}\theta. \end{aligned} \tag{7.13}$$

Substituting $T_1 = \varepsilon\tau$, we get:

$$\begin{aligned} \dot{a}_k &= \frac{\varepsilon}{2\pi}\int_0^{2\pi} f_k(\boldsymbol{y}, \dot{\boldsymbol{y}})\cos\theta_k \,\mathrm{d}\theta, \\ \dot{\phi}_k &= \frac{-\varepsilon}{2\pi a_k}\int_0^{2\pi} f_k(\boldsymbol{y}, \dot{\boldsymbol{y}})\sin\theta_k \,\mathrm{d}\theta. \end{aligned} \tag{7.14}$$

Equation (7.14) is the averaged equation of the weakly nonlinear oscillators in (7.2).

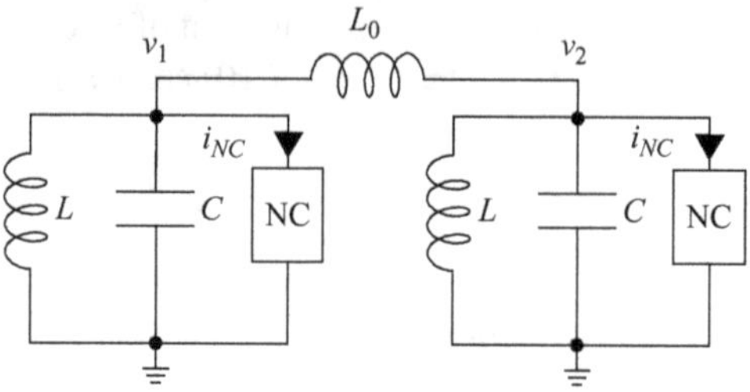

Figure 7.3 Two inductor-coupled hard oscillators

7.2.1.2 Averaged equation of two coupled hard oscillators

In this subsection, the derivation of the averaged equations for the slow variations of two inductor-coupled hard oscillators is explained. Figure 7.3 shows the two hard oscillators connected by inductors (L_0). From Kirchhoff's law, the circuit equation for Figure 7.3 is written as:

$$\frac{d^2 v_k}{dt^2} + \frac{g_1}{C}\left(1 - \frac{3g_3}{g_1}v_k^2 + \frac{5g_5}{g_1}v_k^4\right)\frac{dv_k}{dt} + \left(\frac{1}{LC} + \frac{1}{L_0 C}\right)v_k - \frac{1}{L_0 C}(v_{k+1}) = 0, \tag{7.15}$$

$$k = 1, 2 \quad (v_3 = v_1).$$

By changing each parameter and variable as:

$$t = \frac{\tau}{\sqrt{\left(\frac{1}{LC}\right) + \left(\frac{1}{L_0 C}\right)}}, \quad v_k = \sqrt[4]{\frac{g_1}{5g_5}}\, x_k, \quad \varepsilon \equiv \frac{g_1}{\sqrt{\left(\frac{C}{L}\right) + \left(\frac{C}{L_0}\right)}},$$
$$\alpha \equiv \frac{L}{L + L_0}, \quad \beta \equiv 3\frac{g_3}{\sqrt{5g_1 g_5}}, \tag{7.16}$$

the normalized version of (7.15) in a vector form is written as:

$$\ddot{\boldsymbol{x}} + \varepsilon\left(\dot{\boldsymbol{x}} - \frac{1}{3}\beta\dot{\boldsymbol{x}}_c + \frac{1}{5}\dot{\boldsymbol{x}}_f\right) + \boldsymbol{B}\boldsymbol{x} = 0 \quad \left(\cdot = \frac{d}{d\tau}, \cdot\cdot = \frac{d^2}{d\tau^2}\right), \tag{7.17}$$

where $\boldsymbol{x} = [x_1, x_2]^{\mathrm{T}}$, $\boldsymbol{x}_c = [x_1^3, x_2^3]^{\mathrm{T}}$, $\boldsymbol{x}_f = [x_1^5, x_2^5]^{\mathrm{T}}$, and $\boldsymbol{B} = \begin{bmatrix} 1 & -\alpha \\ -\alpha & 1 \end{bmatrix}$.
The parameter β determines the amplitude of the oscillation, whereas α $(0 \leqq \alpha \leqq 1)$ is the coupling factor. The parameter ε $(\geqq 0)$ is a damping coefficient and indicates the degree of nonlinearity.

The eigenvalues ($=\lambda_k$) of $\boldsymbol{B}$ and the corresponding eigenvectors ($=\boldsymbol{p}_k$) for λ_k are expressed by:

$$\lambda_1 = 1-\alpha, \qquad \lambda_2 = 1+\alpha,$$
$$\boldsymbol{p}_1 = \left[\frac{1}{\sqrt{2}}, \ \frac{1}{\sqrt{2}}\right]^{\mathrm{T}}, \quad \boldsymbol{p}_2 = \left[\frac{1}{\sqrt{2}}, \ \frac{-1}{\sqrt{2}}\right]^{\mathrm{T}}, \quad k = 1,2\ (\lambda_1 < \lambda_2). \tag{7.18}$$

Using a transformation matrix $\boldsymbol{P}$ ($=[\boldsymbol{p}_1, \boldsymbol{p}_2]$), $\boldsymbol{P}^{-1}\boldsymbol{B}\boldsymbol{P}$ becomes a diagonal matrix in which the diagonal elements are λ_k, $k=1,2$. By introducing new variables $\boldsymbol{y}$ ($\equiv \boldsymbol{P}^{-1}\boldsymbol{x}$), (7.17) is written as:

$$\ddot{\boldsymbol{y}} + \boldsymbol{P}^{-1}\boldsymbol{B}\boldsymbol{P}\boldsymbol{y} = \varepsilon\left(-\dot{\boldsymbol{y}} + \frac{1}{3}\beta\boldsymbol{P}^{-1}\dot{\boldsymbol{x}}_c - \frac{1}{5}\boldsymbol{P}^{-1}\dot{\boldsymbol{x}}_f\right). \tag{7.19}$$

Therefore, (7.19) is represented in a scalar form as:

$$\ddot{y}_k + \lambda_k y_k = \varepsilon f_k(\boldsymbol{y}, \dot{\boldsymbol{y}}) \quad (k = 1,2), \tag{7.20}$$

where:

$$\boldsymbol{f}(\boldsymbol{y}, \dot{\boldsymbol{y}}) \left(= [f_1(\boldsymbol{y}, \dot{\boldsymbol{y}}), f_2(\boldsymbol{y}, \dot{\boldsymbol{y}})]^{\mathrm{T}}\right) \equiv -\dot{\boldsymbol{y}} + \frac{1}{3}\beta\boldsymbol{P}^{-1}\dot{\boldsymbol{x}}_c - \frac{1}{5}\boldsymbol{P}^{-1}\dot{\boldsymbol{x}}_f, \tag{7.21}$$

$$\begin{aligned}\boldsymbol{P}^{-1}\dot{\boldsymbol{x}}_c &= \begin{bmatrix} \dfrac{\mathrm{d}}{\mathrm{d}\tau}\left\{\dfrac{1}{\sqrt{2}}\left(\dfrac{1}{\sqrt{2}}y_1 + \dfrac{1}{\sqrt{2}}y_2\right)^3 + \dfrac{1}{\sqrt{2}}\left(\dfrac{1}{\sqrt{2}}y_1 - \dfrac{1}{\sqrt{2}}y_2\right)^3\right\} \\ \dfrac{\mathrm{d}}{\mathrm{d}\tau}\left\{\dfrac{1}{\sqrt{2}}\left(\dfrac{1}{\sqrt{2}}y_1 + \dfrac{1}{\sqrt{2}}y_2\right)^3 - \dfrac{1}{\sqrt{2}}\left(\dfrac{1}{\sqrt{2}}y_1 - \dfrac{1}{\sqrt{2}}y_2\right)^3\right\} \end{bmatrix}, \\ &= \begin{bmatrix} \dfrac{3}{2}y_1^2\dot{y}_1 + \dfrac{3}{2}y_2^2\dot{y}_1 + 3y_1y_2\dot{y}_2 \\ \dfrac{3}{2}y_2^2\dot{y}_2 + \dfrac{3}{2}y_1^2\dot{y}_2 + 3y_1y_2\dot{y}_1 \end{bmatrix},\end{aligned} \tag{7.22}$$

$$\begin{aligned}\boldsymbol{P}^{-1}\dot{\boldsymbol{x}}_f &= \begin{bmatrix} \dfrac{\mathrm{d}}{\mathrm{d}\tau}\left\{\dfrac{1}{\sqrt{2}}\left(\dfrac{1}{\sqrt{2}}y_1 + \dfrac{1}{\sqrt{2}}y_2\right)^5 + \dfrac{1}{\sqrt{2}}\left(\dfrac{1}{\sqrt{2}}y_1 - \dfrac{1}{\sqrt{2}}y_2\right)^5\right\} \\ \dfrac{\mathrm{d}}{\mathrm{d}\tau}\left\{\dfrac{1}{\sqrt{2}}\left(\dfrac{1}{\sqrt{2}}y_1 + \dfrac{1}{\sqrt{2}}y_2\right)^5 - \dfrac{1}{\sqrt{2}}\left(\dfrac{1}{\sqrt{2}}y_1 - \dfrac{1}{\sqrt{2}}y_2\right)^5\right\} \end{bmatrix}, \\ &= \begin{bmatrix} \dfrac{5}{4}y_1^4\dot{y}_1 + \dfrac{15}{2}y_1^2y_2^2\dot{y}_1 + \dfrac{5}{4}y_2^4\dot{y}_1 + 5y_1^3y_2\dot{y}_2 + 5y_1y_2^3\dot{y}_2 \\ \dfrac{5}{4}y_2^4\dot{y}_2 + \dfrac{15}{2}y_2^2y_1^2\dot{y}_2 + \dfrac{5}{4}y_1^4\dot{y}_2 + 5y_2^3y_1\dot{y}_1 + 5y_2y_1^3\dot{y}_1 \end{bmatrix}.\end{aligned} \tag{7.23}$$

By substituting $y_k = a_k \sin(\omega_k \tau + \phi_k)$, $\dot{y}_k = a_k \omega_k \cos(\omega_k \tau + \phi_k)$ with (7.21) into (7.14), the averaged equation for an amplitude a_k of the two inductor-coupled hard oscillators can be written as:

$$\begin{aligned}
\dot{a}_k &= \varepsilon \left\{ -a_k \langle \cos^2\theta_k \rangle + \frac{1}{2}\beta a_k^3 \langle \cos^2\theta_k \sin^2\theta_k \rangle - \frac{1}{4} a_k^5 \langle \cos^2\theta_k \sin^4\theta_k \rangle \right. \\
&\quad + \frac{\beta a_k a_{k+1}^2 \omega_{k+1}}{\omega_k} \langle \cos\theta_k \cos\theta_{k+1} \sin\theta_k \sin\theta_{k+1} \rangle \\
&\quad - \frac{a_k^3 a_{k+1}^2 \omega_{k+1}}{\omega_k} \langle \cos\theta_k \cos\theta_{k+1} \sin^3\theta_k \sin\theta_{k+1} \rangle \\
&\quad + \frac{1}{2}\beta a_k a_{k+1}^2 \langle \cos^2\theta_k \sin^2\theta_{k+1} \rangle - \frac{3}{2} a_k^3 a_{k+1}^2 \langle \cos^2\theta_k \sin^2\theta_k \sin^2\theta_{k+1} \rangle \\
&\quad - \frac{a_k a_{k+1}^4 \omega_{k+1}}{\omega_k} \langle \cos\theta_k \cos\theta_{k+1} \sin\theta_k \sin^3\theta_{k+1} \rangle \\
&\quad \left. - \frac{1}{4} a_k a_{k+1}^4 \langle \cos^2\theta_k \sin^4\theta_{k+1} \rangle \right\} \quad (k = 1, 2, a_3 = a_1, \theta_3 = \theta_1, \omega_3 = \omega_1),
\end{aligned}$$

where the average of a function $h(\theta)$, i.e. $\frac{1}{2\pi}\int_0^{2\pi} h(\theta) d\theta$ is represented by the symbol $\langle h(\theta) \rangle$. Assuming that the ratio of ω_k to ω_{k+1} is a rational number, the averaged terms $\langle \cos^2\theta_k \rangle$, $\langle \cos^2\theta_k \sin^2\theta_k \rangle$, $\langle \cos^2\theta_k \sin^4\theta_k \rangle$, $\langle \cos^2\theta_k \sin^2\theta_{k+1} \rangle$, $\langle \cos^2\theta_k \sin^2\theta_k \sin^2\theta_{k+1} \rangle$, and $\langle \cos^2\theta_k \sin^4\theta_{k+1} \rangle$ can be calculated as $\frac{1}{2}, \frac{1}{8}, \frac{1}{16}, \frac{1}{4}, \frac{1}{16}$, and $\frac{3}{16}$, respectively, and the other terms are zero. Therefore, we arrive at the averaged equation for the slowly varying amplitude:

$$\dot{a_k^2} = \varepsilon \left\{ -a_k^2 + \frac{1}{8}\beta a_k^4 - \frac{1}{32} a_k^6 + \frac{1}{4}\beta a_k^2 a_{k+1}^2 - \frac{3}{16} a_k^4 a_{k+1}^2 - \frac{3}{32} a_k^2 a_{k+1}^4 \right\}. \quad (7.24)$$

In the same way, the averaged equation for the slowly varying phase ϕ_k is calculated using (7.14). The value of the time derivative of ϕ_k eventually becomes zero as follows:

$$\begin{aligned}
\dot{\phi}_k &= \varepsilon \left\{ \langle \cos\theta_k \sin\theta_k \rangle - \frac{1}{2}\beta a_k^2 \langle \cos\theta_k \sin^3\theta_k \rangle + \frac{1}{4} a_k^4 \langle \cos\theta_k \sin^5\theta_k \rangle \right. \\
&\quad - \frac{\beta a_{k+1}^2 \omega_{k+1}}{\omega_k} \langle \cos\theta_{k+1} \sin^2\theta_k \sin\theta_{k+1} \rangle + \frac{a_k^2 a_{k+1}^2 \omega_{k+1}}{\omega_k} \langle \cos\theta_{k+1} \sin^4\theta_k \sin\theta_{k+1} \rangle \\
&\quad - \frac{1}{2}\beta a_{k+1}^2 \langle \cos\theta_k \sin\theta_k \sin^2\theta_{k+1} \rangle + \frac{3}{2} a_k^2 a_{k+1}^2 \langle \cos\theta_k \sin^3\theta_k \sin^2\theta_{k+1} \rangle \\
&\quad \left. + \frac{a_{k+1}^4 \omega_{k+1}}{\omega_k} \langle \cos\theta_{k+1} \sin^2\theta_k \sin^3\theta_{k+1} \rangle + \frac{1}{4} a_{k+1}^4 \langle \cos\theta_k \sin\theta_k \sin^4\theta_{k+1} \rangle \right\} \\
&\qquad (k = 1, 2, \ a_3 = a_1, \theta_3 = \theta_1, \omega_3 = \omega_1) \\
&= 0.
\end{aligned} \quad (7.25)$$

This result allows us to investigate the approximated solutions in (7.17) with the averaged amplitudes $a_k, k = 1, 2$, because θ_k are arbitrary constants determined by the initial condition.

7.2.1.3 Double-mode oscillation in two coupled hard oscillators

The equilibrium points $(\hat{a}_1^2, \hat{a}_2^2)$, which satisfy the conditions $\dot{a_k^2} = 0,\ k = 1, 2$, in (7.24), and correspond to the steady-state solutions of (7.17), are written as:

$$
\begin{aligned}
(\hat{a}_1^2, \hat{a}_2^2) = {} & (0,0), (2\beta \pm 2\sqrt{\beta^2 - 8}, 0), (0, 2\beta \pm 2\sqrt{\beta^2 - 8}), \\
& \left(\frac{3\beta \pm \sqrt{9\beta^2 - 80}}{5}, \frac{3\beta \pm \sqrt{9\beta^2 - 80}}{5} \right), \\
& (\beta \pm \sqrt{16 - \beta^2}, \beta \mp \sqrt{16 - \beta^2}).
\end{aligned}
$$

From (7.9) and the relationship $\boldsymbol{x} = \boldsymbol{P}\boldsymbol{y}$, the steady-state solution in the two inductor-coupled hard oscillator can be represented by the stable equilibrium points $(\hat{a}_1^2, \hat{a}_2^2)$ by:

$$
\begin{bmatrix} x_1 \\ x_2 \end{bmatrix} = \begin{bmatrix} \frac{1}{\sqrt{2}} \hat{a}_1 \sin(\omega_1 \tau + \theta_1) + \frac{1}{\sqrt{2}} \hat{a}_2 \sin(\omega_2 \tau + \theta_2) \\ \frac{1}{\sqrt{2}} \hat{a}_1 \sin(\omega_1 \tau + \theta_1) - \frac{1}{\sqrt{2}} \hat{a}_2 \sin(\omega_2 \tau + \theta_2) \end{bmatrix}. \tag{7.26}
$$

The stability of each solution is determined by the eigenvalues of the Jacobian matrix at the equilibrium point of the linearized system [13]. Then, we can confirm that the following four equilibrium points:

$$
\begin{aligned}
(\hat{a}_1^2, \hat{a}_2^2) = {} & (0,0), \\
& (2\beta + 2\sqrt{\beta^2 - 8}, 0), \\
& (0, 2\beta + 2\sqrt{\beta^2 - 8}), \\
& \left(\frac{3\beta + \sqrt{9\beta^2 - 80}}{5}, \frac{3\beta + \sqrt{9\beta^2 - 80}}{5} \right),
\end{aligned} \tag{7.27}
$$

are stable. The first equilibrium point $(0, 0)$ corresponds to the non-oscillating state. Because $\hat{a}_2(\hat{a}_1) = 0$ and $\hat{a}_1(\hat{a}_2) \neq 0$, the second and third equilibrium points represent the same-phase and reverse-phase solutions where two oscillators oscillate at the single frequency ω_1 (ω_2) and they are synchronized in-phase and anti-phase, respectively. The time series and the trajectories on the phase planes for both solutions with $\varepsilon = 0.01, \alpha = 0.1$, and $\beta = 3.1$ are shown in Figures 7.4 and 7.5.

The fourth equilibrium point in (7.27) corresponds to a double-mode oscillation where the two sinusoidal oscillations are superimposed. Figure 7.6(a) and (b) shows

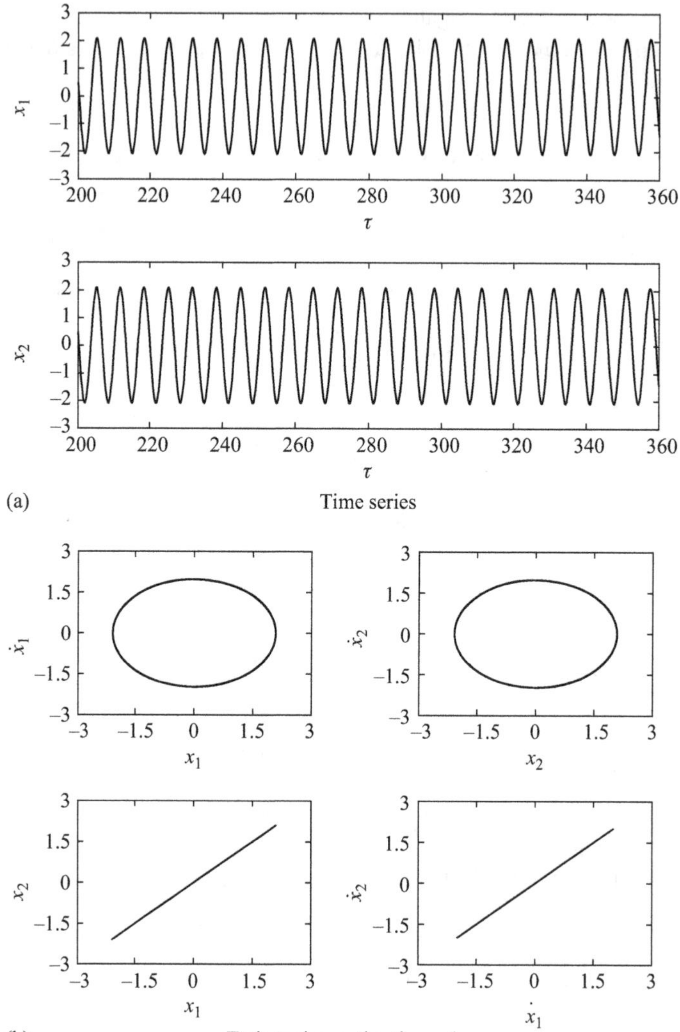

Figure 7.4 Same-phase solution in (7.17) ($\varepsilon = 0.01$, $\alpha = 0.1$, and $\beta = 3.1$)

the time series of x_k, $k = 1, 2$, in (7.17) and the trajectories on the phase planes, respectively. Figure 7.6 shows that x_k undergoes a quasi-periodic oscillation and that the phase relationship between the two oscillators is complicated. The normalized angular frequency component (ω) of x_1 is calculated by applying the fast Fourier transform (FFT) with a normalized angular frequency resolution of 0.0003, as shown in Figure 7.6(c). It is clear that there are two dominant frequency components near $\omega_k = \sqrt{1 \mp \alpha}\,(k = 1, 2)$, $\approx$0.95 and 1.05. It is interesting that the hard nonlinearity for NC as written in (7.1) is necessary for the stable double-mode oscillation in addition

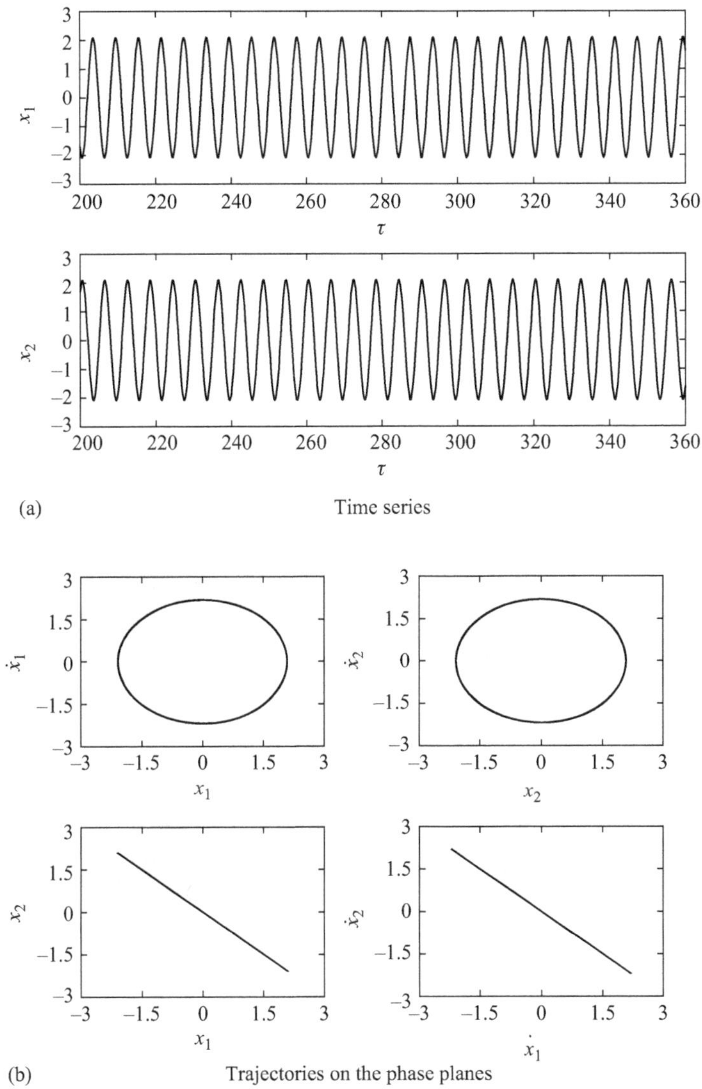

Figure 7.5 Reverse-phase solution in (7.17) ($\varepsilon = 0.01$, $\alpha = 0.1$, and $\beta = 3.1$)

to the single modes even though the double-mode oscillation does not appear when we assume the third-power nonlinearity for NC [14].

Datardina and Linkens derived the averaged equation for the two inductor-coupled oscillators and confirmed the steady states, including the multimode oscillation [11]. Endo and Ohta investigated an arbitrary number of hard oscillator lattices where the oscillators were mutually coupled by an inductor including

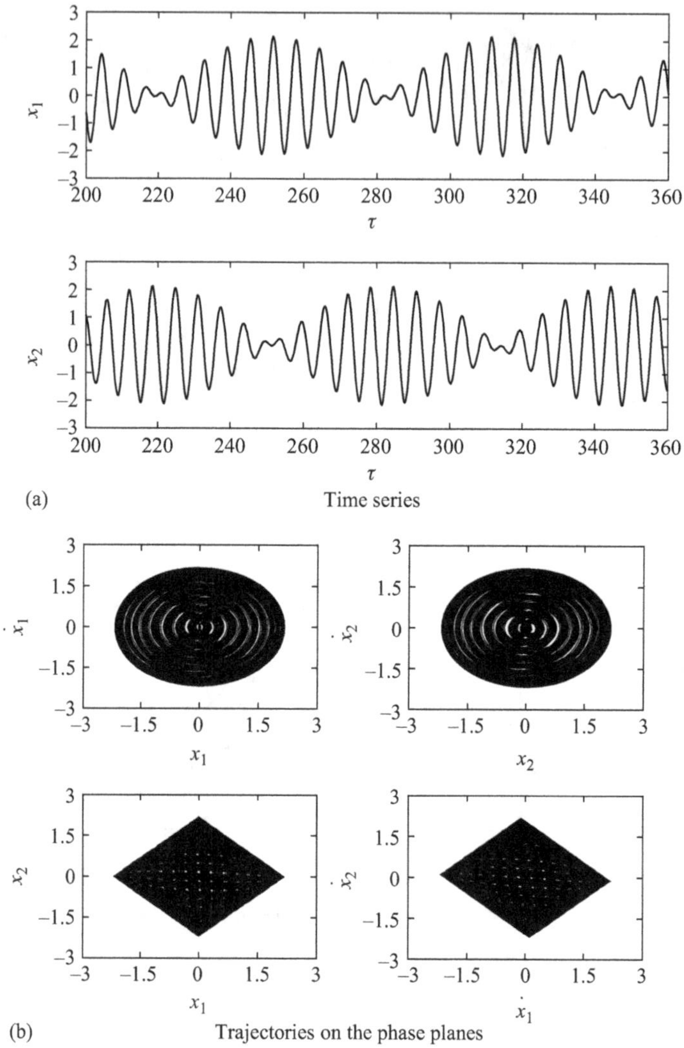

Figure 7.6 Double-mode oscillation in (7.17) ($\varepsilon = 0.01$, $\alpha = 0.1$, and $\beta = 3.1$)

a resistive component with an open boundary condition [12]. They succeeded in deriving the averaged equation for an arbitrary number of coupled systems. They also investigated the stabilities of the single- and double-mode oscillations. It was shown that the existence of a double-mode oscillation depends on the number of oscillators. In addition, the regions of attraction of the single- and double-mode oscillations on the phase planes for the two coupled oscillators were given.

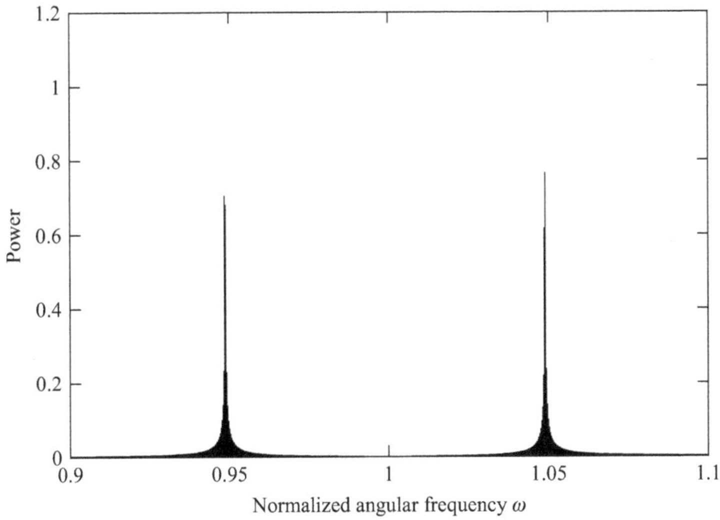

(c) Power spectrum of x_1

Figure 7.6 Continued

7.2.2 Strongly nonlinear oscillators

In strongly nonlinear coupled hard oscillators, a localized phenomenon [2], which cannot be obtained by the averaged equation, is excited. Figure 7.7 shows the localized mode for $\varepsilon = 0.5, \alpha = 0.1$, and $\beta = 3.1$, where x_1 undergoes a large-amplitude oscillation and x_2 oscillates with a small amplitude. Figure 7.8 shows the power spectrum of x_1 corresponding to Figure 7.7. The localized mode corresponds to a periodic oscillation as suggested by the single dominant peak in Figure 7.8. Due to the symmetrical property of this coupled system, another localized mode coexists with the same parameter set as shown in Figure 7.9, where x_1 and x_2 undergo small- and large-amplitude oscillations, respectively.

The two localized modes originate from the nonlinear characteristics of the coupled system. Because the localized mode is a periodic solution and becomes a fixed point on the Poincaré section, the stability of the localized mode can be calculated rigorously [15]. We define the Poincaré section as $\dot{x}_1 = 0$ and map the points when the flow penetrates the section from the positive to the negative side. Figure 7.10 shows the bifurcation structure of the localized modes for $\beta = 3.1$ and $\alpha = 0.1$ when ε is decreased from 0.6. The top and bottom solid curves in Figure 7.10 correspond to the two distinctive localized modes (stable nodes) in Figures 7.7 and 7.9, respectively. In addition, the corresponding unstable saddles are depicted by the dotted curves. The two localized modes simultaneously disappear at $\varepsilon_c \approx 0.448$ via a saddle-node (SN) bifurcation. Moreover, the dominant frequency of the localized mode changes

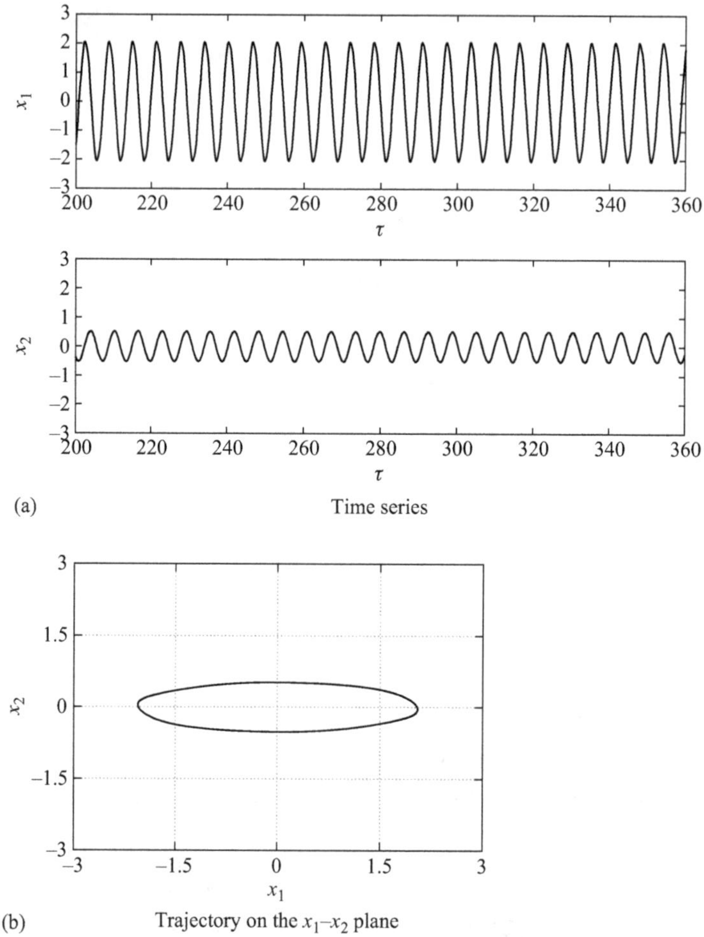

Figure 7.7 Localized mode ($\varepsilon = 0.5$, $\alpha = 0.1$, and $\beta = 3.1$). The initial conditions are $(x_1(0), \dot{x}_1(0), x_2(0), \dot{x}_2(0)) = (2, 0, 0, 0)$

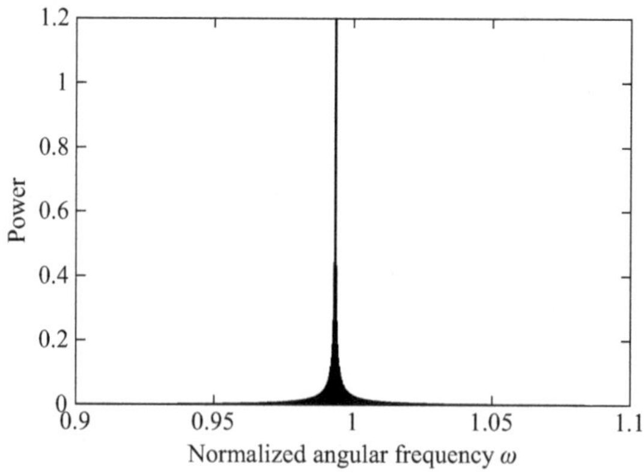

Figure 7.8 Power spectrum of x_1 in Figure 7.7

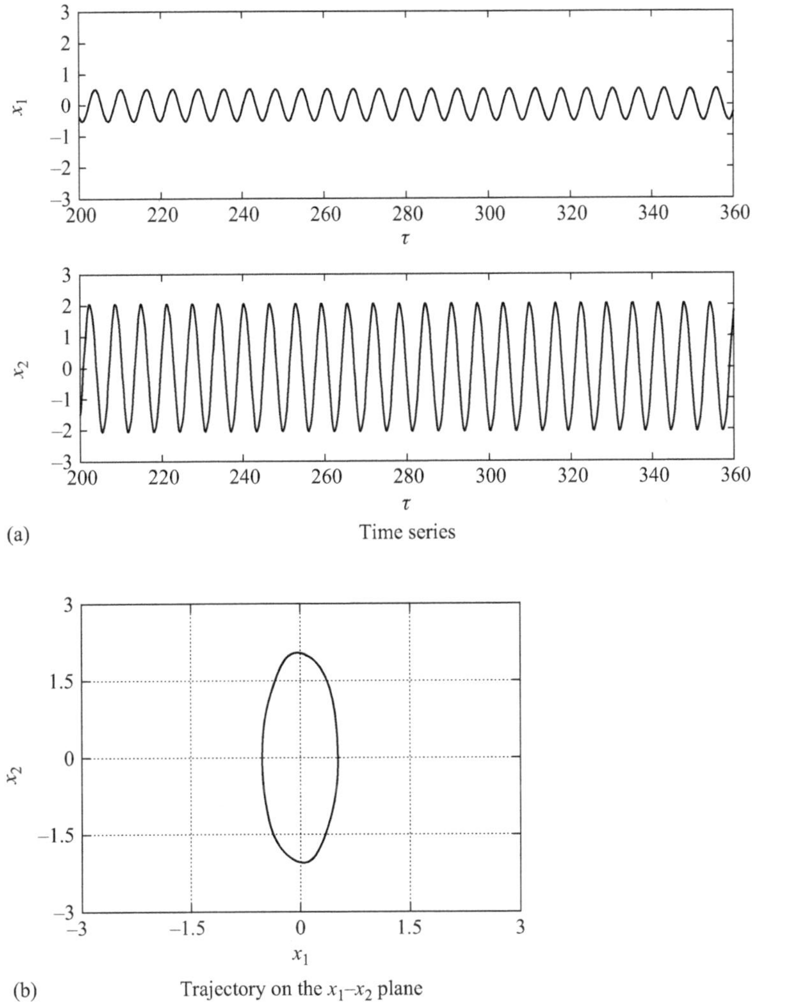

Figure 7.9 Another localized mode ($\varepsilon = 0.5$, $\alpha = 0.1$, and $\beta = 3.1$). The initial conditions are $(x_1(0), \dot{x}_1(0), x_2(0), \dot{x}_2(0)) = (0, 0, 2, 0)$

in terms of the value of ε, as shown in Figure 7.11. Figure 7.11 illustrates that the localized mode has a single angular frequency for $\varepsilon_c \leqq \varepsilon$ as shown in Figure 7.8, whereas it has two dominant angular frequencies for $\varepsilon < \varepsilon_c$.

Immediately after the SN bifurcation, a remarkable transitional phenomenon, called a switching oscillation, emerges in the two coupled hard oscillators [16–18]. Figure 7.12 shows the switching oscillation for $\varepsilon = 0.447, \alpha = 0.1$, and $\beta = 3.1$ where one oscillator oscillates at a larger amplitude while the other oscillates at a smaller amplitude. Note that the solution remains for an extended period of time

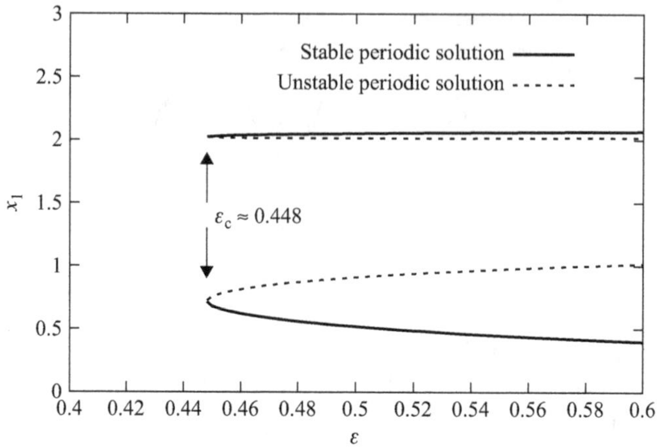

Figure 7.10 Bifurcation structure of the two localized modes in Figures 7.7 and 7.9 near the saddle-node bifurcation point

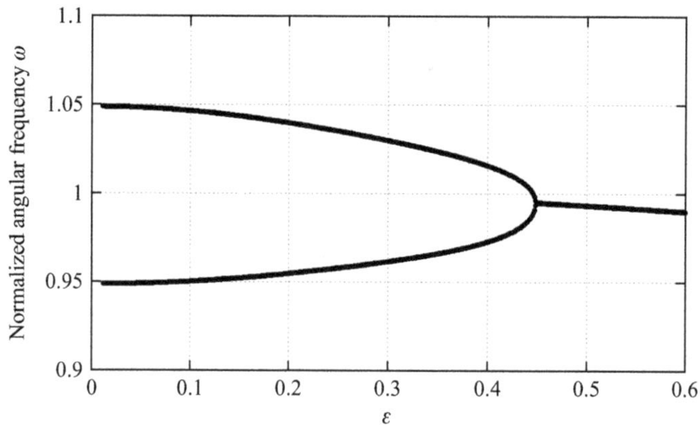

Figure 7.11 Variation of dominant frequencies of x_1 for the localized mode in Figure 7.7

near the trajectories of the two localized modes before the SN bifurcation occurs. Furthermore, the solution has two dominant frequency peaks even though several small peaks also appear, as shown in Figure 7.12(c).

Even though the aligned structure of the two SN bifurcation points is important for observing the switching solution [18], it is not a sufficient condition for the existence of this solution. Figure 7.13 shows the bifurcation structure of the two localized modes for $\beta = 3.2$ and $\alpha = 0.1$, which is qualitatively the same as that in Figure 7.10 except for the value of ε_c. In contrast to the switching oscillation of $\beta = 3.1$, a reverse-phase

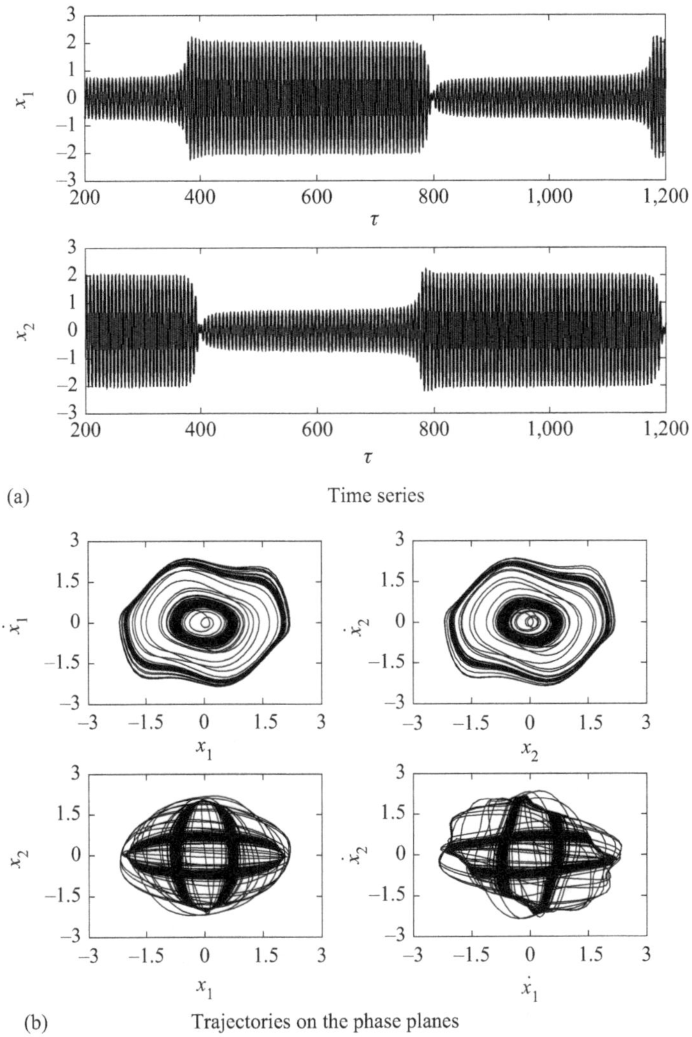

Figure 7.12 Transitional phenomenon immediately after the SN bifurcation ($\varepsilon = 0.447$, $\alpha = 0.1$, and $\beta = 3.1$). The initial conditions are $(x_1(0), \dot{x}_1(0), x_2(0), \dot{x}_2(0)) = (2, 0, 0, 0)$

solution appears immediately after the two localized modes disappear for $\beta = 3.2$. This result suggests that the value of the parameter β, which controls the amplitude of the steady states for the weakly nonlinear oscillators, is also essential to observe a switching oscillation. The initiating mechanisms of the transient phenomena were investigated in detail from a bifurcation theory point of view in References 16–18.

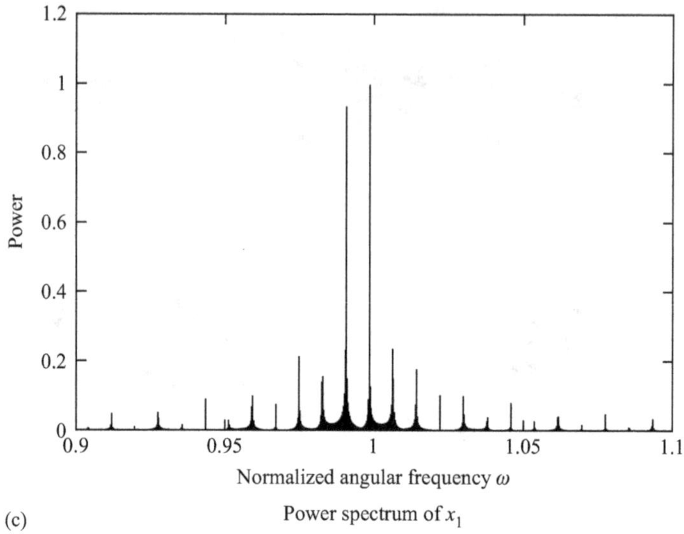

(c)

Figure 7.12 Continued

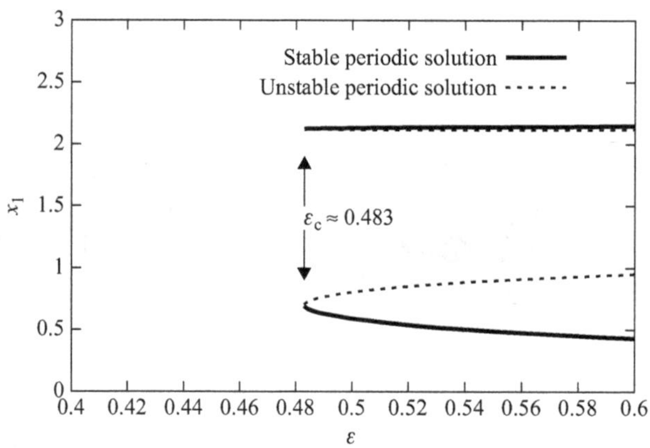

Figure 7.13 Bifurcation structure of the two localized modes near the saddle-node bifurcation point for $\beta = 3.2$ and $\alpha = 0.1$

7.3 Propagating waves in a coupled hard-oscillator ring

In this section, we investigate a localized mode and a propagating wave observed in a coupled hard-oscillator ring. Figure 7.14 shows the one-dimensional array of inductor-coupled hard oscillators. From Kirchhoff's law, the circuit equation of Figure 7.14 is written as

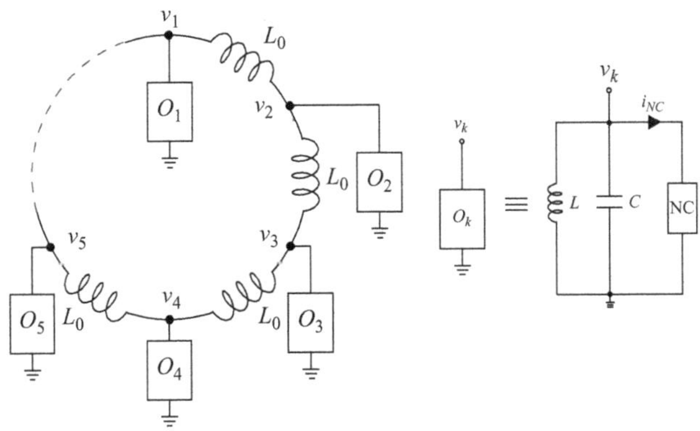

Figure 7.14 Circuit diagrams of coupled hard-oscillator rings

$$\frac{d^2 v_k}{dt^2} + \frac{g_1}{C}\left(1 - \frac{3g_3}{g_1}v_k^2 + \frac{5g_5}{g_1}v_k^4\right)\frac{dv_k}{dt} + \left(\frac{1}{LC} + \frac{1}{L_0 C}\right)v_k - \frac{1}{L_0 C}(v_{k+1} - v_k + v_{k-1}) = 0, \tag{7.28}$$

$$k = 1, 2, \ldots, N \quad (v_0 = v_N, v_{N+1} = v_1),$$

where N is the number of hard oscillators. Substituting the normalized variables and the parameters defined in (7.16) into (7.28), the normalized equation is written as:

$$\ddot{x}_k + \varepsilon(1 - \beta x_k^2 + x_k^4)\dot{x}_k + x_k - \alpha(x_{k-1} - x_k + x_{k+1}) = 0. \tag{7.29}$$

In the following results, the phenomena observed in six inductor-coupled hard-oscillator rings ($N = 6$) are investigated for simplicity.

Figure 7.15 shows the time series and the trajectories on the phase planes for all possible combinations for $\varepsilon = 0.39, \alpha = 0.1$, and $\beta = 3.2$. It is clear that localized excitation appears where x_3 and x_4 are synchronized in anti-phase with the initial conditions denoted in the caption of Figure 7.15. For smaller ε, a propagating wave phenomenon emerges under the same initial conditions as shown in Figure 7.16, where a spatiotemporally localized excitation propagates in one direction with constant speed. That is, a localized large-amplitude excitation in one oscillator propagates to several adjacent oscillators. Figure 7.15(c) shows the power spectrum of x_1 in the propagating wave where several dominant frequency peaks are nearly larger than 1. Furthermore, the stationary localized mode in Figure 7.16 disappears at a pitchfork (PF) bifurcation point ($\varepsilon_{PF} \approx 0.3672$). The propagating wave phenomena are generated via a global bifurcation of maps based on heteroclinic tangle [19,20]. Therefore, the propagating wave in Figure 7.15 emerges before the localized mode disappears, which differs from the condition for the switching oscillations observed in the two inductor-coupled hard oscillators. Note that we set the value of β to 3.2 because this parameter value

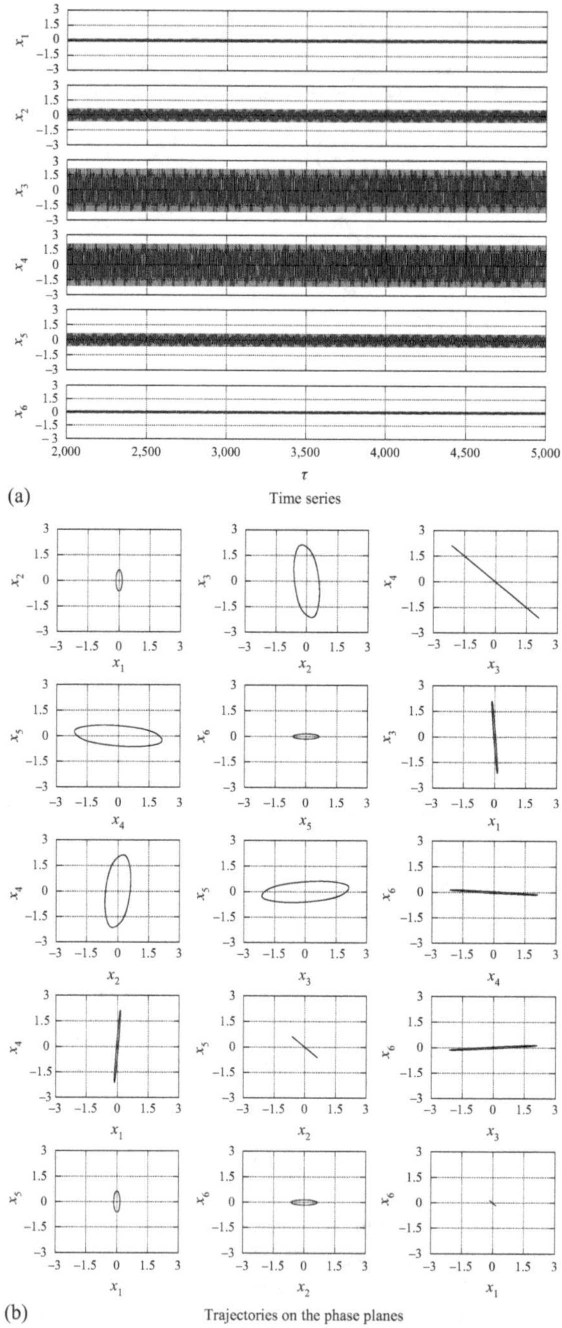

Figure 7.15 Localized mode in six inductor-coupled hard-oscillator rings for $\varepsilon = 0.39$, $\alpha = 0.1$, and $\beta = 3.2$. Initial conditions are $x_1 = 2.0$, $x_2 = 0.5$, $x_3 = 0.1$, $x_4 = 0.2$, $x_5 = 0.9$, $x_6 = -2.2$, $y_1 = 1.8$, $y_2 = -0.3$, $y_3 = y_4 = 0.3$, $y_5 = -2.3$, and $y_6 = 0.4$

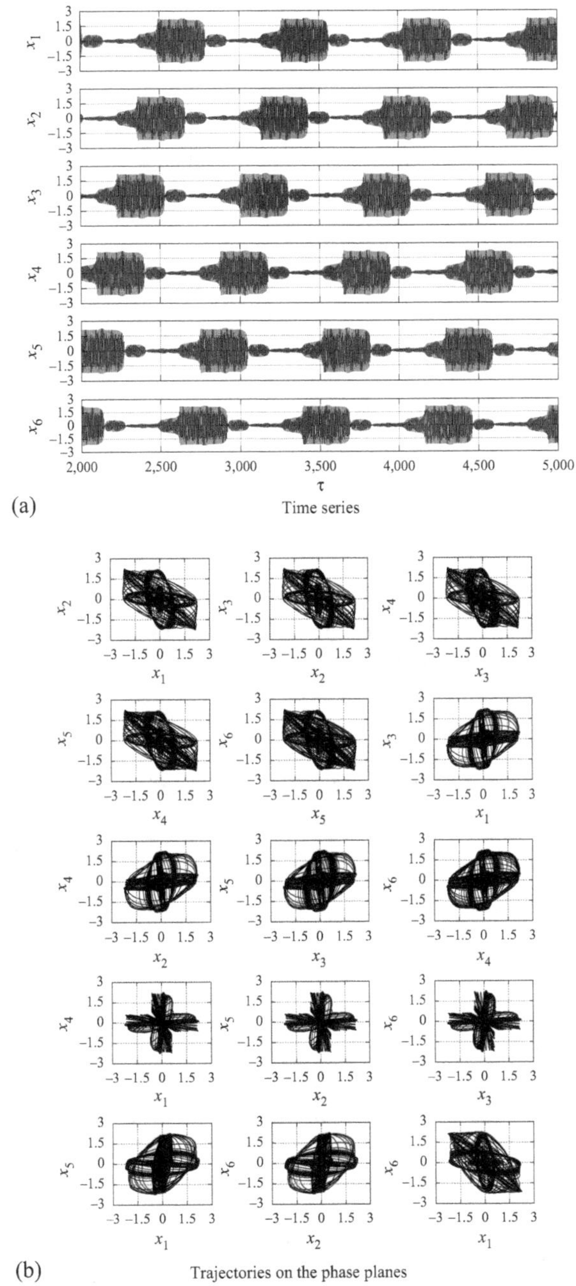

Figure 7.16 Propagating wave in the six inductor-coupled hard oscillators for $\varepsilon = 0.38$, $\alpha = 0.1$, and $\beta = 3.2$. The initial conditions are the same as those in Figure 7.15

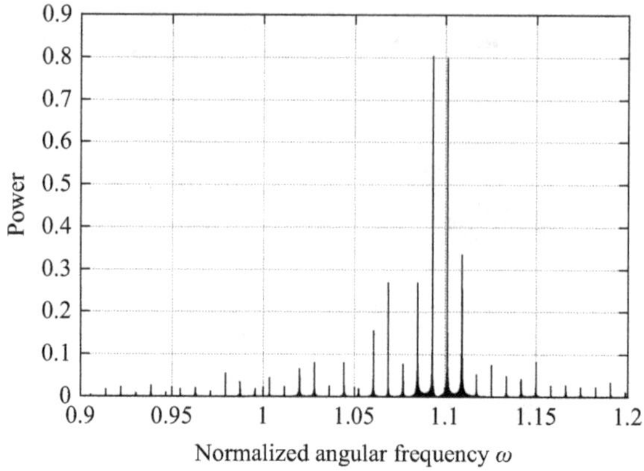

(c) Power spectrum of x_1

Figure 7.16 Continued

is essential to observe the propagating wave, which is consistent with the results for the switching solutions in the two hard oscillators. For $\beta = 3.1$, non-oscillation is observed in the neighbourhood of the bifurcation point of the localized mode.

In addition to the propagating wave, several distinctive propagating waves and different types of switching oscillations occur for different initial conditions with the same parameter set. We have confirmed that one of the switching dynamics corresponds to a double-mode oscillation [8]. In particular, the rotational symmetry of this system (due to its ring coupling structure) plays a role in determining the switching dynamics, which are a type of stationary localized modes for strongly nonlinear oscillators. Therefore, there may exist additional diverse switching oscillations for larger N.

7.4 Conclusions

In this chapter, we investigated the multimode oscillations and the localized modes observed in the inductor-coupled hard oscillators. We surveyed the derivation of the averaged equation for weakly nonlinear oscillators and derived the approximated equation for two inductor-coupled hard oscillators. We confirmed that there exist four distinctive steady states including the double-mode oscillation for weakly nonlinear oscillators. The two distinctive localized modes for strongly nonlinear oscillators simultaneously disappeared via the SN bifurcation. We calculated the variation of the dominant frequency as a function of ε (the degree of nonlinearity). Furthermore, we investigated the propagating wave phenomena and the localized modes observed in six inductor-coupled hard-oscillator rings. The propagating wave originated in the neighbourhood of the local bifurcation point of the localized mode. Note that several propagating waves and switching solutions could be confirmed by circuit experiments [8]. Further detailed results, especially for their initiating mechanisms of the two and six inductor-coupled hard oscillators, have been presented in References 16–18 and 8,19,20, respectively.

References

[1] Flach S., Gorbach A.V. 'Discrete breathers: Advances in theory and applications'. *Physics Reports*. 2008; **467**: 1–116.

[2] Vakakis A.F. (ed.). *Normal Modes and Localization in Nonlinear Systems*. Berlin: Springer-Science+Business Media B.V.; 2001

[3] Nishiura Y., Ueyama D., Yanagita T. 'Chaotic pulses for discrete reaction diffusion systems'. *SIAM Journal of Applied Dynamical Systems*. 2005; **4**: 723–754.

[4] Comte J.C., Morfu S., Marquié P. 'Propagation failure in discrete bistable reaction–diffusion systems: Theory and experiments'. *Physics Reviews E*. 2001; **64**: 4pp.

[5] Binczak S., Bilbault J.M. 'Experimental propagation failure in a nonlinear electrical lattice'. *International Journal of Bifurcation and Chaos*. 2004; **14**: 1819–1830.

[6] Sato M., Hubbard B.E., Sievers A.J. 'Colloquium: Nonlinear energy localization and its manipulation in micro mechanical oscillator arrays'. *Reviews of Modern Physics*. 2006; **78**: 137–157.

[7] Kimura M., Hikihara T. 'Capture and release of traveling intrinsic localized mode in coupled cantilever array'. *Chaos*. 2009; **19**: 7pp.

[8] Shimizu K. 'Experimental observations of propagating waves and switching phenomena in a coupled bistable oscillator system'. *International Journal of Bifurcation and Chaos*. 2014; **24**(12): 21pp.

[9] Nayfeh A.H. *Perturbation Methods*. New York, NY: Wiley; 1981.

[10] Kojima M., Uwate Y., Nishio Y., 'Multimode oscillations in coupled oscillators with high-order-nonlinear characteristics'. *IEEE Transactions on Circuits and Systems I*, 2014; **61**: 2653–2662.

[11] Datardina S.P., Linkens D.A. 'Multimode oscillations in mutually coupled van der Pol type oscillators with fifth-power nonlinear characteristics'. *IEEE Transactions on Circuits and Systems*. 1978; **CAS-25**: 308–315.

[12] Endo T., Ohta T. 'Multimode oscillations in a coupled oscillator system with fifth-power nonlinear characteristics'. *IEEE Transactions on Circuits and Systems*. 1980; **CAS-27**: 277–283.

[13] Strogatz S.H. *Nonlinear Dynamics and Chaos with Applications to Physics, Biology, Chemistry, and Engineering*. Boulder, CO: Westview Press; 2000.

[14] Suezaki T., Mori S. 'Mutual synchronization of two oscillators'. *Transactions of IECE*. 1965; **48**(9): 1551–1557.

[15] Kawakami H. 'Bifurcation of periodic responses in forced dynamic nonlinear circuits: Computation of bifurcation values of the system parameters'. *IEEE Transactions on CAS*. 1984; **CAS-31**(3): 248–260.

[16] Yoshinaga T., Kawakami H. 'Synchronized quasi-periodic oscillations in a ring of coupled oscillators with hard nonlinearities'. *Transactions of IEICE*. 1992; **J75-A**(12): 1811–1818.

[17] Aruga Y., Endo T. 'Transitional dynamics and chaos in coupled oscillator systems'. *IEICE Transactions on Fundamentals*. 2003; **J86-A** (5): 559–568 (in Japanese).

[18] Shimizu K., Endo T., Yoshimura T. 'Transitional dynamics and quasi-periodic solution observed in two asymmetrical coupled oscillators'. *IEICE Transactions on Fundamentals*. 2009; **E92-A** (1): 270–278.

[19] Shimizu K., Komuro M., Endo T. 'Onset of the propagating pulse wave in a ring of coupled bistable oscillators'. *IEICE Nonlinear Theory and Its Applications*. 2011; **2**: 139–151.

[20] Kamiyama K., Komuro M., Endo T. 'Bifurcation analysis of the propagating wave and the switching oscillations in a ring of six-coupled bistable oscillators: Bifurcation starting from type 2 standing wave solution'. *International Journal of Bifurcation and Chaos*. 2012; **22**: 13pp.

Chapter 8
Wave propagation of phase difference in coupled oscillator arrays

Masayuki Yamauchi[1], *Yoshihito Todani*[2] *and Syohei Fujimoto*[3]

There are many synchronization phenomena in coupled oscillator arrays. There is a wave of a phase difference which is one of synchronization phenomena. The wave, which a phase difference between adjacent oscillators propagates, continuously exists. We call the wave a phase-inversion wave. The phase-inversion waves can be observed on a ladder and a cross constructed by van der Pol oscillators. In this section, the phase-inversion waves are introduced.

8.1 Introduction

There are a lot of synchronization phenomena in this world. The synchronization phenomena can be observed in pace maker cells in a heart, in the atomic world, in the moon, and so on [1–7]. Therefore, the living cannot live without synchronization phenomena. Of course, we can observe the synchronization phenomena on electronic circuits [8–17]. When many van der Pol oscillators are coupled by resistors, inductors, or capacitors, we know that all oscillators can synchronize.

We can observe some kinds of waves in many kinds of coupled oscillator arrays of regular structures. For example, there are propagations of voltage amplitudes, phases, and so on. Many of these waves are damped during propagating on the system. However, some waves without damping can be observed. We discovered one of the waves without damping.

In this section, we introduce the wave that the phase state between adjacent oscillators propagates on many van der Pol oscillators which are coupled by inductors as a ladder [12,14], a ring, or a 2D lattice [16]. Furthermore, the wave can propagate on a system that the ladders are connected as a cross [13]. In other words, the phenomenon changes phase states between adjacent oscillators from an in-phase synchronization to an anti-phase synchronization or from the anti-phase synchronization to the in-phase synchronization. We named the phenomenon a phase-inversion

[1]Department of Electric and Computer Engineering, Hiroshima Institute of Technology, Hiroshima, Japan
[2]Nippon Paper Industries Co., Ltd., Tokyo, Japan
[3]TAKAYA Corporation, Okayama, Japan

wave. The phase-inversion waves can be observed in all oscillators which are in the in-phase synchronization or a phase state that the in-phase synchronization and the anti-phase synchronization alternately exist. The mechanisms of phase-inversion waves are explained by using phase differences between adjacent oscillators and an instantaneous frequency of each oscillator circuit.

8.2 Circuit model

In this section, special synchronization phenomena which are observed on many van der Pol oscillators coupled by inductors are introduced. Each van der Pol oscillator is shown as a ball (see Figure 8.1(a)). A nonlinear negative resistor of each van der Pol oscillator is constructed by an op-amp and three resistors in actual circuit experiments (see Figure 8.1(b)). The v–i characteristics of nonlinear negative resistors are shown by the following approximate equation:

$$f(v) = -g_1 v + g_3 v^3. \tag{8.1}$$

An equivalent figure in which van der Pol oscillators are coupled by inductors is written in Figure 8.1(c). Circuit models of a ring, a ladder, and a 2D lattice are shown by using the equivalent figure in Figures 8.2–8.4. As a special model, we introduce a cross system of which four ladders are coupled by four inductors and an oscillator as a cross (see Figure 8.5). In this cross system, values of the four coupling inductors between the center oscillator (OSC_c) and ladders are changed from values of the coupling inductors in the ladders. The circuit equations of each model are normalized, and they are calculated by using fourth-order Runge–Kutta method in simulations. The "k" is an integer number in this chapter.

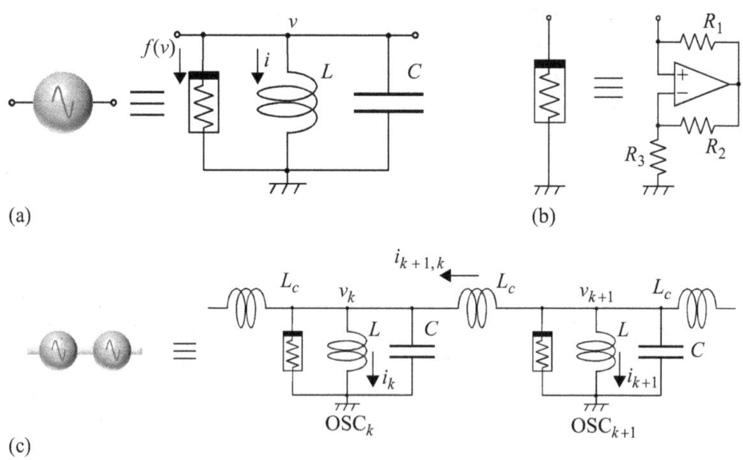

Figure 8.1 Equivalent models. (a) A van der Pol oscillator, (b) nonlinear negative resistor, and (c) two van der Pol oscillators coupled by an inductor

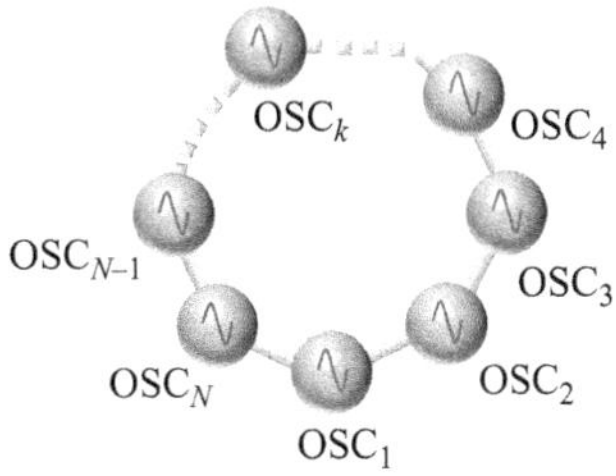

Figure 8.2 Ring system

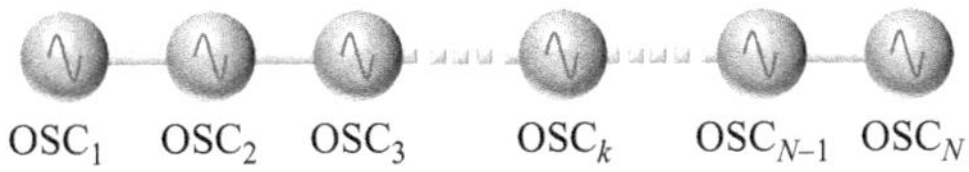

Figure 8.3 Ladder system

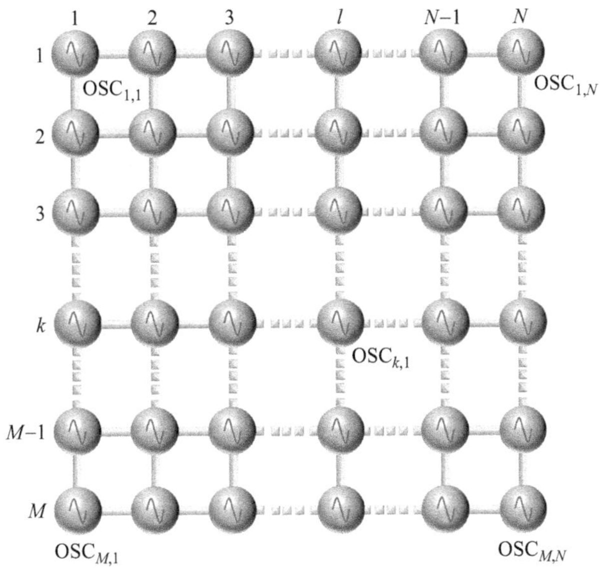

Figure 8.4 2D lattice system

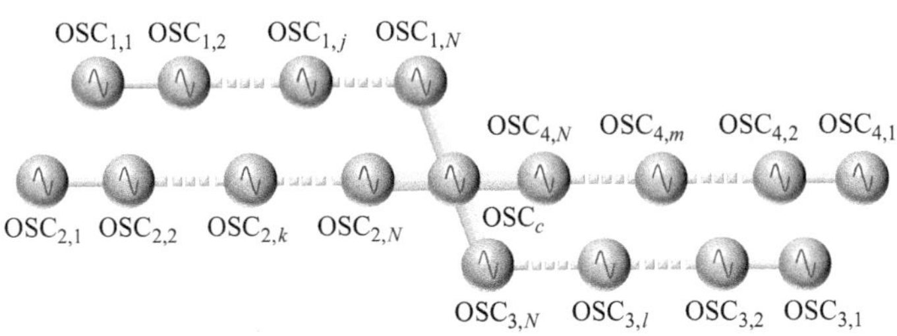

Figure 8.5 Cross system

⟨Circuit equations of a ring system⟩

When N van der Pol oscillators are coupled by inductors as a ring (see Figure 8.2), circuit equations of OSC_k are set up (see (8.2)). A current of an inductor of the OSC_k is set as i_k, and a voltage of the inductor is set as v_k (see Figure 8.1(c)). A current of a coupling inductor between OSC_k and OSC_{k+1} is shown as $i_{k+1,k}$ (see Figure 8.1(c)).

$$\begin{cases} L\dfrac{di_k}{dt} = v_k, \\ C\dfrac{dv_k}{dt} + i_k + f(v_k) + i_{k,a} - i_{b,k} = 0, \\ ((k, a, b) = (1, N, 2), (k, a, b) = (N, N-1, 1), \text{ or} \\ 2 \le k \le N-1 \text{ and } (a, b) = (k-1, k+1)). \end{cases} \tag{8.2}$$

These circuit equations are normalized by following equations:

$$v_k = \sqrt{\frac{g_1}{3g_3}}y_k, \quad i_k = \sqrt{\frac{Cg_1}{3Lg_3}}x_k, \quad t = \sqrt{LC}\tau, \quad \alpha = \frac{L}{L_c}, \quad \varepsilon = g_1\sqrt{\frac{L}{C}}. \tag{8.3}$$

The α expresses coupling strength and the ε shows nonlinearity. Normalized equations are shown as follows:

$$\begin{cases} \dfrac{dx_k}{d\tau} = y_k, \\ \dfrac{dy_k}{d\tau} = -x_k + \varepsilon\left(y_k - \dfrac{1}{3}y_k^3\right) + \alpha(x_a + x_b - 2x_k), \\ ((k, a, b) = (1, 2, N), (k, a, b) = (N, 1, N-1), \text{ or} \\ 2 \le k \le N-1 \text{ and } (a, b) = (k+1, k-1)). \end{cases} \tag{8.4}$$

⟨Circuit equations of a ladder system⟩

Circuit equations of OSC_k are obtained when N van der Pol oscillators are coupled by inductors as a ladder. The circuit equations are normalized by the same method as previously described and are shown as follows.

— Middle oscillators ($2 \le k \le N-1$) —

$$\begin{cases} \dfrac{dx_k}{d\tau} = y_k, \\ \dfrac{dy_k}{d\tau} = -x_k + \varepsilon\left(y_k - \dfrac{1}{3}y_k^3\right) + \alpha(x_{k+1} + x_{k-1} - 2x_k). \end{cases} \tag{8.5}$$

— Edge oscillators ($(k, n) = (1, 2)$ or $(N, N-1)$) —

$$\begin{cases} \dfrac{dx_k}{d\tau} = y_k, \\ \dfrac{dy_k}{d\tau} = -x_k + \varepsilon\left(y_k - \dfrac{1}{3}y_k^3\right) + \alpha(x_n - x_k). \end{cases} \tag{8.6}$$

⟨Circuit equations of a 2D lattice system⟩

Circuit equations of $\mathrm{OSC}_{k,l}$ are obtained when $M \times N$ van der Pol oscillators are coupled by inductors as a 2D lattice. The circuit equations are normalized by the same method.

— Inside oscillators ($2 \le k \le M-1$ and $2 \le l \le N-1$) —

$$\begin{cases} \dfrac{dx_{k,l}}{d\tau} = y_{k,l}, \\ \dfrac{dy_{k,l}}{d\tau} = -x_{k,l} + \varepsilon\left(y_{k,l} - \dfrac{1}{3}y_{k,l}^3\right) + \alpha\left(x_{k+1,l} + x_{k-1,l} + x_{k,l+1} + x_{k,l-1} - 4x_{k,l}\right). \end{cases} \tag{8.7}$$

— Side oscillators (Top: $(k, a, b) = (1, 0, 1)$ and $2 \le l \le N-1$,
Bottom: $(k, a, b) = (M, 0, -1)$ and $2 \le l \le N-1$,
Left: $(l, a, b) = (1, 1, 0)$ and $2 \le k \le M-1$,
Right: $(l, a, b) = (N, -1, 0)$ and $2 \le k \le M-1$) —

$$\begin{cases} \dfrac{dx_{k,l}}{d\tau} = y_{k,l}, \\ \dfrac{dy_{k,l}}{d\tau} = -x_{k,l} + \varepsilon\left(y_{k,l} - \dfrac{1}{3}y_{k,l}^3\right) + \alpha\left(x_{k+a,l+b} + x_{k-a,l-b} + x_{k+b,l+a} - 3x_{k,l}\right). \end{cases} \tag{8.8}$$

— Corner oscillators (Left-top: $(k, l, a, b) = (1, 1, 1, 1)$,
Right-top: $(k, l, a, b) = (1, N, 1, -1)$, Left-bottom: $(k, l, a, b) = (M, 1, -1, 1)$,
Right-bottom: $(k, l, a, b) = (M, N, -1, -1)$ —

$$\begin{cases} \dfrac{dx_{k,l}}{d\tau} = y_{k,l}, \\ \dfrac{dy_{k,l}}{d\tau} = -x_{k,l} + \varepsilon\left(y_{k,l} - \dfrac{1}{3}y_{k,l}^3\right) + \alpha\left(x_{k+a,l} + x_{k,l+b} - 2x_{k,l}\right). \end{cases} \tag{8.9}$$

⟨Circuit equations of a cross system⟩

When four ladders, which N van der Pol oscillators are coupled by inductors, are coupled to a center oscillator OSC_c by inductors as a cross, the normalized circuit equations of $\mathrm{OSC}_{k,l}$ and OSC_c are developed (see (8.10)–(8.13)). The four coupling inductors between ladders and the OSC_c are set as L_{cc}. Therefore, circumjacent coupling parameter α of OSC_c is set as α_c.

— Middle oscillators ($1 \le k \le 4$ and $2 \le l \le N-1$) —

$$\begin{cases} \dfrac{dx_{k,l}}{d\tau} = y_{k,l}, \\ \dfrac{dy_{k,l}}{d\tau} = -x_{k,l} + \varepsilon\left(y_{k,l} - \dfrac{1}{3}y_{k,l}^3\right) + \alpha\left(x_{k,l+1} + x_{k,l-1} - 2x_{k,l}\right). \end{cases} \tag{8.10}$$

— Edge oscillators ($1 \leq k \leq 4$) —

$$\begin{cases} \dfrac{dx_{k,1}}{d\tau} = y_{k,1}, \\ \dfrac{dy_{k,1}}{d\tau} = -x_{k,1} + \varepsilon\left(y_{k,1} - \dfrac{1}{3}y_{k,1}^3\right) + \alpha\left(x_{k,2} - x_{k,1}\right). \end{cases} \tag{8.11}$$

— Adjacent oscillators to the center oscillator OSC$_c$ ($1 \leq k \leq 4$) —

$$\begin{cases} \dfrac{dx_{k,N}}{d\tau} = y_{k,N}, \\ \dfrac{dy_{k,N}}{d\tau} = -x_{k,N} + \varepsilon\left(y_{k,N} - \dfrac{1}{3}y_{k,N}^3\right) + \alpha\left(x_{k,N-1} - x_{k,N}\right) + \alpha_c\left(x_c - x_{k,N}\right). \end{cases} \tag{8.12}$$

— Center oscillator OSC$_c$ —

$$\begin{cases} \dfrac{dx_c}{d\tau} = y_c, \\ \dfrac{dy_c}{d\tau} = -x_c + \varepsilon\left(y_c - \dfrac{1}{3}y_c^3\right) + \alpha_c\left(x_{1,N} + x_{2,N} + x_{3,N} + x_{4,N} - 4x_c\right), \end{cases} \tag{8.13}$$

where $\alpha_c = \dfrac{L}{L_{cc}}$.

8.3 Phase-inversion waves

The phase-inversion wave which is changing phase states between adjacent oscillators can be observed on many kinds of synchronization phenomena. The synchronization phenomena can be observed on the ring systems, the ladder systems, and 2D lattice systems and the cross systems.

8.3.1 Basic synchronization phenomena

An in-phase synchronization is a synchronization phenomenon of which phase differences between adjacent oscillators are zero. If the phase differences are π, a phase state is called an anti-phase synchronization. An in-and-anti-phase synchronization is a synchronization phenomenon in which the in-phase synchronizations and the anti-phase synchronizations are alternatively existing (see Figure 8.6). In Figure 8.6, the "/" means the in-phase synchronization and "\" shows the anti-phase synchronization. If phase states between an edge oscillator and the next edge oscillator are the anti-phase synchronization like in Figure 8.6 when the system is the ladder systems, the in-and-anti-phase synchronization is stable. Therefore, if there are not phase-inversion waves on the system, the in-and-anti-phase synchronization can be observed when the number of oscillators is an even number. The in-and-anti-phase synchronization can be observed on the ring system and the 2D lattice system too.

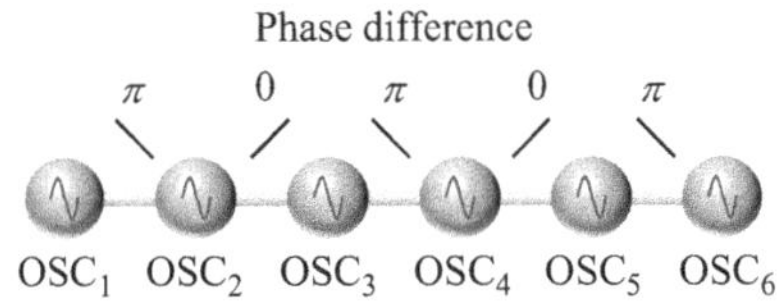

Figure 8.6 In-and-anti-phase synchronization

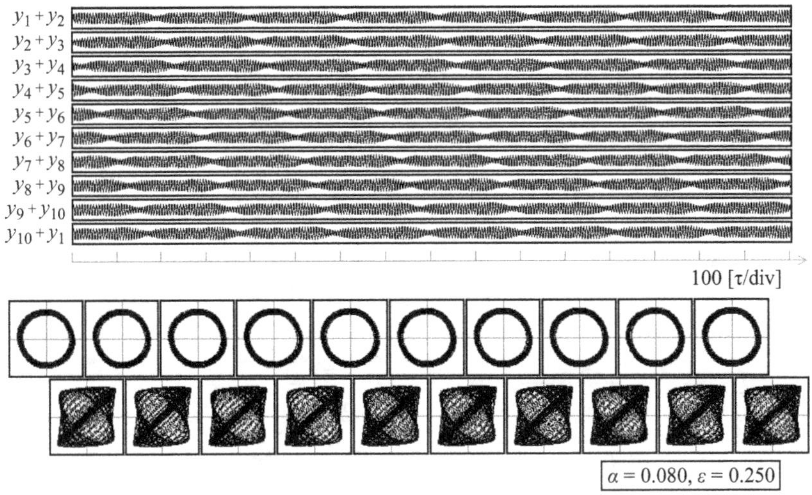

Figure 8.7 Collaterally propagating two phase-inversion waves in the in-phase synchronization on a ring system (N = 10)

8.3.2 Sample of phase-inversion waves

Samples of the phase-inversion waves, which are observed on the ring system, the ladder system, the 2D lattice system, and the cross system, are shown in Figures 8.7–8.17.

<Ring>

A sample of two phase-inversion waves on the ring system which is constructed by ten oscillators is shown in Figures 8.7–8.9. The upper side of Figure 8.7 is constructed by stacked ten thin rectangle boxes. Horizontal axes show the time, and vertical axes express sum of voltages between adjacent oscillators ($y_k + y_{k+1}$, or $v_k + v_{k+1}$). Therefore, if a phase state between the adjacent oscillators is the in-phase synchronization, an amplitude of each graph becomes large. If the phase state is the anti-phase synchronization, the amplitude becomes zero. Black regions show around the in-phase synchronization, and white regions show around the anti-phase synchronization. In Figure 8.7, we can see that two phase-inversion waves are collaterally propagating to same direction. The first phase-inversion wave changes from the in-phase

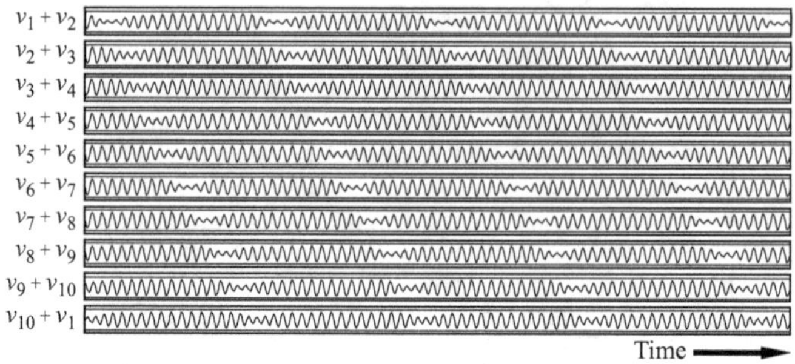

Figure 8.8 Actual circuit experimental result of collaterally propagating two phase-inversion waves in the in-phase synchronization on a ring system ($N = 10$, $L \simeq 52.2$ mH, $C \simeq 52.2$ nF, $R_1 \simeq R_2 \simeq 8.2$ kΩ, $R_3 \simeq 32.0$ kΩ, and $L_c \simeq 305$ mH)

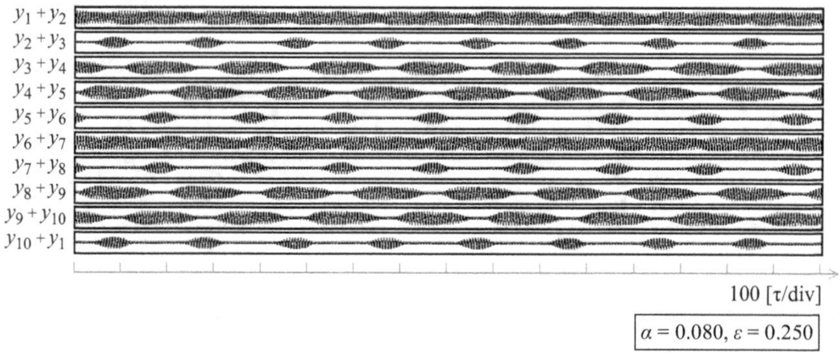

Figure 8.9 Propagation and reflection between two phase-inversion waves in the in-and-anti-phase synchronization on a ring system ($N = 10$)

synchronization to the anti-phase synchronization, and the second phase-inversion wave changes from the anti-phase synchronization to the in-phase synchronization.

An attractor of each oscillator is shown in the lower side of Figure 8.7. A horizontal axis is a voltage y_k of each oscillator, and a vertical axis shows a current x_k. We can understand that the nonlinearity of this system is low, because the shape of attractor is almost a round shape. A left edge attractor is an attractor of OSC_1 and a right edge attractor is an attractor of OSC_{10}.

Boxes under the attractors show phase states between adjacent oscillators. Shape of "/" expresses the in-phase synchronization, and "\" shows the anti-phase synchronization. These figures are almost black rectangle shape, respectively. Therefore, we can understand that the phase states are changing from the in-phase synchronization to the anti-phase synchronization or the anti-phase synchronization to the in-phase synchronization continuously. The left edge box and the right edge box show the phase states between OSC_1 and OSC_2 and between OSC_{10} and OSC_1, respectively.

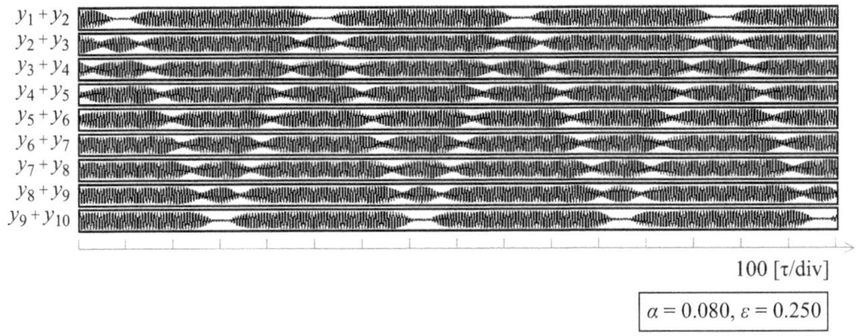

Figure 8.10 Collaterally propagating two phase-inversion waves in the in-phase synchronization on a ladder system ($N = 10$)

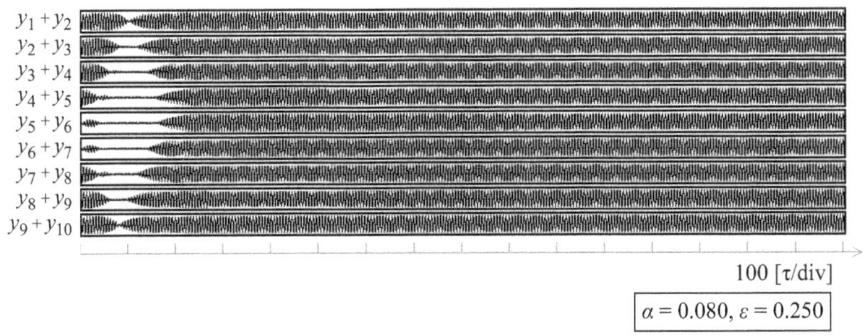

Figure 8.11 Reflection at the edges and disappearance of the two phase-inversion waves in the in-phase synchronization on a ladder system ($N = 10$)

Figure 8.8 shows an actual circuit experimental result. We can see that the actual circuit experimental result is the same shape as the simulation result in Figure 8.7.

Figure 8.9 shows that two phase-inversion waves are continuously existing in the in-and-anti-phase synchronization. A reflection between two phase-inversion waves occurs and is continuously propagating.

<Ladder>

The phase-inversion waves on a ladder system are shown in Figures 8.10–8.14 when N is ten. The graphs are constructed by stacked nine thin rectangle boxes, because the number between oscillators is nine.

Figure 8.10: Two phase-inversion waves collaterally propagate and reflect at the edges.

Figure 8.11: Two phase-inversion waves propagate to each different direction, the phase-inversion waves reflect at the edges, and the phase-inversion waves collide and disappear.

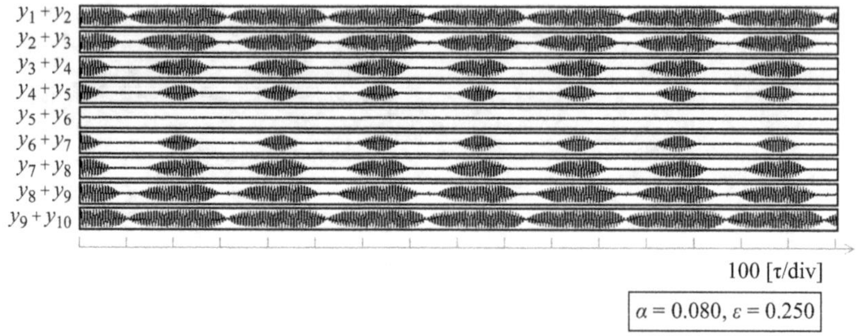

Figure 8.12 Reflection between two phase-inversion waves in the in-phase synchronization on a ladder system (N = 10)

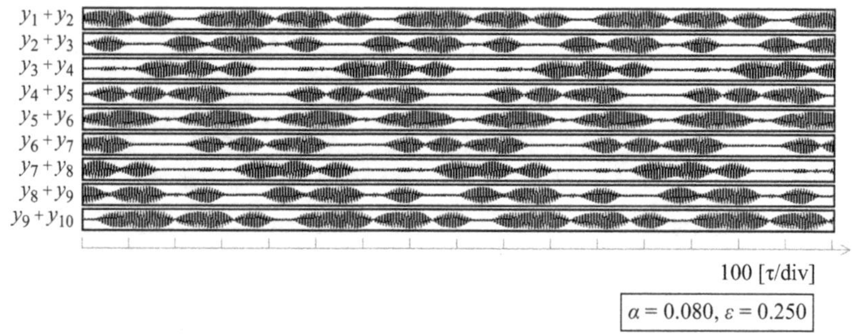

Figure 8.13 Independent propagation of four phase-inversion waves in the in-phase synchronization on a ladder system (N = 10)

Figure 8.12: Two phase-inversion waves propagate to each different direction, reflect at the edges, and the phase-inversion waves collide and reflect. The phase-inversion waves continuously exist.

Figure 8.13: Four phase-inversion waves propagate, reflect at the edges, and collide and reflect. The phase-inversion waves continuously exist.

Figure 8.14: Two phase-inversion waves propagate, reflect at the edges, and collide and reflect in the in-and-anti-phase synchronization. The phase-inversion waves continuously exist.

<2D lattice>

A sample of the phase-inversion waves on a 2D lattice system is shown in Figure 8.15. Each circle (Figure 8.15-A) shows an attractor of each oscillator. If each circle is assumed as an oscillator, we think that this figure is easily understood. Each rectangle figure between the circles shows an itinerancy of the phase state between

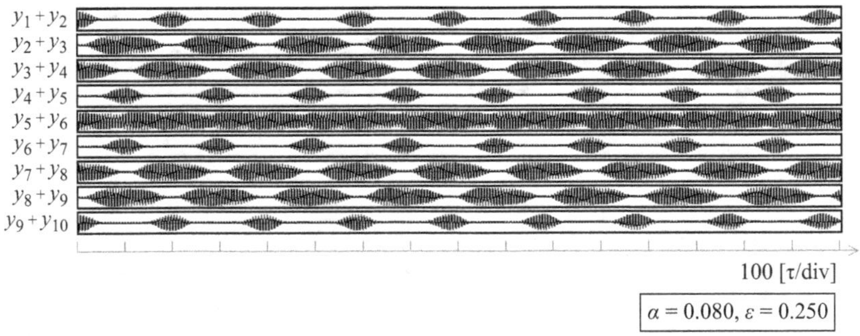

Figure 8.14 Propagation and reflection between two phase-inversion waves in the in-and-anti-phase synchronization on a ladder system (N = 10)

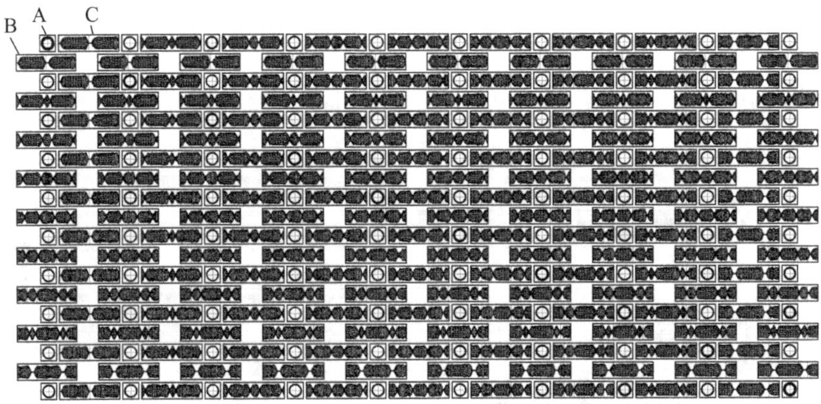

Figure 8.15 The phase-inversion waves in the in-phase synchronization on 2D lattice ((M,N) = (10,10), α = 0.02, and ε = 0.10)

adjacent oscillators. Therefore, Figure 8.15-B shows the itinerancy of the phase state between $OSC_{1,1}$ and $OSC_{2,1}$, and Figure 8.15-C shows the itinerancy of the phase state between $OSC_{1,1}$ and $OSC_{1,2}$.

Two phase-inversion waves in each column and in each row are collaterally propagating to same direction. Therefore, 50 horizontal phase-inversion waves are continuously existing in rows, and 50 vertical phase-inversion waves are continuously existing in columns.

<Cross system>

Figure 8.16 shows a simulation result of four phase-inversion waves in the in-phase synchronization on a cross system ($N = 4$). Figure 8.16 is separated into four groups. Each group is constructed by what four thin rectangles are piled up. The

Figure 8.16 A simulation result of continuously propagating four phase-inversion waves in the in-phase synchronization on a cross system ($N = 4$, $\alpha = 0.06$, $\alpha_c = 0.10$, and $\varepsilon = 0.25$)

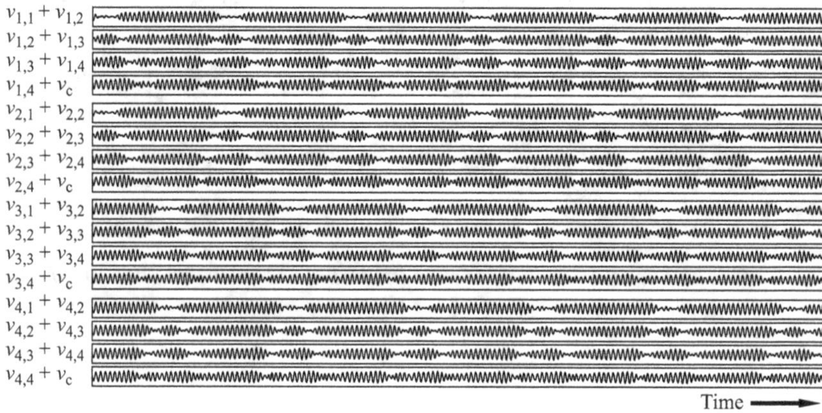

Figure 8.17 An actual experimental result of continuously propagating four phase-inversion waves in the in-phase synchronization on a cross system ($N = 4$, $L \simeq 51.4$ mH, $C \simeq 10.7$ nF, $R_1 \simeq R_2 \simeq 8.2$ kΩ, $R_3 \simeq 6.8$ kΩ, and $L_c \simeq 305$ mH)

bottom thin rectangle in each group shows an itinerancy of a phase state between $\mathrm{OSC}_{1,4}$, and OSC_c ($k = 1, 2, 3$, and 4). The phase-inversion waves penetrate the center oscillator OSC_c and are continuously existing on two ladders. An actual circuit experimental result is shown in Figure 8.17. We can see that the phase-inversion waves of the actual circuit experimental result are the same shape as the phase-inversion waves of the simulation result.

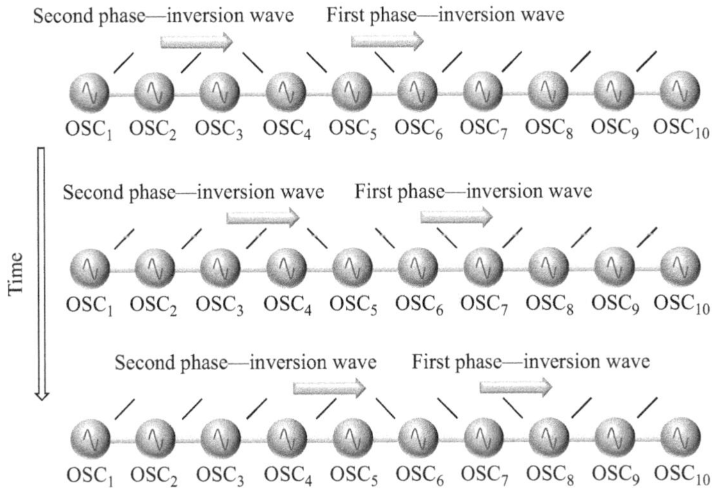

Figure 8.18 Shift of phase states by propagation of the phase-inversion waves

8.3.3 Characteristics of the phase-inversion waves

The phase-inversion waves have some characteristics. On the ring system, three characteristics can be observed. Four characteristics can be observed on the ladder system. The number of characteristic is six on the 2D lattice system. We introduce four basic characteristics.

1. Propagation (ring system, ladder system, and 2D lattice system)
   A model of shift of phase states by the propagation which is the most basic characteristic is shown in Figure 8.18. Two phase-inversion waves are propagating from left to right on ten oscillators coupled as a ladder. The "/" and the "\" show the in-phase synchronization and the anti-phase synchronization, respectively. The first phase-inversion wave changes phase states between adjacent oscillators from the in-phase synchronization to the anti-phase synchronization. The phase states are changed from the anti-phase synchronization to the in-phase synchronization by the second phase-inversion wave. The propagation is such phenomenon.
   On 2D lattice system, the phase-inversion waves can propagate to a vertical direction in each column and to a horizontal direction in each row. In other words, the vertical phase-inversion waves and the horizontal phase-inversion waves exist independently.
2. Reflection between two phase-inversion waves (ring system, ladder system, and 2D lattice system)
   When two phase-inversion waves which come from each direction arrive at adjacent oscillators at same time, the refection can be observed.
3. Disappearance (ring system, ladder system, and 2D lattice system)
   When two phase-inversion waves which come from each direction arrive at an oscillator at same time, the disappearance can be observed.

4. Reflection at an edge (ladder system and 2D lattice system)
   In the ladder system, the phase-inversion wave arrives at an edge, and reflects. In the 2D lattice system, a reflection at a side and a reflection at a corner can be observed. A mechanism of the reflection at a corner differs from a mechanism of reflection at a side.

8.3.4 Propagation velocity of phase-inversion waves

In the simulations, a propagation velocity of the phase-inversion waves depends on the coupling parameter α which is obtained by using the inductor L in each oscillator and the coupling inductor L_c. Therefore, the propagation velocity depends on the coupling inductor L_c in the experimental results when the inductor L is not changed in each actual circuit experiment. The simulation results and the actual circuit experimental results are shown in Figures 8.19 and 8.20, respectively. The propagation velocity of the phase-inversion waves accelerates as the α is increasing or the L_c is decreasing.

8.3.5 Mechanisms

Mechanisms of the propagation, reflection, and disappearance of the phase-inversion waves can be explained by using phase differences between adjacent oscillators and instantaneous frequencies of each oscillator.

The phase differences and the instantaneous frequencies are obtained by the following method. Waveforms of OSC_k and OSC_{k+1} are shown in Figure 8.21. The $\tau_k(a)$ means time of ath maximum value of a voltage of OSC_k. Therefore, a phase difference $\Phi_{k,k+1}(a)$ between OSC_k and OSC_{k+1} and an instantaneous frequency $f_k(a)$ of OSC_k are calculated by using (8.14) and (8.15), respectively:

$$\Phi_{k,k+1}(a) = \frac{\tau_k(a) - \tau_{k+1}(a)}{\tau_k(a) - \tau_k(a-1)} \times 2\pi, \tag{8.14}$$

$$f_k(a) = \frac{1}{\tau_k(a) - \tau_k(a-1)}. \tag{8.15}$$

8.3.5.1 Theoretical frequencies

The in-phase synchronization, the anti-phase synchronization, and the in-and-anti-phase synchronization can be observed on the ring systems. The phase-inversion waves change instantaneous frequencies among frequencies of these steady states. A frequency of the each steady state can be calculated theoretically when the oscillation waveform can be assumed as a sinusoidal wave. The frequencies can be obtained by using the averaging method and so on [8–10]. However, in this section, we easily obtain these frequencies by calculating instantaneous reactive powers of the ring system. An instantaneous power of the inductance L, an instantaneous power of the

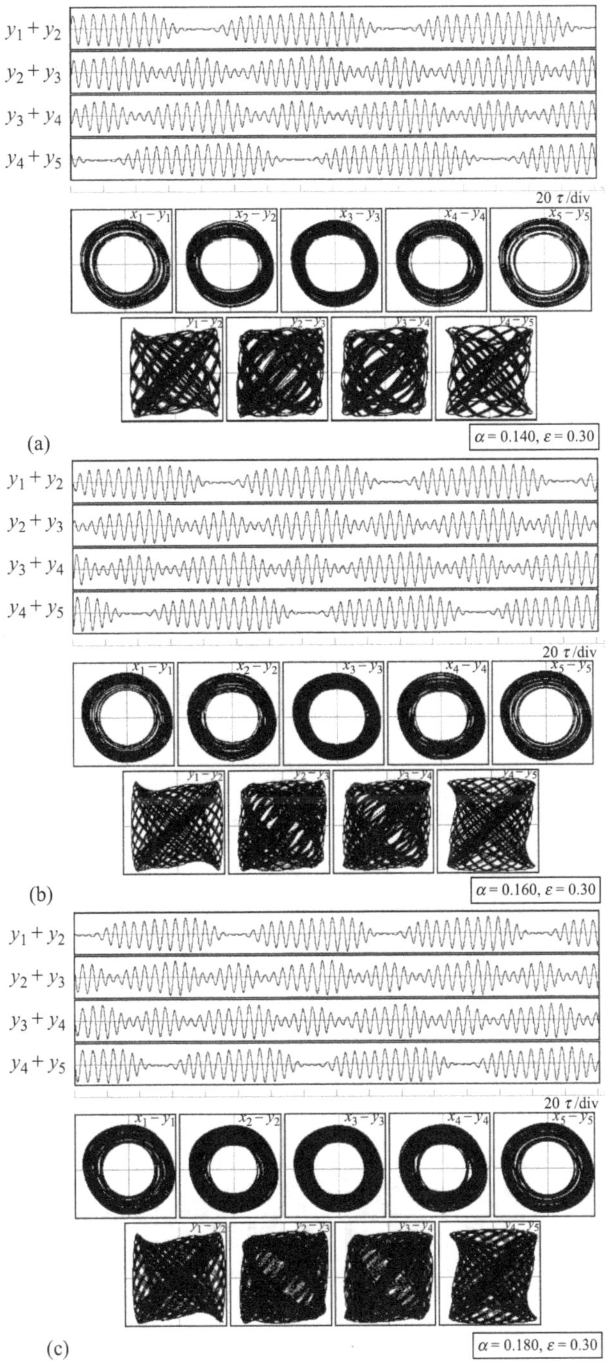

Figure 8.19 Comparison of propagation velocity in simulation results ($N = 5$ and $\varepsilon = 0.30$). (a) $\alpha = 0.14$, (b) $\alpha = 0.16$, and (c) $\alpha = 0.18$

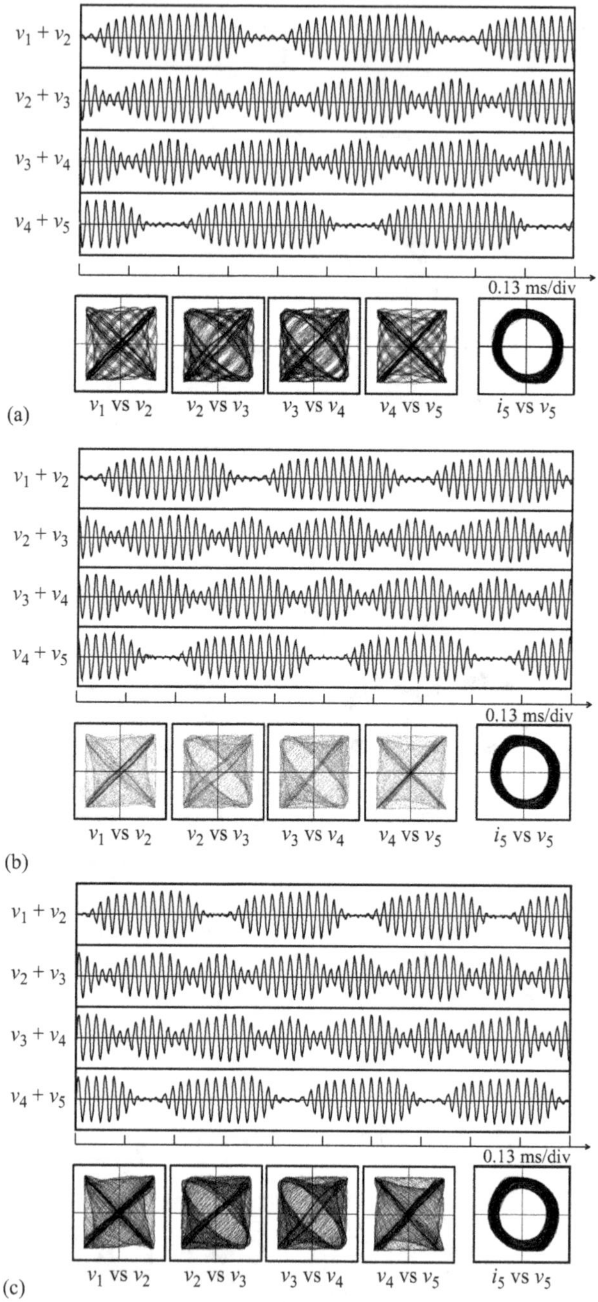

Figure 8.20 Comparison of propagation velocity in actual circuit experiments ($N = 5$, $L \simeq 51.2$ mH, $C \simeq 6.7$ nH, $R_1 \simeq R_2 \simeq 8.2$ kΩ, and $R_3 \simeq 1.8$ kΩ). (a) $L_c \simeq 405$ mH, (b) $L_c \simeq 355$ mH, and (c) $L_c \simeq 305$ mH

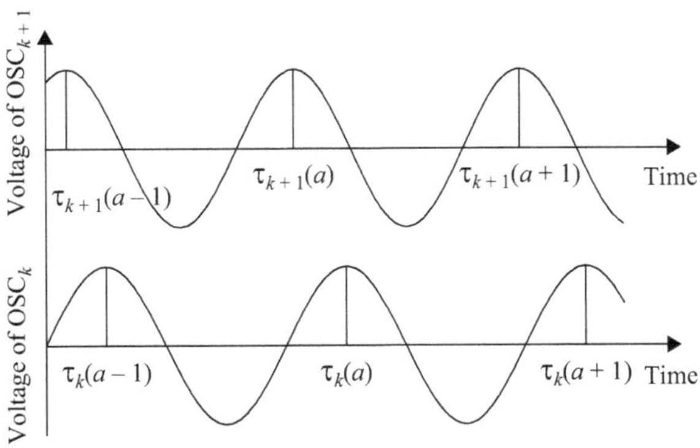

Figure 8.21 Waveform of each oscillator

coupling inductance L_c, and an instantaneous power of the capacitance C are named p_{Lk}, $p_{Lck+1,k}$, and p_{Ck}, respectively:

$$\begin{cases} L\dfrac{di_k}{dt} = v_k, \\ L_c\dfrac{di_{k+1,k}}{dt} = v_{k+1} - v_k, \\ p_{Lk} = v_k i_k, \\ p_{Ck} = Cv_k\dfrac{dv_k}{dt}, \\ p_{Lck+1,k} = (v_{k+1} - v_k)\, i_{k+1,k}. \end{cases} \tag{8.16}$$

Equation (8.16) are normalized by using (8.3).

$$\begin{cases} p_{Lk} = \dfrac{\delta}{\varepsilon} x_k \dfrac{dx_k}{d\tau}, \\ p_{Ck} = \dfrac{\delta}{\varepsilon}\dfrac{dx_k}{d\tau}\dfrac{d^2x_k}{d\tau^2}, \\ p_{Lck+1,k} = \alpha\dfrac{\delta}{\varepsilon}(x_{k+1} - x_k)\left(\dfrac{dx_{k+1}}{d\tau} - \dfrac{dx_k}{d\tau}\right), \end{cases} \tag{8.17}$$

where $\delta = g_1^2/3g_3$. Three synchronization states of the ring system with four oscillators are calculated. The x_1, x_2, x_3, and x_4 are assumed as same amplitude A and as same angular frequency ω, and are expressed as follows:

$$x_k = A\sin(\omega\tau + \theta_k) \quad (k = 1, 2, 3, \text{ or } 4)\,. \tag{8.18}$$

Total instantaneous reactive power p_r of the ring system is obtained by using (8.17) and (8.18) as follows:

$$
\begin{aligned}
p_r &= \sum_{k=1}^{4}(p_{Lk} + p_{Ck}) + p_{Lc2,1} + p_{Lc3,2} + p_{Lc4,3} + p_{Lc1,4} \\
&= \frac{\omega\delta A^2}{2\varepsilon}(1-\omega^2)(\sin(2\omega\tau + 2\theta_1) + \sin(2\omega\tau + 2\theta_2) + \sin(2\omega\tau + 2\theta_3) \\
&\quad + \sin(2\omega\tau + 2\theta_4)) + \frac{\omega\alpha\delta A^2}{\varepsilon}(\sin(2\omega\tau + 2\theta_1) + \sin(2\omega\tau + 2\theta_2) \\
&\quad + \sin(2\omega\tau + 2\theta_3) + \sin(2\omega\tau + 2\theta_4) - \sin(\omega\tau + \theta_1)\cos(\omega\tau + \theta_4) \\
&\quad - \sin(\omega\tau + \theta_2)\cos(\omega\tau + \theta_1) - \sin(\omega\tau + \theta_3)\cos(\omega\tau + \theta_2) \\
&\quad - \sin(\omega\tau + \theta_4)\cos(\omega\tau + \theta_3) - \sin(\omega\tau + \theta_4)\cos(\omega\tau + \theta_1) \\
&\quad - \sin(\omega\tau + \theta_1)\cos(\omega\tau + \theta_2) - \sin(\omega\tau + \theta_2)\cos(\omega\tau + \theta_3) \\
&\quad - \sin(\omega\tau + \theta_3)\cos(\omega\tau + \theta_4)).
\end{aligned} \tag{8.19}
$$

<Frequency of the in-phase synchronization>

First, a frequency of the in-phase synchronization is obtained. The frequency is called f_{in}. We plug zero into θ_1, θ_2, θ_3, and θ_4 of (8.19). We assume the p_r is zero, and calculate the ω_{in} and the frequency f_{in}

$$
\begin{aligned}
p_r &= 2(1-\omega_{in}^2)\sin(2\omega_{in}\tau) \\
&\qquad + \alpha(4\sin(2\omega_{in}\tau) - 8\sin(\omega_{in}\tau)\cos(\omega_{in}\tau)) = 0, \\
2(1&-\omega_{in}^2)\sin(2\omega_{in}\tau) = 0.
\end{aligned} \tag{8.20}
$$

Even if τ is changed, this equation has to become zero. The ω_{in} is larger than zero. Therefore, the following equation is obtained:

$$
\begin{aligned}
\omega_{in} &= 1, \\
f_{in} &= \frac{1}{2\pi}.
\end{aligned} \tag{8.21}
$$

<Frequency of the anti-phase synchronization>

Next, a frequency of the anti-phase synchronization (f_{anti}) is obtained. We plug zero into θ_1 and θ_3, and plug π into θ_2 and θ_4 in (8.19). We assume the p_r is zero, and calculate the ω_{anti}

$$
\begin{aligned}
p_r &= 2(1-\omega_{anti}^2)\sin(2\omega_{anti}\tau) + 8\alpha\sin(2\omega_{anti}\tau) = 0, \\
1 &- \omega_{anti}^2 + 4\alpha = 0.
\end{aligned} \tag{8.22}
$$

Even if τ is changed, this equation has to become zero. The ω_{anti} is larger than zero. Therefore, the following equation is obtained:

$$\begin{aligned} \omega_{anti} &= \sqrt{1+4\alpha}, \\ f_{anti} &= \frac{\sqrt{1+4\alpha}}{2\pi}. \end{aligned} \tag{8.23}$$

<Frequency of the in-and-anti-phase synchronization>

Finally, a frequency of the in-and-anti-phase synchronization ($f_{in\text{-}anti}$) is obtained. We plug zero into θ_1, and θ_2, and plug π into θ_3 and θ_4 in (8.19). We assume the p_r is zero, and calculate the $\omega_{in\text{-}anti}$

$$\begin{aligned} &p_r = 2\left(1-\omega_{in\text{-}anti}^2\right)\sin(2\omega_{in\text{-}anti}\tau) + 4\alpha\sin(2\omega_{in\text{-}anti}\tau), \\ &1-\omega_{in\text{-}anti}^2 + 2\alpha = 0. \end{aligned} \tag{8.24}$$

Even if τ is changed, this equation has to become zero. The $\omega_{in\text{-}anti}$ is larger than zero. Therefore, the following equation is obtained:

$$\begin{aligned} \omega_{in\text{-}anti} &= \sqrt{1+2\alpha}, \\ f_{in\text{-}anti} &= \frac{\sqrt{1+2\alpha}}{2\pi}. \end{aligned} \tag{8.25}$$

<Comparison between simulation results and theoretical results>

In Section 8.3.2, the phase-inversion waves are observed on the ring system and the ladder system when α is 0.08 and ε is 0.25. This value is assigned to α of (8.21), (8.23), and (8.25). The theoretical values of frequencies are compared from the simulation values. Theoretical values of the f_{in}, the $f_{in\text{-}anti}$, and the f_{anti} are almost 0.159, 0.171, and 0.183, respectively. The simulation results of the f_{in}, the $f_{in\text{-}anti}$, and the f_{anti} are almost 0.159, 0.171, and 0.182, respectively, when ε is 0.25. The errors of the f_{in}, the $f_{in\text{-}anti}$, and the f_{anti} are almost 0.000%, 0.000%, and 0.564%. We can understand that the errors are small enough. These theoretical values are used for analyzing mechanisms.

8.3.5.2 Propagation mechanism on the ladder system and the ring system

Mechanisms of propagations which are observed in the in-phase synchronization and in the in-and-anti-phase synchronization on the ladder system and on the ring system are expressed.

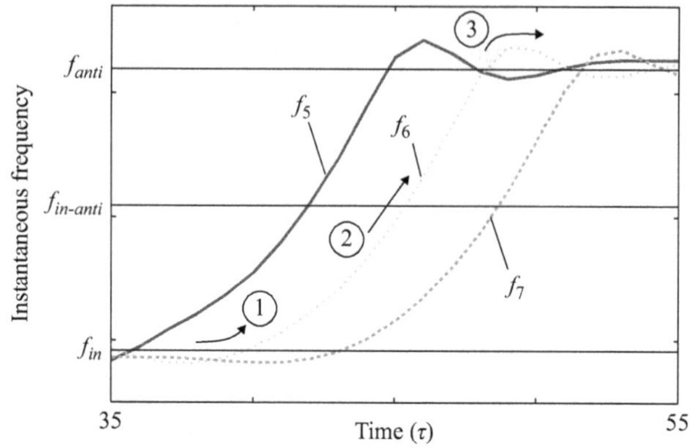

Figure 8.22 Instantaneous frequencies of the propagation in the in-phase synchronization

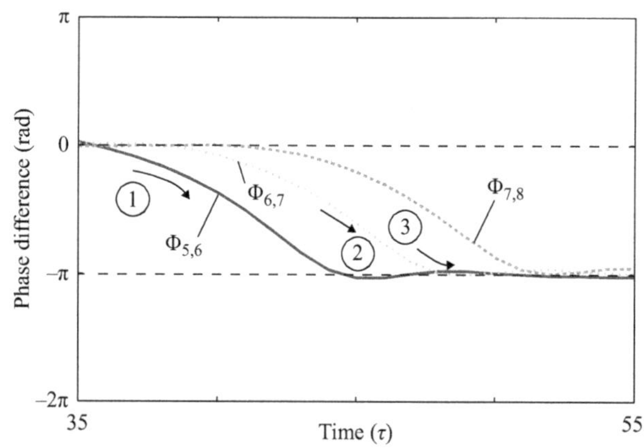

Figure 8.23 Phase differences of the propagation in the in-phase synchronization

<In the in-phase synchronization>

Figure 8.22 shows itinerancies of instantaneous frequencies and Figure 8.23 shows itinerancies of phase differences between adjacent oscillators.

1. ① in Figures 8.22 and 8.23: Let us assume that all oscillators are the in-phase synchronization, and that a phase-inversion wave, which is changing from the in-phase synchronization to the anti-phase synchronization, comes from the direction of OSC_1 and arrives at OSC_6.
2. ① in Figures 8.22 and 8.23: An instantaneous frequency f_6 of OSC_6 changes from the f_{in} to the f_{anti} as $\Phi_{5,6}$ is changing to $-\pi$.
3. ② in Figures 8.22 and 8.23: The phase difference $\Phi_{6,7}$ between OSC_6 and OSC_7 decreases from 0 to $-\pi$ by shift of f_6. Changing speeds of the phase differences and the instantaneous frequencies are decided by the difference between the f_{in}

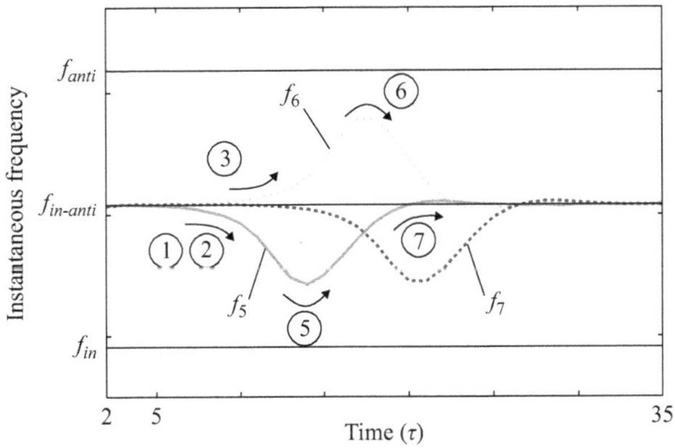

Figure 8.24 Instantaneous frequencies of the propagation in the in-and-anti-phase synchronization

and the f_{anti}. This means that the difference between the f_{in} and the f_{anti} decides the propagation speed of the phase-inversion wave. In other words, the propagation speed depends on the coupling parameter α.

4. ③ in Figures 8.22 and 8.23: f_6 becomes to the f_{anti}, and $\Phi_{6,7}$ arrives at almost $-\pi$.

The mechanism of changing from the anti-phase synchronization to the in-phase synchronization by the phase-inversion wave can be explained in the similar manner.

<In the in-and-anti-phase synchronization>

Figure 8.24 shows itinerancies of instantaneous frequencies and Figure 8.25 shows itinerancies of phase differences between adjacent oscillators.

1. ① in Figures 8.24 and 8.25: Let us assume that the oscillators of the system synchronize at the in-and-anti-phase and that the phase-inversion wave is arriving at OSC_5 from the direction of OSC_1. Phase states between OSC_4 and OSC_5, between OSC_5 and OSC_6, and between OSC_6 and OSC_7 are the anti-phase synchronization, the in-phase synchronization, and the anti-phase synchronization, respectively.
2. ① in Figures 8.24 and 8.25: f_5 changes from the $f_{in\text{-}anti}$ toward the f_{in}, because the phase state between OSC_4 and OSC_5 changes from the anti-phase synchronization $-\pi$ to the in-phase synchronization -2π.
3. ② in Figures 8.24 and 8.25: The phase state of OSC_5 is late to OSC_6, because f_5 decreases from the $f_{in\text{-}anti}$ toward the f_{in}. Therefore, $\Phi_{5,6}$ starts to increase from 0 to π.
4. ③ in Figure 8.24: f_6 starts to change from the $f_{in\text{-}anti}$ toward the f_{anti}, because $\Phi_{5,6}$ changes from 0 into π.
5. ④ in Figure 8.25: $\Phi_{6,7}$ changes from π to 0, because f_6 changes toward the f_{anti}.
6. ⑤ in Figure 8.24: f_5 starts to return to the $f_{in\text{-}anti}$ before f_5 arrives at the f_{in}, because the phase state between OSC_4 and OSC_5 is changing to the in-phase

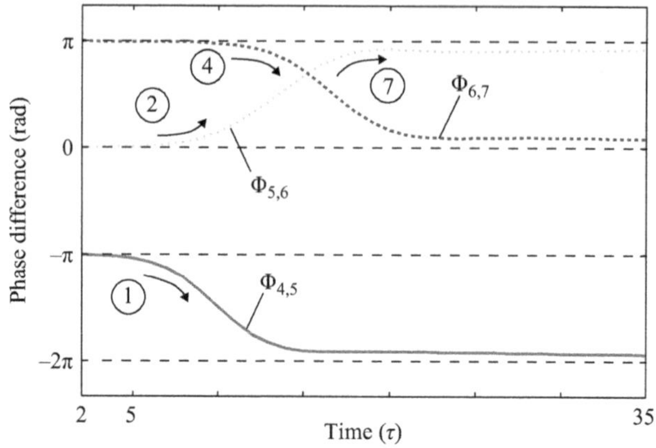

Figure 8.25 Phase differences of the propagation in the in-and-anti-phase synchronization

synchronization and the phase state between OSC_5 and OSC_6 is changing to the anti-phase synchronization.

7. ⑥ in Figure 8.24: f_6 starts to return to the $f_{in\text{-}anti}$ before f_6 arrives at the f_{anti}, because the phase states between OSC_5 and OSC_6 is changing to the anti-phase synchronization and the phase state between OSC_6 and OSC_7 is changing to the in-phase synchronization.
8. ⑦ in Figures 8.24 and 8.25: f_5 becomes to the $f_{in\text{-}anti}$, when the $\Phi_{4,5}$ arrives at -2π and the $\Phi_{5,6}$ becomes to almost π.

The phase-inversion wave propagates in the in-and-anti-phase synchronization by the above mechanism.

8.3.5.3 Reflection mechanism between the phase-inversion waves on the ladder system and the ring system

<In the in-phase synchronization>

Figure 8.26 shows itinerancies of instantaneous frequencies and Figure 8.27 shows itinerancies of phase differences between adjacent oscillators.

1. ① in Figures 8.26 and 8.27: Let us assume that OSC_4–OSC_7 are the anti-phase synchronization and that two phase-inversion waves, which switch from the anti-phase synchronization to the in-phase synchronization, are arriving at OSC_4 and OSC_7 at the almost same time from the directions of OSC_1 and OSC_N, respectively.
2. ① in Figures 8.26 and 8.27: $\Phi_{4,5}$ and $\Phi_{6,7}$ start to change to 2π and -2π at almost the same time, respectively.
3. ② in Figure 8.26: Instantaneous frequencies of OSC_5 f_5 and OSC_6 f_6 start to change from the f_{anti} toward the f_{in}, because $\Phi_{4,5}$ and $\Phi_{6,7}$ are changing to 2π and -2π, respectively.

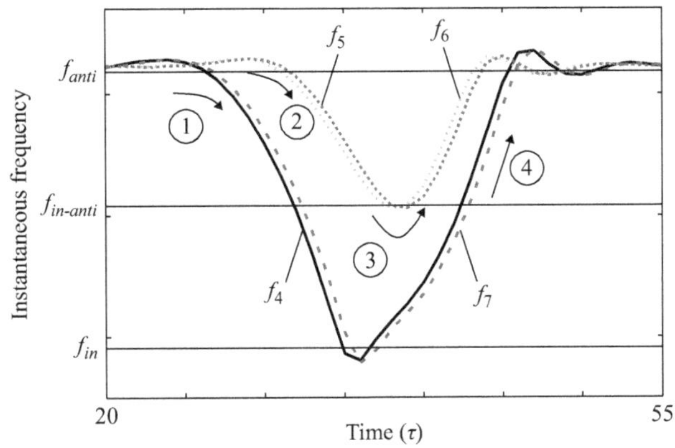

Figure 8.26 Instantaneous frequencies of the reflection between two phase-inversion waves in the in-phase synchronization

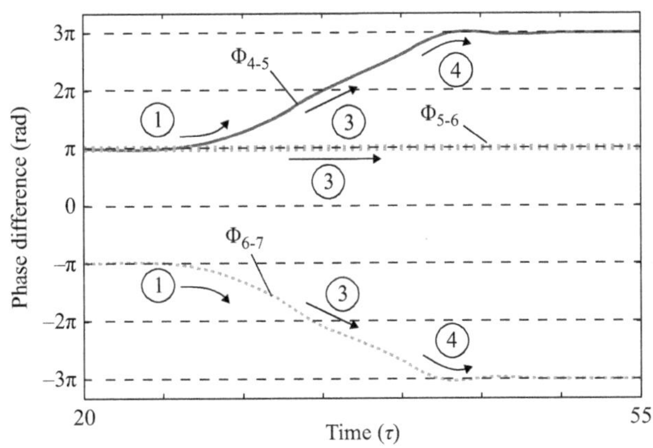

Figure 8.27 Phase differences of the reflection between two phase-inversion waves in the in-phase synchronization

4. ③ in Figures 8.26 and 8.27: However, the anti-phase synchronization between OSC_5 and OSC_6 does not break, because f_5 and f_6 change almost simultaneously. Namely, $\Phi_{5,6}$ does not change from π. Hence, f_5 and f_6 do not arrive at the f_{in} and start to return toward the f_{anti}, and $\Phi_{4,5}$ and $\Phi_{6,7}$ are continuously changing until reaching 3π and -3π, respectively.
5. ④ in Figures 8.26 and 8.27: $\Phi_{4,5}$ and $\Phi_{6,7}$, respectively, arrive at 3π and -3π, and f_4 and f_7 become to the f_{anti}. Two phase-inversion waves are reflected.

The phase-inversion waves changing from the in-phase synchronization to the anti-phase synchronization can be explained in the similar manner.

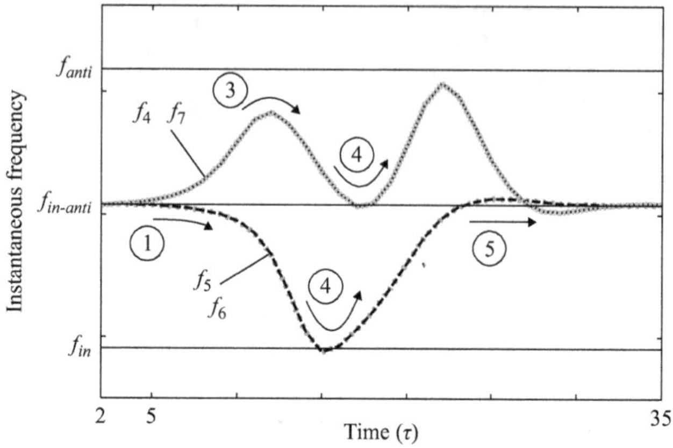

Figure 8.28 Instantaneous frequencies of the reflection between two phase-inversion waves in the in-and-anti-phase synchronization

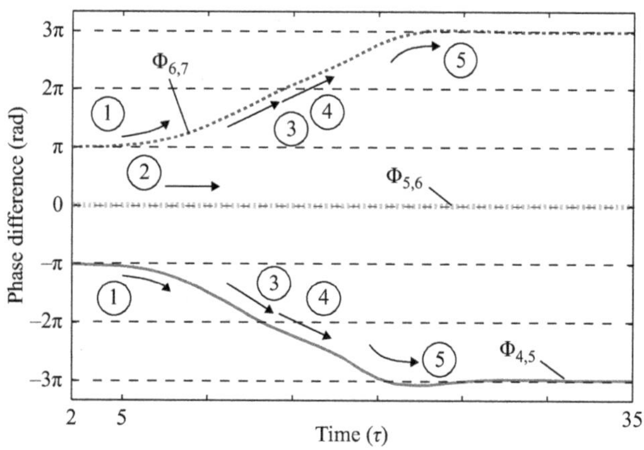

Figure 8.29 Phase differences of the reflection between two phase-inversion waves in the in-and-anti-phase synchronization

<In the in-and-anti-phase synchronization>

Figure 8.28 shows itinerancies of instantaneous frequencies and Figure 8.29 shows itinerancies of phase differences between adjacent oscillators.

1. ① in Figures 8.28 and 8.29: Let us assume that the oscillators of the system synchronize at the in-and-anti-phase and that two phase-inversion waves are arriving at OSC_4 and OSC_7 at the same time from the direction of OSC_1 and OSC_N, respectively. The phase states between OSC_4 and OSC_5 and between OSC_6 and OSC_7 are the anti-phase synchronization and the phase state between OSC_5 and OSC_6 is the in-phase synchronization.

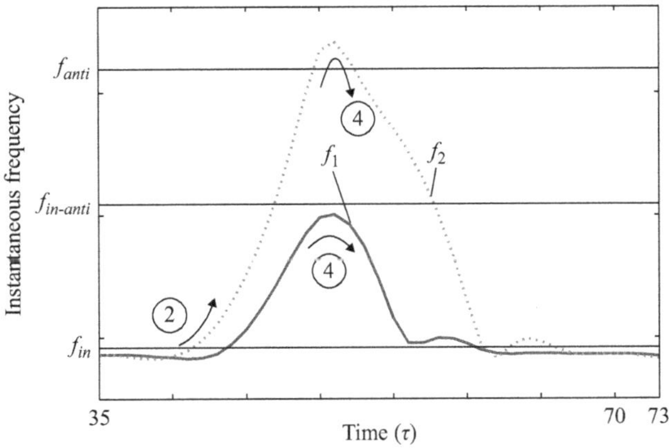

Figure 8.30 Instantaneous frequencies of the reflection at an edge in the in-phase synchronization

2. ① in Figures 8.28 and 8.29: f_5 and f_6 start to change from the $f_{in\text{-}anti}$ toward the f_{in}, because $\Phi_{4,5}$ and $\Phi_{6,7}$ change to -2π and 2π, respectively.
3. ② in Figure 8.29: $\Phi_{5,6}$ does not change, because f_5 and f_6 start to change at the same time.
4. ③ in Figures 8.28 and 8.29: f_4 and f_7 start to change to the $f_{in\text{-}anti}$ again before they reach the f_{anti}, because the phase states between OSC_3 and OSC_4 and between OSC_4 and OSC_5 are interchanging and the phase states between OSC_6 and OSC_7 and between OSC_7 and OSC_8 are also interchanging.
5. ④ in Figures 8.28 and 8.29: When $\Phi_{4,5}$ and $\Phi_{6,7}$ reach -2π and 2π, f_5 and f_6 arrive at the f_{in}. However, $\Phi_{4,5}$ and $\Phi_{6,7}$ are continuously changing to -3π and 3π, because f_4 and f_7 are the $f_{in\text{-}anti}$. Therefore, f_5 and f_6 change from the f_{in} toward the $f_{in\text{-}anti}$, and f_4 and f_7 change from the $f_{in\text{-}anti}$ to the f_{anti}
6. ⑤ in Figures 8.28 and 8.29: $\Phi_{4,5}$ and $\Phi_{6,7}$ become -3π and 3π, and f_4–f_7 arrive at the $f_{in\text{-}anti}$.

Reflections between two phase-inversion waves are occurred in the in-and-anti-phase synchronization by this mechanism.

8.3.5.4 Reflection mechanism at the edge on the ladder system

<In the in-phase synchronization>

Figure 8.30 shows itinerancies of instantaneous frequencies and Figure 8.31 shows itinerancies of phase differences between adjacent oscillators.

1. ① in Figure 8.31: Let us assume that only OSC_1 and OSC_2 are the in-phase synchronization and that the phase-inversion wave changing from the in-phase synchronization to the anti-phase synchronization is arriving at this edge.

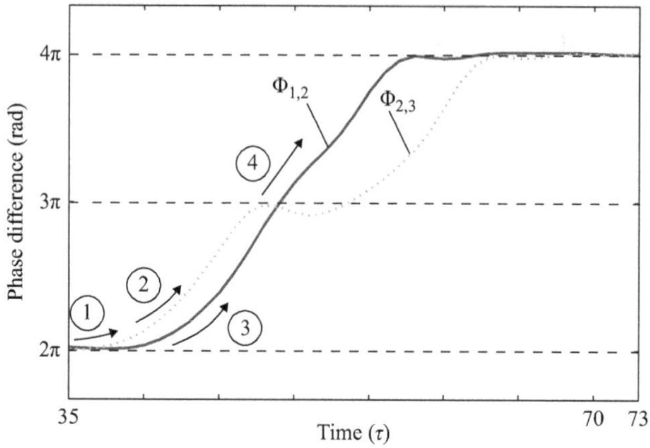

Figure 8.31 Phase differences of the reflection at an edge in the in-phase synchronization

2. ② in Figures 8.30 and 8.31: The instantaneous frequency f_2 of OSC_2 starts to change from the f_{in} toward the f_{anti}, because the in-phase synchronization between OSC_2 and OSC_3 is changed by the phase-inversion wave.
3. ③ in Figure 8.31: $\Phi_{1,2}$ starts to increase from 2π because the f_2 changes.
4. ④ in Figures 8.30 and 8.31: f_1 changes from the f_{in} to the f_{anti}. However, f_1 cannot arrive at the f_{anti}, because OSC_1 is an oscillator of an edge and there are not effect from the other side. $\Phi_{1,2}$ is continuously increasing until arriving at 4π.
5. ④ in Figures 8.30 and 8.31: f_2 starts to decrease again toward the f_{in} because $\Phi_{1,2}$ is changing to 4π. Then, $\Phi_{2,3}$ starts to change toward 4π.

The phase-inversion wave reflects at an edge by this mechanism.

<In the in-and-anti-phase synchronization>

Figure 8.32 shows itinerancies of instantaneous frequencies and Figure 8.33 shows itinerancies of phase differences between adjacent oscillators.

1. ① in Figures 8.32 and 8.33: Let us assume that the oscillators of the system synchronize at the in-and-anti-phase and that the phase-inversion wave reaches OSC_3 from the direction of OSC_N. The phase state between OSC_1 and OSC_2 is the anti-phase synchronization and the phase state between OSC_2 and OSC_3 is the in-phase synchronization.
2. ① in Figures 8.32 and 8.33: f_2 starts to change from the $f_{in\text{-}anti}$ to the f_{anti}, because the phase state between OSC_2 and OSC_3 changes from the in-phase synchronization to the anti-phase synchronization by the phase-inversion wave.
3. ② in Figure 8.33: $\Phi_{1,2}$ starts to change from π to 2π, because f_2 is changing toward the f_{anti}.

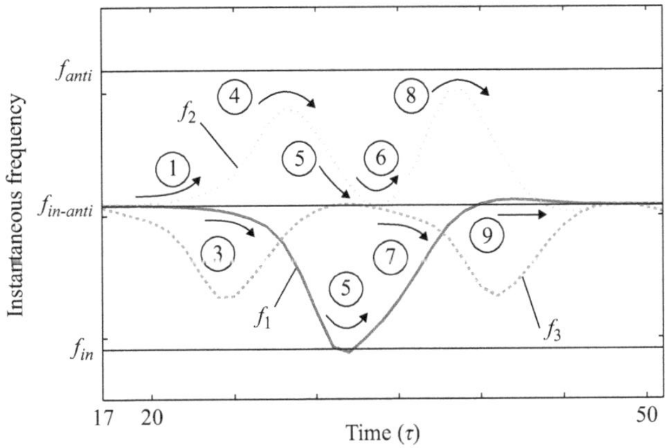

Figure 8.32 Instantaneous frequencies of the reflection at an edge in the in-and-anti-phase synchronization

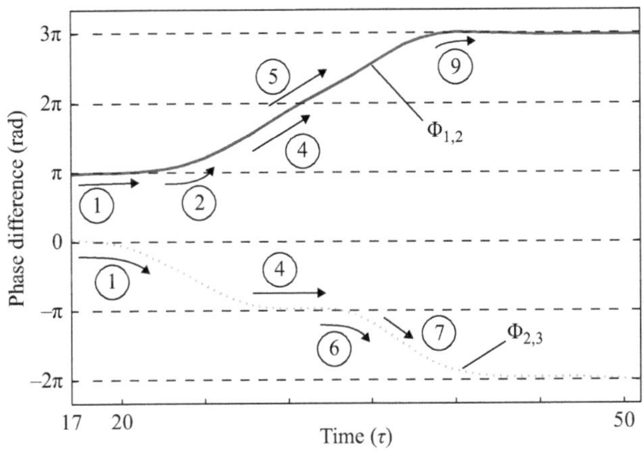

Figure 8.33 Phase differences of the reflection at an edge in the in-and-anti-phase synchronization

4. ③ in Figure 8.32: f_1 starts to decrease toward the f_{in}, because $\Phi_{1,2}$ is changing toward 2π.
5. ④ in Figures 8.32 and 8.33: f_2 starts to decrease to the $f_{in\text{-}anti}$ again before f_2 arrives at the f_{anti}, because the phase states between OSC_1 and OSC_2 and between OSC_2 and OSC_3 are interchanging.
6. ⑤ in Figures 8.32 and 8.33: f_1 and f_2 arrive at the f_{in} and $f_{in\text{-}anti}$, respectively, when $\Phi_{1,2}$ reaches at 2π. However, the in-phase synchronization between OSC_1 and OSC_2 is not stable, because f_1 and f_2 are not a same value. $\Phi_{1,2}$ is continuously changing to 3π, because f_1 is smaller than f_2. Further, f_1 and f_2 start to increase toward the f_{anti}, because $\Phi_{1,2}$ is changing toward 3π.

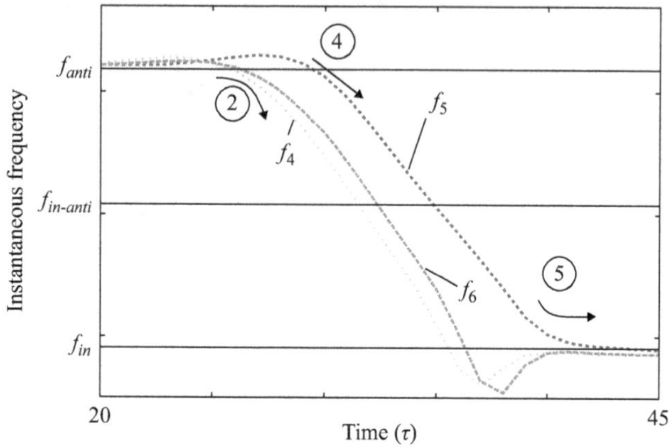

Figure 8.34 Instantaneous frequencies of the disappearance between two phase-inversion waves in the in-phase synchronization

7. ⑥ in Figures 8.32 and 8.33: $\Phi_{2,3}$ starts to change to -2π, because f_2 is larger than f_3.
8. ⑦ in Figures 8.32 and 8.33: f_3 changes toward the f_{in}, because $\Phi_{2,3}$ is changing toward -2π.
9. ⑧ in Figure 8.32: f_2 changes to the $f_{in\text{-}anti}$ again before f_2 arrives at the f_{anti}, because the phase states between OSC_1 and OSC_2 and between OSC_2 and OSC_3 are interchanging.
10. ⑨ in Figures 8.32 and 8.33: f_1 and f_2 arrive at the $f_{in\text{-}anti}$ again, when $\Phi_{1,2}$ arrives at 3π.

The phase-inversion wave reflects at an edge in the in-and-anti-phase synchronization by above mechanism.

8.3.5.5 Disappearance mechanism between the phase-inversion waves on the ladder system and the ring system

Figure 8.34 shows itinerancies of instantaneous frequencies and Figure 8.35 shows itinerancies of phase differences between adjacent oscillators.

1. ① in Figures 8.34 and 8.35: Let us assume that OSC_4–OSC_6 synchronize at the anti-phase synchronization and that two phase-inversion waves switching from the anti-phase synchronization to the in-phase synchronization are arriving at OSC_4 and OSC_6 from the directions of OSC_1 and OSC_N at the almost same time, respectively.
2. ② in Figure 8.34: Oscillation frequencies of OSC_4 f_4 and OSC_6 f_6 change from the f_{anti} to the f_{in} because the phase-inversion waves arrive at OSC_4 and OSC_6.
3. ③ in Figure 8.35: $\Phi_{4,5}$ and $\Phi_{5,6}$ change from π to 2π and to 0 at the almost same time, respectively.
4. ④ in Figures 8.34 and 8.35: The instantaneous frequency f_5 of OSC_5 starts to decrease from the f_{anti} to the f_{in}, because $\Phi_{4,5}$ and $\Phi_{5,6}$ are changing.

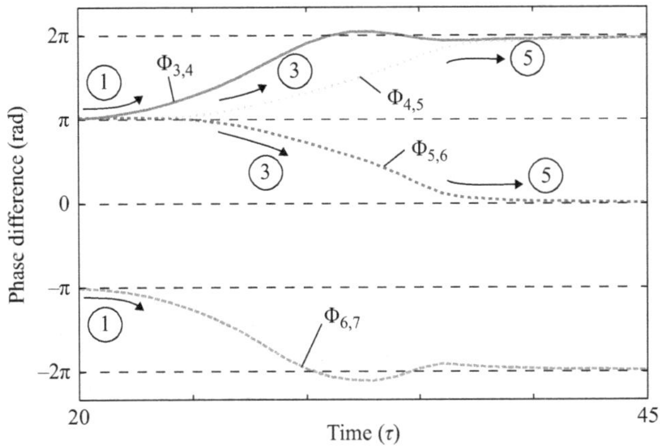

Figure 8.35 Phase differences of the disappearance between two phase-inversion waves in the in-phase synchronization

5. ⑤ in Figures 8.34 and 8.35: f_5 arrives at the f_{in} when $\Phi_{4,5}$ and $\Phi_{5,6}$ reach at 2π and 0, respectively. Two phase-inversion waves disappear.

The phase-inversion waves, which switch from the in-phase synchronization to the anti-phase synchronization, can be explained in the similar manner.

8.4 Conclusion

In this section, we introduced the phase-inversion waves which continuously propagate on the coupled oscillator systems. Furthermore, mechanisms of basic characteristic of the phase-inversion waves are explained by using the instantaneous frequencies and the phase differences between adjacent oscillators. Mechanisms of other characteristics can be explained by the same method.

Acknowledgment

This work was supported by JSPS KAKENHI Grant Number 27530152.

References

[1] Y. Kuramoto and T. Yamada, "Pattern formation in oscillatory chemical reactions," *Prog. Theor. Phys.*, vol. 56, no. 3, pp. 724–739, Sep. 1976.

[2] D. C. Michaels, E. P. Matyas and J. Jalife, "Dynamic interactions and mutual synchronization of sinoatrial node pacemaker cells," *Circ. Res. J. Am. Heart Assoc.*, 58, pp. 706–720,1986.

[3] J. Buck, "Synchronous rhythmic flashing of fireflies. II," *Quart. Rev. Biol.*, vol. 63, no. 3, pp. 265–289, Sep. 1988.

[4] K. Nomura, Y. Takei and Y. Yanagida, "Comparison of cardio-locomotor synchronization during running and cycling," *Eur. J. Appl. Phesiol.*, vol. 89, no. 3–4, pp. 221–229, May 2003.

[5] I. V. Belykh, E. Lange, M. Hasler, "Synchronization of bursting neurons: what matters in the network topology," *Phys. Rev. Lett.*, vol. 94, no. 18, p. 188101, May 2005.

[6] M. J. Moore and T. Nakano, "Oscillation and synchronization of molecular machines by the diffusion of inhibitory molecules," *IEEE Trans. Nanotechnol.*, vol. 12, no. 4, pp. 601–608, May 2013.

[7] S. Strogatz, *Sync: The Emerging Science of Spontaneous Order*, Hachette Books, New York, NY, Mar. 2003, ISBN 9780786868445.

[8] T. Endo and S. Mori, "Mode analysis of a multimode ladder oscillator," *IEEE Trans. Circuits Syst.*, vol. 23, no. 2, pp. 100–113, Feb. 1976.

[9] T. Endo and S. Mori, "Mode analysis of a two-dimensional low-pass multimode oscillator," *IEEE Trans. Circuits Syst.*, vol. 23, no. 9, pp. 517–530, Sep. 1976.

[10] T. Endo and S. Mori, "Mode analysis of a ring of a large number of mutually coupled van der Pol oscillators," *IEEE Trans. Circuits Syst.*, vol. 25, pp. 7–18, Jan. 1978.

[11] H. Kanasugi, S. Moro and S. Mori, "Van der Pol oscillators coupled by piecewise-linear negative resistor – asynchronous oscillations by self-modulation effect," *IEICE Trans. Fundamentals*, vol. E79-A, no. 10, pp. 1551–1562, Oct. 1996.

[12] M. Yamauchi, M. Wada, Y. Nishio and A. Ushida, "Wave propagation phenomena of phase states in oscillators coupled by inductors as a ladder," *IEICE Trans. Fundamentals*, vol. E82-A, no. 11, pp. 2592–2598, Nov. 1999.

[13] M. Akazawa, M. Yamauchi, Y. Nishio and A. Ushida, "Phase-inversion-waves on coupled oscillators by inductors as a cross," in *Proceedings of the 2002 International Symposium on Nonlinear Theory and Its Applications, NOLTA 2002*, Xian, China, pp. 575–578, Oct. 2002.

[14] M. Yamauchi, M. Okuda, Y. Nishio and A. Ushida, "Analysis of phase-inversion waves in coupled oscillators synchronizing at in-and-anti-phase," *IEICE Trans. Fundamentals*, vol. E86-A, no. 7, pp. 1799–1806, Jul. 2003.

[15] K. Shimizu, M. Komuro and T. Endo, "Onset of the propagating pulse wave in a ring of coupled bistable oscillators," *IEICE Nonlinear Theory and Its Applications*, vol. 2, no. 1, pp. 139–151, 2011.

[16] S. Kunihiro, R. Okamoto, H. Aburatani, M. Yamauchi and Y. Nishio, "Actual circuit experiments and computer simulations of phase-inversion waves in in-and-anti-phase synchronization on 2D lattice oscillators," in *Proceedings of the 2011 International, Symposium on Nonlinear Theory and Its Applications, NOLTA2011*, Kobe, Japan, pp. 480–483, Sep. 2011.

[17] K. Suzuki, T. Tsubone, "In-phase and anti-phase synchronization phenomena in coupled systems of piecewise constant oscillators," *IEICE Trans. Fundamentals*, vol. E98-A, no. 1, pp. 340–353, Jun. 2015.

Chapter 9
Coupled oscillator networks with frustration

Yoko Uwate[1] *and Yoshifumi Nishio*[1]

Coupled oscillatory circuits provide simple models for describing high-dimensional nonlinear phenomena occurring in our everyday world. We often feel that the coupled oscillatory networks behave like human, especially, when we carry out circuit experiments. In this chapter, we pay our attention to frustration influence for synchronization phenomena of coupled oscillatory networks. We consider two types of coupled oscillatory networks with frustration. First one is that the oscillator in a ring-coupled van der Pol oscillator has different frequency to the others. Second one is that the coupling structure affects the synchronization state in coupled polygonal oscillatory networks.

9.1 Introduction

The synchronization phenomena observed from coupled oscillators are suitable model to analyze the natural phenomena. Therefore, many researchers have proposed different coupled oscillatory networks and have discovered many interesting synchronization phenomena [1–4]. Such oscillatory systems can produce the interesting phase pattern including: wave propagation, clustering, and complex phase pattern [5–8].

In our research group, we have focused on synchronization phenomena of coupled oscillators under a difficult situation for the circuit. Setou *et al.* have reported the synchronization phenomena in N oscillators coupled by resistors as a star. The oscillation stop in some range of the coupling resistors was confirmed [9]. Such synchronization phenomena are very interesting. Because each oscillator seems to be understanding their condition or situation in the system. However, it is difficult to investigate synchronization phenomena of large-scale circuits, namely, van der Pol oscillators coupled in a star topology do not synchronize when the number of coupling oscillators is larger than 4. In order to arrange such system for large-scale circuits, the ring topology network with frequency error is required.

[1]Department of Electrical and Electronic Engineering, Tokushima University, Tokushima, Japan

On the other hand, Endo and Mori have presented the details of a theoretical analysis and corresponding circuit experiments on electrical circuits oscillators arranged in a ladder, a ring, and in a two-dimensional array topology [5,10,11]. Recently, grid cells, which are the type of neurons in the entorhinal cortex, show remarkable hexagonal activity patterns [12,13]. The polygonal structure of neuronal firing pattern has possible important role for emerging unified computational framework [14]. Namely, it is important to study synchronization phenomena observed from the polygonal oscillatory networks with the several characteristics in the human brain as we described above, for modeling neurobiological systems and applying its high-functional information processing to engineering applications.

In this chapter, synchronization phenomena in coupled polygonal oscillatory networks with frustration are investigated. First, we consider a ring consisting of van der Pol oscillators with different oscillation frequencies. It is possible to add the frequency error to the van der Pol oscillators by setting the different value of a capacitor of oscillators from the others. By using computer simulations, we observe that the van der Pol oscillators with frequency error can produce synchronization phenomena such as oscillation death, independent oscillation, and double-mode oscillation for large-scale oscillatory systems.

Second, synchronization phenomena in coupled polygonal oscillatory networks with frustration are investigated. In this system, van der Pol oscillators are connected to every corner of polygonal network. First, we consider two identical oscillatory networks are coupled by sharing a branch as a symmetrical system. In this oscillatory system, frustration is occurred by the shared branch. We observe the different synchronization state than the coupled oscillatory system without frustration. We confirm that the phase difference between the shared oscillators is shifted, then other oscillators synchronize to compensate this phase shift. How do the shared oscillators determine this phase shift? What effects will be occurred when such oscillatory system is applied for the large-scale network? In order to make clear these questions, we focus on the power consumption of the coupling resistors in the whole system and investigate the phase difference between adjacent oscillators. By using computer simulations and theoretical analysis, we confirm that coupled oscillators tend to synchronize to minimize the power consumption of the whole system. Furthermore, the phase difference of the shared oscillators is solved by finding the minimum value of the power consumption function.

Finally, we consider the case of that the van der Pol oscillators are coupled strongly by increasing the coupling strength. We propose new circuit model which is including the earth resistance in all ground parts. The synchronization phenomena in coupled polygonal oscillatory networks with strong frustration are investigated. We confirm that the amplitude of the oscillators decreases by increasing the value of the coupling strength and oscillation death is occurred at un-frustrated oscillators. We expect that the results of this study contribute to understanding of synchronization phenomena observed in general complex networks.

9.2 Frustration in ring van der Pol oscillators with different frequencies

9.2.1 Circuit model

The basic circuit model and the concept of circuit system are shown in Figure 9.1. In the system, only the Nth oscillator has different oscillation frequencies from the others. We realize the frequency error of the Nth van der Pol oscillator by setting the different values of the capacitor of this oscillator in comparison to the others.

First, we assume that the v_k–i_{Rk} characteristics of the nonlinear resistor in each oscillator is represented by the following third-order polynomial equation:

$$i_{Rk} = -g_1 v_k + g_3 {v_k}^3. \tag{9.1}$$

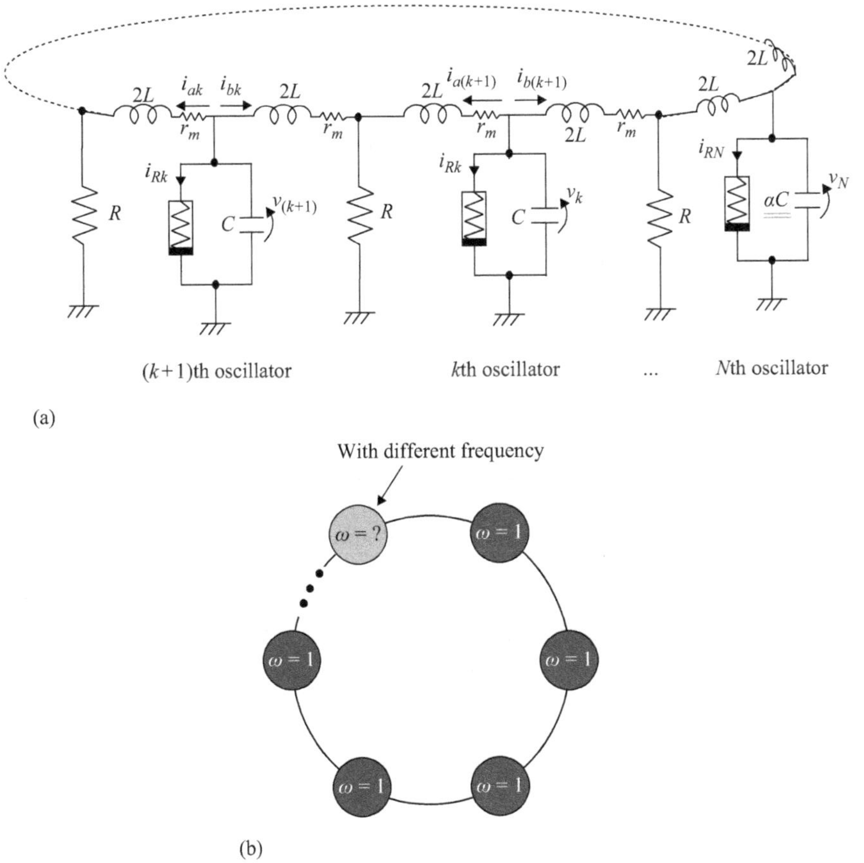

Figure 9.1 Circuit model and concept of circuit system. (a) Ring of van der Pol oscillators and (b) conceptual circuit model

By changing the variables and the parameters,

$$v_k = \sqrt{\frac{g_1}{3g_3}}x_k, \quad i_k = \sqrt{\frac{g_1}{3g_3}}\sqrt{\frac{C}{L}}y_k, \quad t = \sqrt{LC}\tau,$$

$$\varepsilon = g_1\sqrt{\frac{L}{C}}, \quad \gamma = r\sqrt{\frac{C}{L}}, \quad \alpha = \frac{1}{\omega^2}, \quad \eta = r_m\sqrt{\frac{C}{L}},$$

the normalized circuit equations of the ring of oscillators are given as:

$$\begin{cases} \dfrac{dx_k}{d\tau} = \omega_k^2\varepsilon x_k(1 - x_N{}^2) - \omega_k^2(y_{aN} + y_{bN}), \\ \dfrac{dy_{ak}}{d\tau} = \dfrac{1}{2}x_k - \eta y_{ak} - \gamma(y_{ak)} + y_{b(k+1)}), \\ \dfrac{dy_{bk}}{d\tau} = \dfrac{1}{2}x_k - \eta y_{bk} - \gamma(y_{a(k-1)} + y_{bk)}), \\ \qquad\qquad (k = 1, 2, \ldots, N), \end{cases} \tag{9.2}$$

where:

$$y_{a0} = y_{aN}, \qquad y_{b(N+1)} = y_{b1}. \tag{9.3}$$

It should be noted that ω_k denotes the frequency of kth oscillators, γ corresponds to the coupling strength and that ε corresponds to the nonlinearity of oscillators. Equation (9.2) is calculated by using the fourth-order Runge–Kutta method.

9.2.2 Synchronization phenomena

Figure 9.2 shows the observed phenomena for coupling oscillator number $N = 3$. For the case that all of these oscillations have the same frequency, we can observe that the system is synchronized at the three phases (Figure 9.2(a)). When the frequency of the third oscillator is varied, we confirm that oscillation of the third oscillator stops, namely oscillation death appears (Figure 9.2(b)). As increasing the frequency of the third oscillator, oscillation of the third oscillator starts again (Figure 9.2(c)). However, the third oscillator is not synchronized to the others. Namely, the third oscillator oscillates alone and the others keep anti-phase synchronization. Furthermore, we observe the double-mode oscillation as shown in Figure 9.2(d). We observe the similar synchronization states when the number of oscillator is set to $N = 4$ as shown in Figure 9.3.

Next, we calculate the relationship between the amplitude of the Nth oscillator and the frequency ω_N when the nonlinearity parameter is changed from $\varepsilon = 0.06$ to 0.2. The simulated results for $N = 3, 4, 9, 10, 15$, and 16 are shown in Figure 9.4. The graph form of amplitude is different between odd and even number of coupling oscillators. In the case of odd number, the amplitude takes the peak on $\omega_N = 1.0$ and around $\omega_N = 1.2$ the oscillation death appears when the nonlinearity parameter ε_N is set to 0.06. While, in the case of even number, although the amplitude takes the peak on $\omega = 1.0$, the oscillation death cannot be observed.

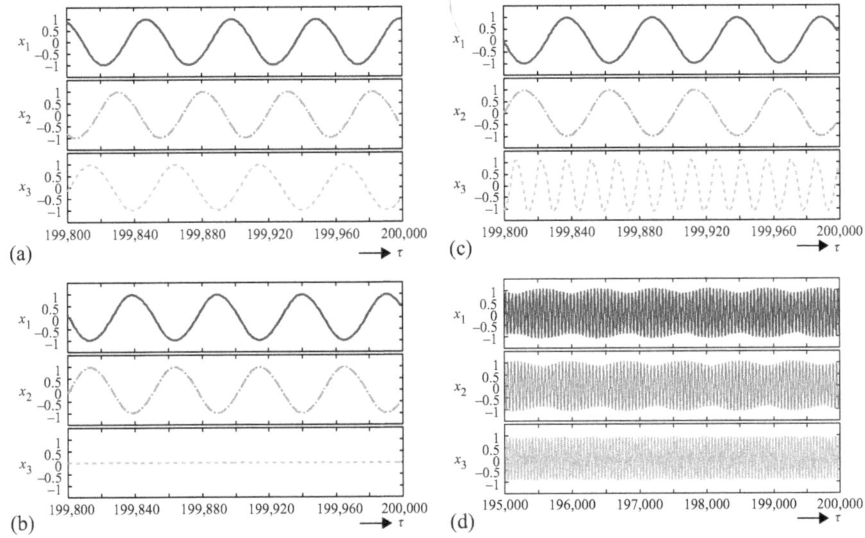

Figure 9.2 *Computer calculation for $N = 3$ ($\varepsilon = 0.2$, $\gamma = 0.2$, $\eta = 0.01$). (a) Three-phase synchronization ($\omega_1 = \omega_3 = \omega_3 = 1.0$). (b) Anti-phase and oscillation death ($\omega_1 = 1.0$, $\omega_2 = 1.0$, $\omega_3 = 0.64$). (c) Anti-phase and independent oscillation ($\omega_1 = 1.0$, $\omega_2 = 1.0$, $\omega_3 = 3.47$). (d) Double-mode oscillation ($\omega_1 = 1.0$, $\omega_2 = 1.0$, $\omega_3 = 1.07$)*

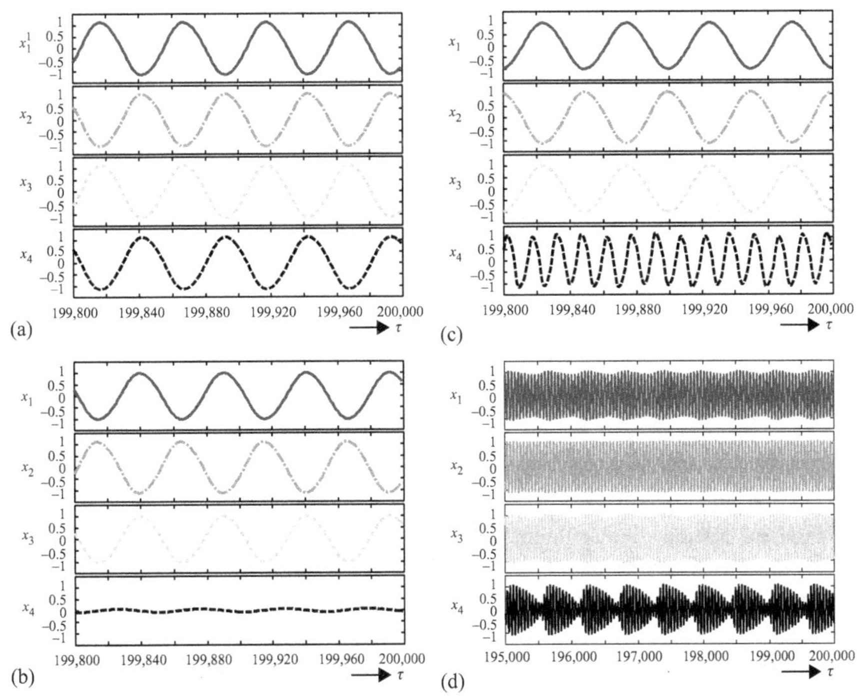

Figure 9.3 *Computer calculation for $N = 4$ ($\varepsilon = 0.2$, $\gamma = 0.2$). (a) Anti-phase synchronization ($\omega_\alpha = \omega_0 = 1.0$). (b) Anti-phase and oscillation death ($\omega_\alpha = 0.64$). (c) Anti-phase and independent oscillation ($\omega_\alpha = 3.47$). (d) Double-mode oscillation ($\omega_\alpha = 1.12$)*

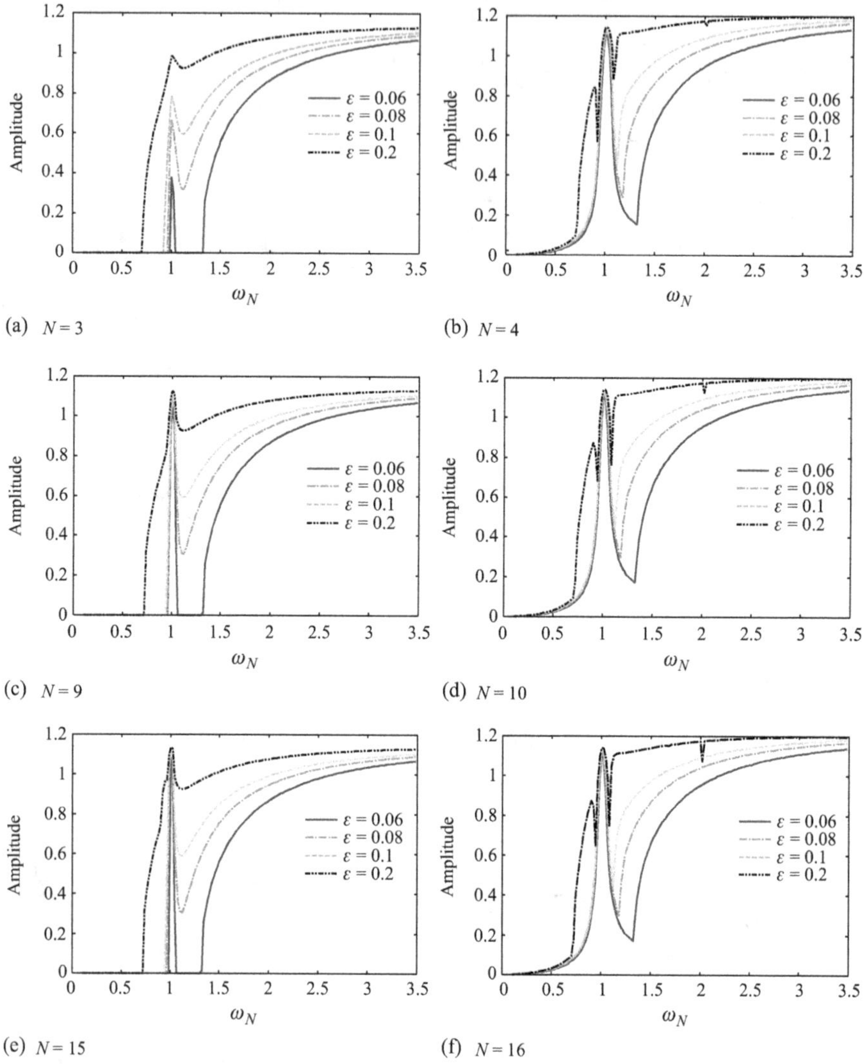

Figure 9.4 Relation between amplitude and ω_N

Figure 9.5 shows the frequency at the breakdown of the anti-phase or N-phase synchronization by changing the number of oscillators. From these results, we confirm that the even number coupling is more stable than the odd number coupling systems for the frequency error.

In this section, we have investigated a ring coupling van der Pol oscillators with different oscillation frequencies. By computer simulations, we observe that the van der Pol oscillators with frequency error can produce the interesting nonlinear phenomena such as oscillation death, independent oscillation, and double-mode oscillation.

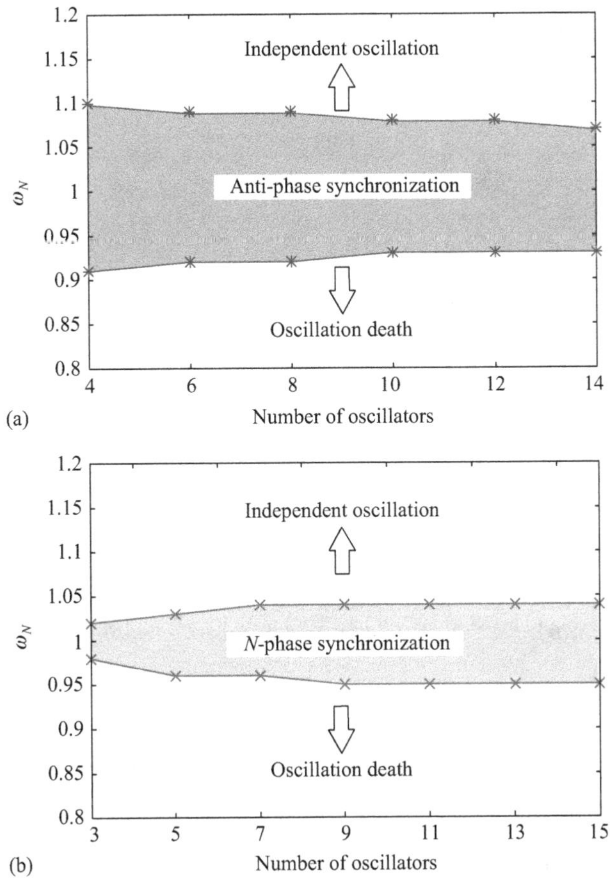

Figure 9.5 *Frequency for breakdown of synchronization ($\varepsilon = 0.2$, $\gamma = 0.2$, $\eta = 0.01$). (a) Breakdown for anti-phase synchronization. (b) Breakdown for N-phase synchronization*

9.3 Frustration in coupled polygonal oscillatory networks

In this section, we focus on the structure of the circuit networks. Here, we investigate two types (strongly and weakly couplings) of polygonal oscillatory networks.

9.3.1 Weakly coupled oscillators

First, we investigate the synchronization phenomena in the symmetrical coupled oscillatory systems. Two identical odd-number-node polygonal oscillatory networks are coupled by sharing a branch as shown in Figure 9.6. We call the first and

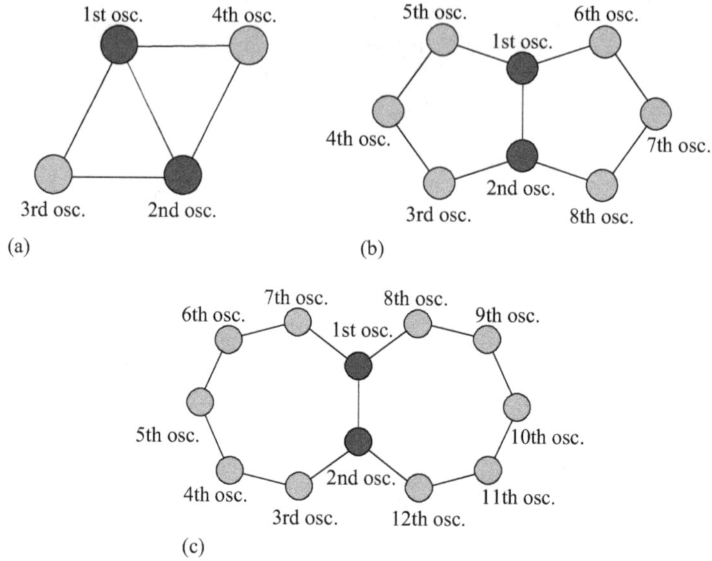

Figure 9.6 Two coupled oscillatory networks (symmetric model). (a) 3-3 coupling network. (b) 5-5 coupling network. (c) 7-7 coupling network

the second oscillators which are connected to the both polygonal networks "shared oscillators."

Figure 9.7 shows a circuit model of the 3-3 coupled networks. We develop the expression of the circuit equations of this model. The v_k–i_{Rk} characteristics of the nonlinear resistor are assumed to be the following third-order polynomial equation:

$$i_{Rk} = -g_1 v_k + g_3 v_k^3 \quad (g_1, g_3 > 0), \quad (k = 1, 2, 3, 4). \tag{9.4}$$

Then the circuit dynamics is described by the following ordinary differential equation:

$$\begin{cases} C\dfrac{dv_k}{dt} = -i_{Rk} - i_{ak} - i_{bk} - i_{ck}, \\ 3L\dfrac{di_{ak}}{dt} = v_k - r_m i_{ak} - R(i_{ak} + i_n), \\ 3L\dfrac{di_{bk}}{dt} = v_k - r_m i_{bk} - R(i_{bk} + i_n), \\ 3L\dfrac{di_{ck}}{dt} = v_k - r_m i_{ck} - R(i_{ck} + i_n), \\ \qquad\qquad\qquad (k = 1, 2, 3, 4). \end{cases} \tag{9.5}$$

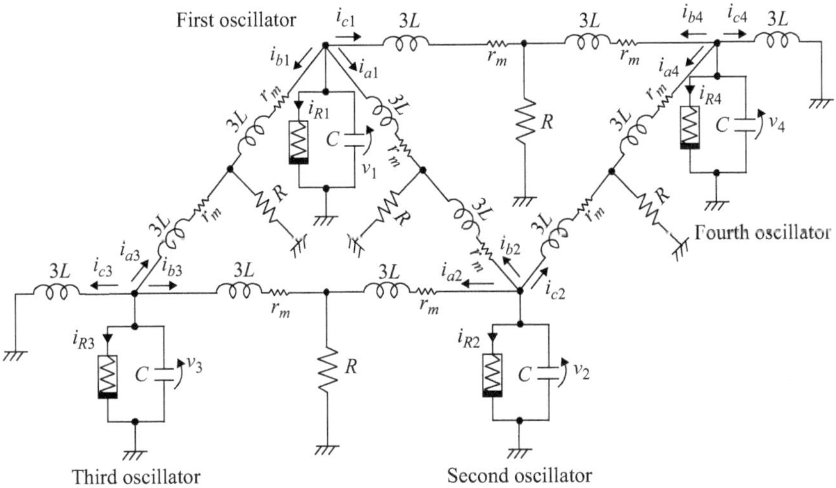

Figure 9.7 Circuit model of 3-3 coupled networks

where i_n denotes the current from the neighbor oscillator over the corresponding coupling resistor. By using the variables and the parameters,

$$t = \sqrt{LC}\tau, \quad v_k = \sqrt{\frac{g_1}{3g_3}}x_k, \quad i_{ak} = \sqrt{\frac{g_1}{3g_3}}\sqrt{\frac{C}{L}}y_{ak},$$

$$i_{bk} = \sqrt{\frac{g_1}{3g_3}}\sqrt{\frac{C}{L}}y_{bk}, \quad i_{ck} = \sqrt{\frac{g_1}{3g_3}}\sqrt{\frac{C}{L}}y_{ck}, \quad i_n = \sqrt{\frac{g_1}{3g_3}}\sqrt{\frac{C}{L}}y_n,$$

$$\varepsilon = g_1\sqrt{\frac{L}{C}}, \quad \gamma = R\sqrt{\frac{C}{L}}, \quad \eta = r_m\sqrt{\frac{C}{L}},$$

$$(k = 1, 2, 3, 4).$$

The normalized circuit equations governing the circuit are expressed as:

$$\begin{cases} \dfrac{dx_k}{d\tau} = \varepsilon\left(1 - \dfrac{1}{3}{x_k}^2\right)x_k - (y_{ak} + y_{bk} + y_{ck}), \\ \dfrac{dy_{ak}}{d\tau} = \dfrac{1}{3}\{x_k - \eta y_{ak} - \gamma(y_{ak} + y_n)\}, \\ \dfrac{dy_{bk}}{d\tau} = \dfrac{1}{3}\{x_k - \eta y_{bk} - \gamma(y_{bk} + y_n)\}, \\ \dfrac{dy_{ck}}{d\tau} = \dfrac{1}{3}\{x_k - \eta y_{ck} - \gamma(y_{ck} + y_n)\}, \\ \qquad\qquad (k = 1, 2, 3, 4). \end{cases} \tag{9.6}$$

In these equations, γ is the coupling strength, and ε denotes the nonlinearity of the oscillators. For the computer simulations, we calculate (9.6) using the fourth-order Runge–Kutta method with the step size $h = 0.005$. The parameters of this circuit model are fixed as $\varepsilon = 0.1$, $\gamma = 0.1$, and $\eta = 0.0001$.

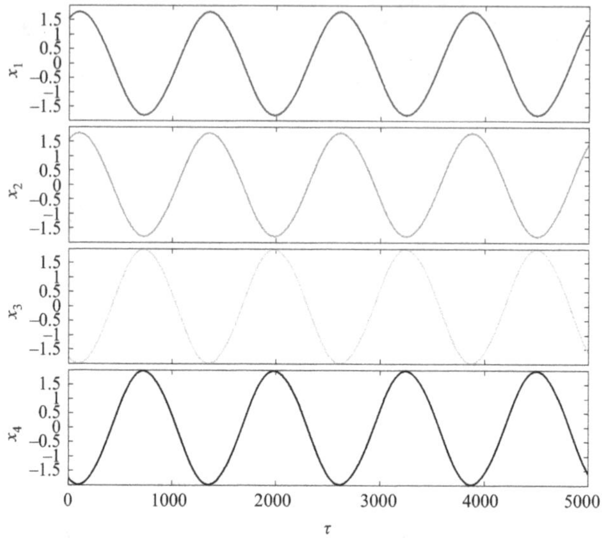

(a) Computer simulation

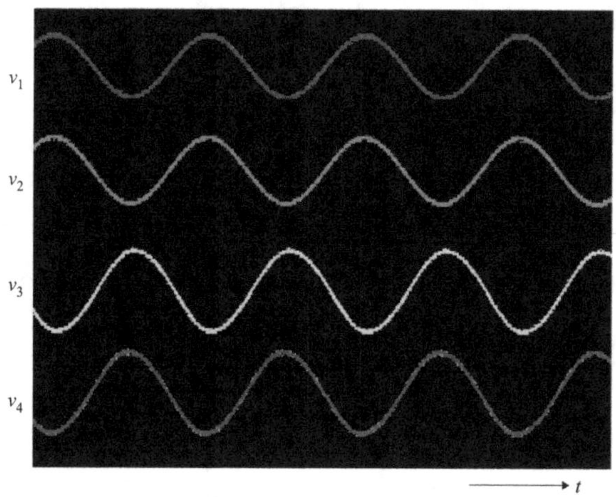

(b) Circuit experiment

Figure 9.8 *Time wave form of the voltage charged at the capacitance of each oscillator. (a) $\varepsilon = 0.1$, $\gamma = 0.1$, and $\eta = 0.0001$. (b) $L = 30$ mH and $C = 33$ nF*

9.3.1.1 Synchronization phenomena

Figure 9.8 shows the time wave form of the voltage charged at the capacitance of each oscillator obtained from the computer simulation and the circuit experiment. From Figure 9.8, we can see that the first and the second oscillators are synchronized at in-phase (phase difference: 0°). While, the other pairs of the oscillators synchronize with anti-phase (phase difference: 180°).

Table 9.1 Phase difference of odd–odd symmetrical coupled networks (computer simulation)

Symmetrical coupling *N-N* ($O_N = 2N - 2$)	Phase difference	
	Shared (first–second) oscillators	Other combinations
3-3 ($O_N = 4$)	0°	180°
5-5 ($O_N = 8$)	106°	153°–154°
7-7 ($O_N = 12$)	130°	158°–159°

We did not expect such synchronization states. Because, in the case of three coupled van der Pol oscillators as a ring topology, we can observe three-phase synchronization state (phase difference: 120°) [15]. Hence, we assumed that a pair of three-phase synchronizations could be observed. It is interesting to observe this unexpected synchronization state.

Next, we investigate the phase difference of two coupled odd-number-nodes polygonal oscillatory networks when the number of nodes of the both networks is changed from 5 to 11. The simulation results are summarized in Table 9.1. O_N denotes the total number of nodes in the whole system. From Table 9.1, we can see that the coupled oscillators do not synchronize with N-phase[2] in each case.

How is the phase difference between the shared oscillators determined?

9.3.1.2 Power consumption

In order to make clear the mechanism of that N-phase synchronization could not be observed in this coupled networks, we focus on the power consumption of the coupling resistors in the whole system. We assume the current of the inductor ($3L$) as follows:

$$i_k(t) = \sqrt{2}I_e \sin(\omega t + \varphi_k) \quad (k = 1, 2, 3, \ldots, O_N), \tag{9.7}$$

where ω is the natural frequency of van der Pol oscillator as $\omega = \frac{1}{\sqrt{LC}}$. When the coupling resistance is assumed to $R = 1$, the average power consumption of the coupling resistor between kth and $(k+1)$st oscillators is described as follows:

$$P = \frac{1}{2\pi}\int_0^{2\pi} \{i_k(t) + i_{k+1}(t)\}^2 dt. \tag{9.8}$$

We calculate the power consumption of the whole system by using the phase difference obtained from the computer simulations. For comparison, the power consumption of N-phase synchronization state is also calculated. Table 9.2 summarizes

[2] In Reference 16, synchronization phenomena in a ring of van der Pol oscillators are investigated. For even-number-node rings, anti-phase synchronization (π-phase shift between adjacent oscillators) is observed. For odd-number-node rings, N-phase synchronization ($(N-1)\pi/N$ phase shift between adjacent oscillators) is observed.

Table 9.2 Power consumption of odd–odd symmetrical coupled networks

Symmetrical coupling N-N($O_N = 2N - 2$)	Power consumption	
	Simulation	**N-phase presumption**
3-3 ($O_N = 4$)	$4I_e^2$	$5I_e^2$
5-5 ($O_N = 8$)	$3.132I_e^2$	$3.437I_e^2$
7-7 ($O_N = 12$)	$2.413I_e^2$	$2.580I_e^2$

the calculation results when N is changed from 3 to 11. We confirm that the total power consumption of N-phase synchronization state is larger than the computer simulation results. In the case of 3-3 coupled networks, the total number of coupling resistors is five. When the coupled oscillators are synchronized with three phase, the power consumption of each coupling resistor is I_e^2. Then, the power consumption of the whole system becomes $5I_e^2$ ($=5 \times I_e^2$). While, when the shared oscillators synchronize with in-phase, the power consumption of the shared oscillators is $4I_e^2$ ($=(I_e + I_e)^2$). Other power consumptions become zero because of anti-phase synchronizations. Hence, N-phase synchronization cannot be obtained in this type of two coupled networks.

9.3.1.3 Theoretical analysis

Here, we try to derive the phase differences between oscillators theoretically.

First, let the phase difference between the shared oscillators be θ. Second, we assume that the phase difference of the other adjacent oscillators is described by the following equation:

$$\varphi_N = \pi - \frac{\theta}{N-1}. \tag{9.9}$$

If the network has no frustration, then the two adjacent oscillators tend to synchronize with anti-phase (π phase shift). If the phase difference of the shared oscillators is zero, then the other oscillators have no frustration. However, the phase difference of the shared oscillators is defined as θ. Hence, the phase difference of the other oscillators is shifted $\theta/(N_o - 1)$ from π, where $(N_o - 1)$ is the number of the branches except the shared branch.

Under this assumption, the power consumption of the whole system is expressed by the following equation:

$$P = \frac{1}{2\pi}\int_0^{2\pi} \{\sin \omega t + \sin(\omega t + \theta)\}^2 dt$$
$$+ \frac{2(N-1)}{2\pi}\int_0^{2\pi} \left\{\sin \omega t \sin\left(\omega t + \pi - \frac{\theta}{N-1}\right)\right\}^2 dt, \tag{9.10}$$

where the amplitude of the current is normalized to $\sqrt{2}I_e$, and the coupling resistance is assumed to $R = 1$ without loss of generality. By solving (9.10):

$$P = 1 + \cos\theta + 2(N-1)\left\{1 + \cos\left(\pi - \frac{\theta}{N-1}\right)\right\}. \tag{9.11}$$

At the minimum value of P, $\frac{dP}{d\theta} = 0$ is satisfied:

$$\frac{dP}{d\theta} = -\sin\theta + 2\sin\left(\pi - \frac{\theta}{N-1}\right) = 0. \tag{9.12}$$

The phase difference between the shared oscillators θ can be obtained by solving (9.12) numerically.

Table 9.3 summarizes the theoretical results of the phase differences. We confirm that the theoretical results agree well with the computer simulation results (see Table 9.1).

As an example, Figure 9.9 shows the graphs of (9.11) and (9.12) and the polar representation of the phase differences in the whole system for 5-5 coupled networks. From Figure 9.9(a), we can see that extreme value of (9.8) corresponds to 0 (zero)

Table 9.3 Phase difference of odd–odd symmetrical coupled networks (theoretical analysis)

Symmetrical coupling N-N ($O_N = 2N - 2$)	**Phase difference**	
	Shared oscillators (θ)	**Other combinations (φ_N)**
3-3 ($O_N = 4$)	0°	180°
5-5 ($O_N = 8$)	110°	152.5°
7-7 ($O_N = 12$)	131°	158.2°

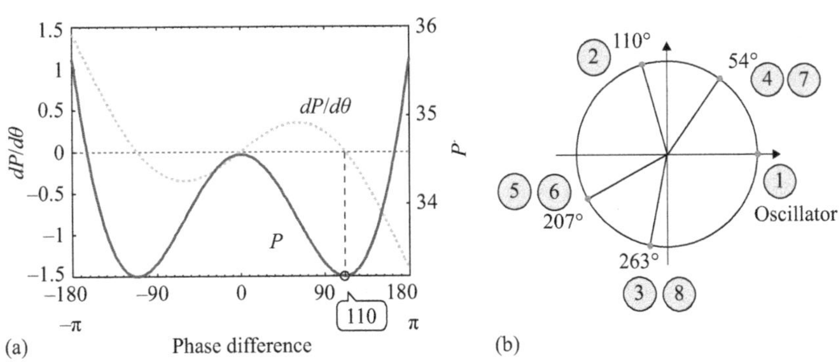

Figure 9.9 Functions of (a) power consumption and (b) polar representation for 5-5 coupled network

of (9.11). We confirm the angle $(4\pi - \theta)$ is divided equally by all oscillators except the shared oscillators (Figure 9.9(b)).

In this analysis, we assume in (9.7) that the inductor currents do not contain the subharmonics, and that their amplitudes are the same all over the network. The theoretical results with these assumptions agree well with the simulated results and the experimental results at least qualitatively. Furthermore, for the standard averaging method, the solution of van der Pol oscillator is assumed to be sinusoidal wave when the nonlinearity is small. We consider that it is an advantage of the proposed theoretical analysis to explain the synchronization phenomena by using only the fundamental frequency and the same amplitude as a simple theoretical approach.

9.3.2 Strongly coupled oscillators

9.3.2.1 Circuit model

Next, we investigate the synchronization phenomena when van der Pol oscillators are strongly coupled. Figure 9.10 shows the circuit model of the 3-3 coupling networks. The novel part of this system is that the earth resistances are inserted to the third and the fourth oscillators.

9.3.2.2 Synchronization phenomena

Figure 9.11 shows the observed attractors of 3-3 coupling network by changing the coupling strength. When the coupling strength is weak, the amplitude of all oscillators are almost same as shown in Figure 9.11(A). By increasing the value of the coupling strength, the amplitude of the shared oscillators becomes smaller than the

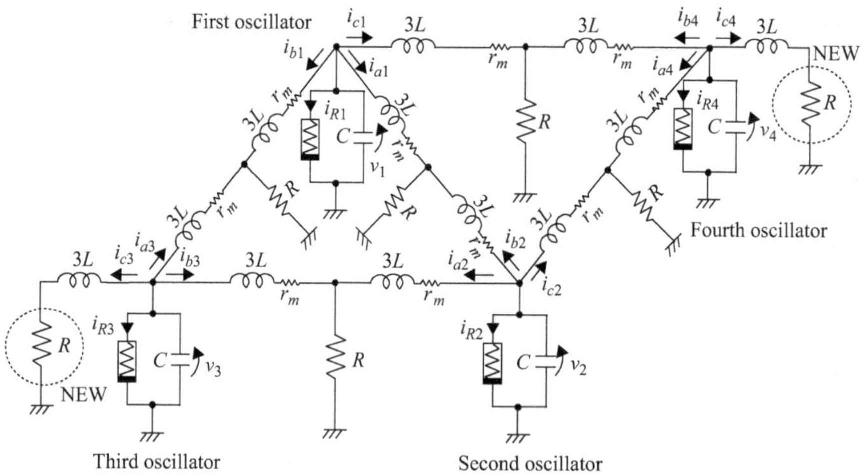

Figure 9.10 Coupling model (3-3 strongly coupling networks)

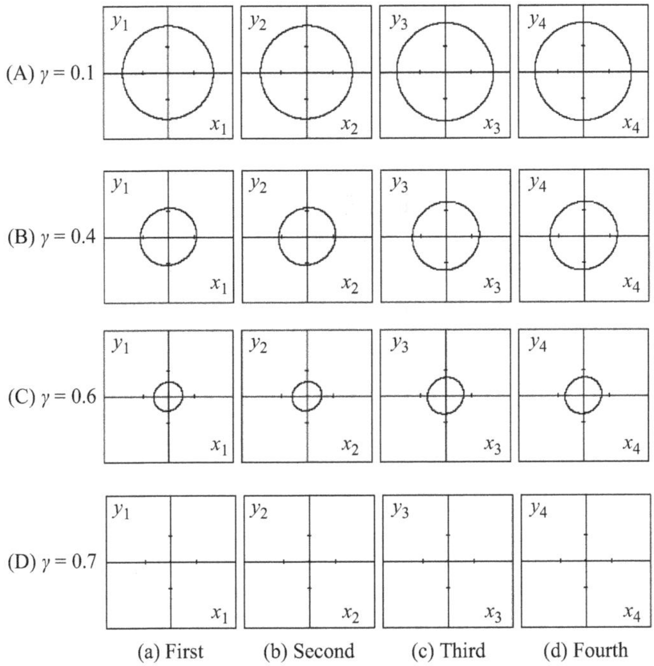

Figure 9.11 Attractors (3-3 coupling network)

others (Figure 9.11(B) and (C)). When the coupling strength is larger than $\gamma = 0.8$, oscillation death of all oscillators is observed as shown in Figure 9.11(D).

Next, Figure 9.12 shows the observed attractors of 5-5 coupling network by changing the coupling strength. In this circuit model, the amplitude of fourth oscillator (which is located farthest place from the shared oscillators) decreases with the coupling strength. We observe the oscillation death of the fourth oscillator when the coupling strength is set to $\gamma = 1.0$. In the case of 7-7 coupling network, we observe similar results with 5-5 coupling network as shown in Figure 9.13.

Figure 9.14 shows the change of the amplitude observed from each network. In the case of 3-3 coupling network, amplitude death of all oscillators is occurred at the same time. While, in the cases of 5-5 and 7-7 coupling networks, first, the amplitude death of the oscillators located farthest place from the shared oscillators is occurred. After that, the other oscillators stop to oscillate at the same time.

We explain amplitude death as physical phenomena. The earth resistance is not inserted in two shared oscillators, namely the shared oscillators do not tend to oscillation death by controlling the phase difference to minimize energy. Then next oscillators from the shared oscillators try to oscillate to synchronize with the shared oscillators. Furthermore, after amplitude death is occurred, the network topology is changed from ring to two ladders.

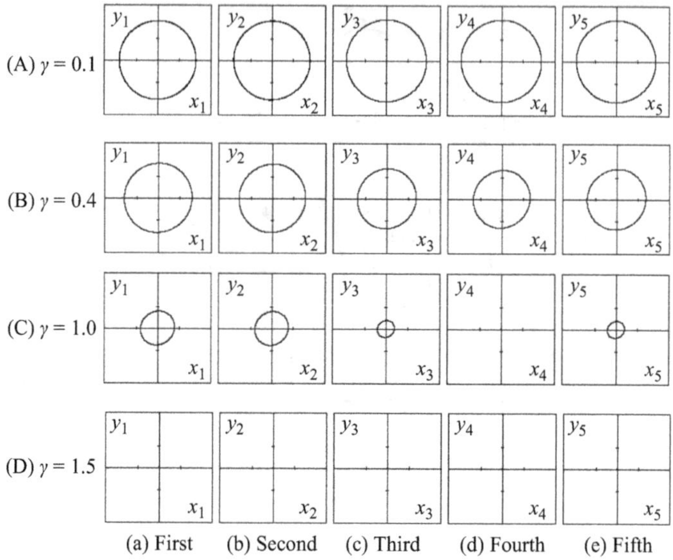

Figure 9.12 Attractors (5-5 coupling network)

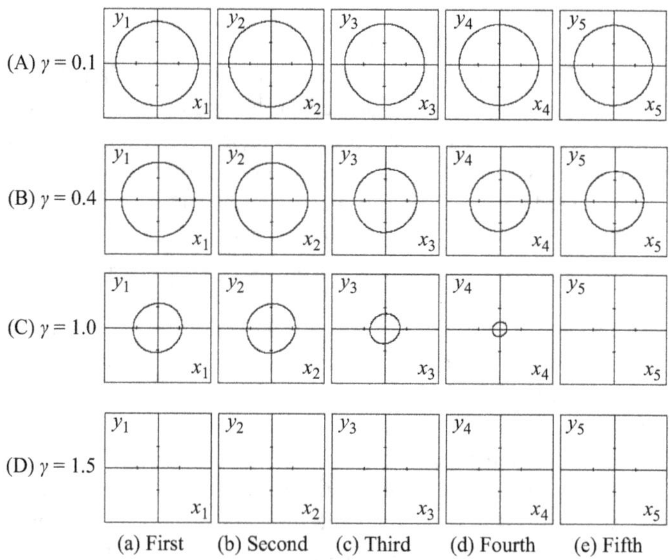

Figure 9.13 Attractors (7-7 coupling network)

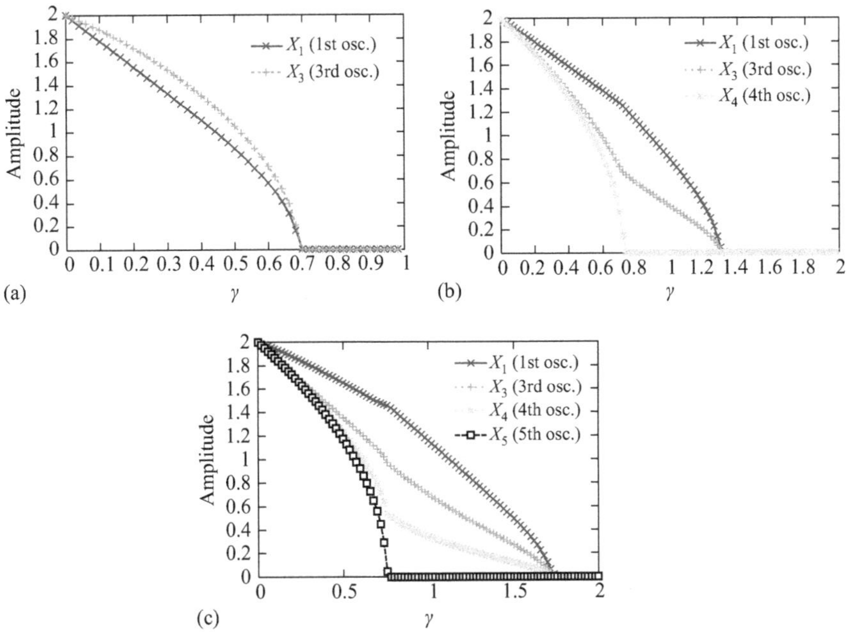

Figure 9.14 Amplitudes (symmetric model). (a) 3-3 coupling network. (b) 5-5 coupling network. (c) 7-7 coupling network

9.4 Conclusions

We introduced two types of frustration for coupled oscillatory systems.

First, we consider a ring consisting of van der Pol oscillators with different oscillation frequencies. By using computer simulations, we have observed that the van der Pol oscillators with frequency error can produce synchronization phenomena such as oscillation death, independent oscillation, and double-mode oscillation for large-scale oscillatory systems.

Second, we have investigated synchronization phenomena in coupled polygonal oscillatory networks with frustration. We focused on phase difference between adjacent oscillators and power consumption of the coupling resistors in the whole system. By using computer simulations and theoretical analysis, we have confirmed that coupled oscillators tend to synchronize to minimize the power consumption of the whole system. Furthermore, we have proposed the theoretical analysis to solve the phase difference by finding the minimum value of the power function. We have confirmed that the proposed theoretical approach matched well with the computer simulation results.

Finally, we consider the case that the van der Pol oscillators are coupled strongly by increasing the coupling strength. We proposed new circuit model which is including the earth resistance in all ground parts. The synchronization phenomena in coupled polygonal oscillatory networks with strong frustration were investigated. We confirmed that the amplitude of the oscillators decreases by increasing the value of the coupling strength and oscillation death was occurred at un-frustrated oscillators.

For the future work, we would like to consider the influence of the power of subharmonics of the inductor currents for the proposed theoretical analysis. And, we develop an efficient method to find the minimum value of the power function for the large-scale complex networks with many variables. Furthermore, in order to fill the gap between the proposed polygonal networks and real networks, we would like to analyze the coupled polygonal oscillatory networks including additional noises, frequency, and parameter errors theoretically.

References

[1] M.G. Rosenblum, A.S. Pikovsky and J. Kurths, "Phase synchronization of chaotic oscillators," *Physical Review Letters*, vol. 76, no. 11, pp. 1804–1807, Mar. 1996.

[2] W. Wang, I.Z. Kiss and J.L. Hudson, "Experiments on arrays of globally coupled chaotic electrochemical oscillators: Synchronization and clustering," *Chaos*, vol. 10, no. 1, pp. 248–256, Mar. 2000.

[3] M. Yamauchi, Y. Nishio and A. Ushida, "Phase-waves in a ladder of oscillators," *IEICE Transactions on Fundamentals*, vol. E86-A, no. 4, pp. 891–899, Apr. 2003.

[4] H.B. Fotsina and J. Daafouza, "Adaptive synchronization of Uncertain chaotic Colpitts oscillators based on parameter identification," *Physics Letters A*, vol. 339, pp. 304–315, May 2005.

[5] T. Endo and S. Mori, "Mode analysis of a ring of a large number of mutually coupled van der Pol oscillators," *IEEE Transactions on Circuits and Systems*, vol. 25, no. 1, pp. 7–18, Jan. 1978.

[6] Y. Nishio and S. Mori, "Mutually coupled oscillators with an extremely large number of steady states," in *Proceedings of ISCAS'92*, vol. 2, pp. 819–822, San Diego, May 1992.

[7] M. Yamauchi, Y. Nishio and A. Ushida, "Phase-waves in a ladder of oscillators," *IEICE Transactions on Fundamentals*, vol. E86-A, no. 4, pp. 891–899, Apr. 2003.

[8] Y. Uwate and Y. Nishio, "Complex phase synchronization in an array of oscillators coupled by time-varying resistor," in *Proceedings of IJCNN'06*, pp. 8345–8350, Vancouver, Jul. 2006.

[9] Y. Setou, Y. Nishio and A. Ushida, "Synchronization phenomena in resistively coupled oscillators with different frequencies," *IEICE Transactions on Fundamentals*, vol. E79-A, no. 10, pp. 1575–1580, Oct. 1996.

[10] T. Endo and S. Mori, "Mode analysis of a multimode ladder oscillator," *IEEE Transactions on Circuits and Systems*, vol. 23, pp. 100–113, Feb. 1976.

[11] T. Endo and S. Mori, "Mode analysis of two-dimensional low-pass multimode oscillator," *IEEE Transactions on Circuits and Systems*, vol. 23, pp. 517–530, Sep. 1976.

[12] T. Hafting, M. Fyhn, S. Molden, M.B. Moser and E.I. Moser, "Microstructure of a spatial map in the entorhinal cortex," *Nature*, Vol. 436, no. 11, pp. 801–806, Aug. 2005.

[13] F. Sargolini, M. Fyhn, T. Hafting, *et al.*, "Conjunctive representation of position, direction, and velocity in entorhinal cortex," *Science*, vol. 312, pp. 758–762, May 2006.

[14] H. Mhatre, A. Gorchetchnikov and S. Grossberg, "Grid cell hexagonal patterns formed by fast self-organized learning within entorhinal cortex," *Hippocampus*, vol. 22, pp. 320–334, 2012.

[15] Y. Uwate, Y. Nishio and R. Stoop, "Synchronization in three coupled van der Pol oscillators with different coupling strength," in *Proceedings of NCSP'10*, pp. 109–112, Honolulu, Mar. 2010.

[16] Y. Setou, Y. Nishio and A. Ushida, "Synchronization phenomena in many oscillators coupled by resistors as a ring," in *Proceedings of APCCAS'94*, pp. 570–575, Taipei, Dec. 1994.

Chapter 10

Graph comparison and synchronization in complex networks*

Hui Liu[1], *Ming Cao*[2] *and Chai Wah Wu*[3]

We study bounds on the coupling strengths required to synchronize dynamical systems coupled via undirected and directed complex networks. In particular, we leverage tools from spectral graph theory to derive conditions based on the topological structures of the networks.

10.1 Introduction

Synchronization phenomena in various complex networks have attracted great attention in the past decades [1–7]. The synchronization phenomena in complex networks have been useful in the study of many diverse fields, such as flocking behavior in birds and social insects, social dynamics in populations, and coordination of generators in electronic power grids. During the past 20 years, significant progress has been made in this research area. In general, there have been three fundamental methods to study the synchronization of complex dynamical networks, which are briefly reviewed below. First, the master stability function method, proposed in Reference 1, studies the *local* synchronization problem for linearly coupled chaotic systems through the calculation of the Lyapunov exponents of the linearized systems. Second, a general systematic framework based on the Lyapunov function method was presented in Reference 7 to study the synchronization of nonlinear dynamical systems with diffusive couplings. A key contribution of References 7,8 is that analyzing the magnitude of the second smallest eigenvalue (λ_2) of the Laplacian matrix of the underlying network can provide insights on the synchronization behavior. Third, along a different

[1]School of Automation, Key Laboratory of Image Processing and Intelligent Control of Education Ministry of China, Huazhong University of Science and Technology, Wuhan, China
[2]Faculty of Mathematics and Natural Sciences, ENTEG, University of Groningen, Groningen, The Netherlands
[3]IBM T. J. Watson Research Center, Yorktown Heights, NY, USA

line, the connection graph method [9–11] uses the topological information of the underlying graph and determines a lower bound of weight for each edge to synchronize the whole network. The lower bound is constructed based on the total length of all the paths passing through a chosen edge in the graph. In summary, the first method addresses local synchronization in complex networks, and the latter two tackle global synchronization.

In this chapter, we introduce some recent progress in our work [12,13] on global synchronization in complex networks. The work bridges the main results developed separately by the two different approaches, i.e., the Lyapunov function method and the connection graph method. We propose a set of new methodologies to allocate coupling strengths in order to guarantee global complete synchronization in complex networks.

We introduce new tools of graph comparison to explore graphical conditions for network synchronization [12,13]. Graph comparison techniques have been developed in the past to bound the second smallest eigenvalues of Laplacian matrices of *undirected* graphs [14–16], where the bounds are obtained by embedding complete graphs into the graph under study. More general ideas for graph comparison have been reported in References 17,18, where the comparison of combinatorial features can be carried out between two arbitrary graphs with the same vertex set for the purpose of bounding any eigenvalues of Laplacian matrices of the graphs.

In this chapter, we follow the approach delineated in References 17,18 to study conditions for synchronization in complex networks based on graph comparison. By doing so, we prove that the synchronization condition given in Reference 9 for allocating coupling strengths can be explained by comparing the network graph with the corresponding complete graph. We then propose different coupling strength allocation strategies by comparing the network graphs with other typical network structures. Our method is especially useful in large, time-varying, complex networks, since adaptive strength allocations can be carried out using only information about local network topologies.

We further develop our methodologies for general networks associated with *directed* graphs. We prove that the synchronization conditions for allocating coupling strengths given in References 10,11 can be explained by comparing the network graphs with the corresponding complete graphs. We construct an algorithm that incorporates graph comparison procedures to find candidate sets of coupling strengths for synchronization. To keep the computation tractable, for a large network that can be decomposed into smaller strongly connected subnetworks, we run the algorithm locally for each subnetwork. The topological conditions proposed are typically easier to check.

The rest of this chapter is organized as follows. In Section 10.2, we review a classical complex dynamical network model and an existing relevant result on synchronization. In Section 10.3, we introduce the graph comparison tools from spectral graph theory. Then we give several synchronization criteria using graph comparison and propose methods to allocate coupling strengths for undirected networks in Section 10.4. We extend this analysis to directed networks in Section 10.5. Finally, concluding remarks are presented in Section 10.6.

10.2 Network model and preliminaries

We consider a network of $n > 1$ coupled identical oscillators whose dynamics are described by:

$$\dot{x}_i = f(x_i) + \sum_{j=1}^{n} \varepsilon_{ij}(t) P x_j, \qquad i = 1, \ldots, n, \tag{10.1}$$

where $x_i \in \mathbb{R}^d$ is the state of the ith oscillator, $f(\cdot) : \mathbb{R}^d \to \mathbb{R}^d$ denotes the identical self-dynamics of each oscillator, the diagonal (0, 1)-matrix $P \in \mathbb{R}^{d \times d}$ determines through which components of the states that the oscillators are coupled together, and $\varepsilon_{ij}(t) \geq 0$ $(i \neq j)$ describes the time-varying strength of the coupling from oscillator j to i at time t, $\varepsilon_{ii}(t)$ are set to be $-\sum_{j=1, j \neq i}^{n} \varepsilon_{ij}(t)$ for $i = 1, \ldots, n$. The couplings between the oscillators can be conveniently described by a weighted graph $\mathbb{G}(t) = (\mathscr{V}, \mathscr{E}, \varepsilon(t))$ with the vertex set $\mathscr{V} = \{1, \ldots, n\}$, the edge set $\mathscr{E} \subseteq \mathscr{V} \times \mathscr{V}$, and the weight function $\varepsilon : \mathscr{E} \to \mathbb{R}$. There is an edge from vertex j to i if and only if $\varepsilon_{ij}(t) > 0$ and the weights ε_{ij} may change with time. Let $L_{\mathbb{G}(t)} \in \mathbb{R}^{n \times n}$ be the Laplacian matrix [19] of the graph $\mathbb{G}(t)$. Then the ijth entry of $L_{\mathbb{G}(t)}$ is $-\varepsilon_{ij}(t)$ for $1 \leq i, j \leq n$.

Let $\otimes$ denote Kronecker product. We can rewrite system (10.1) in the compact form as follows:

$$\dot{x} = \begin{pmatrix} f(x_1) \\ f(x_2) \\ \vdots \\ f(x_N) \end{pmatrix} - (L_{\mathbb{G}(t)} \otimes P) x, \tag{10.2}$$

where $x = (x_1^\top, x_2^\top, \ldots, x_n^\top)^\top$.

Definition 10.1 ([7]). *The synchronization manifold $\mathscr{M}$ of system (10.1) is defined as the linear subspace $\{x : x_i = x_j, \forall i, j\}$. If x approaches the set $\mathscr{M}$, i.e., $\|x_i - x_j\| \to 0$ for any i and j, as $t \to +\infty$, then we say the network of coupled oscillators synchronizes asymptotically.*

It is easy to see that each point in the manifold $\mathscr{M}$ is indeed an equilibrium point of system (10.1).

System (10.1) has been used widely to study under what conditions the coupled oscillators can achieve global and complete synchronization, where for any initial condition, $\|x_i(t) - x_j(t)\| \to 0$ as $t \to \infty$ for all i, j [7]. In this chapter, we explore synchronization conditions in complex dynamical networks using tools of graph comparison from spectral graph theory. Toward this end, we make the following standard technical assumption about the individual dynamics $f(\cdot)$.

Assumption 10.1. *For any two vectors $y, z \in \mathbb{R}^d$, there exists a sufficiently large positive constant a, such that the inequality:*

$$(y - z)^\top (f(y) - f(z) - a P (y - z)) \leq -c \|y - z\|^2 \tag{10.3}$$

holds for some positive constant c.

Here, the constant a can be determined by both the function f and the inner coupling matrix P. Assumption 10.1 implies that two coupled oscillators are always able to get synchronized when their coupling is sufficiently strong. An equivalent assumption has been made in References 9–11, which guarantees that the whole network of oscillators can get synchronized when the coupling strengths between oscillators are sufficiently large. More explanations about this assumption and about how to determine the threshold a are provided in Appendix A.

We first define some notations. Let A and B be n-dimensional real square matrices. We say $A \succ 0$ ($A \succeq 0$) if A is positive definite (positive semidefinite). We say $A \succ B$ if $A - B \succ 0$, i.e., $A - B$ is positive definite; in other words, all eigenvalues of $(A-B)+(A-B)^{\mathrm{T}}$ are positive. Similarly, we say $A \succeq B$ if $A - B$ is positive semidefinite. We also apply this notation to undirected graphs.

In the following, we introduce a general synchronization criterion for networks with time-varying dynamics. We use $\mathscr{W}_s$ to denote the set of irreducible, symmetric matrices that have zero row sums and non-positive off-diagonal elements.

Theorem 10.1. *Let $Y(t)$ be a $d \times d$ time-varying matrix and V a $d \times d$ symmetric, positive definite matrix such that $(y-z)^{\mathrm{T}} V(f(y,t) + Y(t)y - f(z,t) - Y(t)z) \leq -c\|y-z\|^2$ for some $c > 0$ and all y, z, t. Then system (10.1) synchronizes globally if there exists an $n \times n$ matrix $U \in \mathscr{W}_s$ such that:*

$$(U \otimes V)\left(L_{\mathbb{G}(t)} \otimes (-P) - I_n \otimes Y(t)\right) \preceq 0 \tag{10.4}$$

for all t.

Proof. Follows from Theorem 2 from Reference *8* and Chapter 4 in Reference *7*. □

Now we present a synchronization criterion directly using graph inequalities. Here, graph $\mathbb{G}(t)$ can be undirected or directed.

Theorem 10.2. *Under Assumption 10.1, the synchronization manifold of system (10.1) is globally asymptotically stable if there exists a connected undirected graph $\mathbb{G}_0$ with the same vertex set of $\mathbb{G}(t)$ such that:*

$$L_{\mathbb{G}_0} L_{\mathbb{G}(t)} - a L_{\mathbb{G}_0} \succeq 0, \quad \text{for all } t. \tag{10.5}$$

Proof. Assumption 10.1 on the self-dynamics $f(\cdot)$ is equivalent to the condition that $(y-z)^{\mathrm{T}} V(f(y,t) + Y(t)y - f(z,t) - Y(t)z) \leq -c\|y-z\|^2$ when we set $Y(t) = -aP$, $V = I_d$. To apply Theorem 10.1, we choose $Y(t) = -aP$, $V = I_d$, and $U = L_{\mathbb{G}_0}$. Then from (10.4), we have $(L_{\mathbb{G}_0} \otimes I_d)(L_{\mathbb{G}(t)} \otimes (-P) - I_n \otimes (-aP)) \preceq 0$, i.e., $L_{\mathbb{G}_0} L_{\mathbb{G}(t)} \otimes (-P) - aL_{\mathbb{G}_0} \otimes (-P) \preceq 0$. Since $-P \preceq 0$, this is satisfied if $L_{\mathbb{G}_0} L_{\mathbb{G}(t)} - aL_{\mathbb{G}_0} \succeq 0$. Therefore, the complete synchronization of system (10.1) is guaranteed if $L_{\mathbb{G}_0} L_{\mathbb{G}(t)} \succeq a L_{\mathbb{G}_0}$ for all t. □

10.3 Tools of graph comparison

The use of graphical properties in (10.5) inspires us to introduce tools of graph comparison to study the synchronization problem.

Definition 10.2. *For two undirected graphs $\mathbb{H}_1$ and $\mathbb{H}_2$ with the same vertex set $\mathscr{V} = \{1, \ldots, n\}$, we say:*

$$\mathbb{H}_1 \succeq \mathbb{H}_2$$

if their Laplacian matrices satisfy $L_{\mathbb{H}_1} \succeq L_{\mathbb{H}_2}$.

For a graph $\mathbb{G}$ with vertex set $\mathscr{V}$, we use λ_k, $1 \leq k \leq n$, to denote the kth smallest eigenvalue of $L_{\mathbb{G}}$. For graphs $\mathbb{G}$ and $\mathbb{H}$ with the same vertex set, we consider some multiple $c\,\mathbb{G}$ of graph $\mathbb{G}$. Using Courant–Fischer Theorem [18], one can easily prove the following result.

Lemma 10.1. *If $\mathbb{G}$ and $\mathbb{H}$ are the graphs with the same vertex set $\mathscr{V}$ satisfying $c\mathbb{G} \succeq \mathbb{H}$, then:*

$$c\lambda_k(\mathbb{G}) \geq \lambda_k(\mathbb{H})$$

for all $1 \leq k \leq n$.

To apply additional tools from spectral graph theory, we need to introduce another equivalent definition of the Laplacian matrix of graphs. Following Reference 17, we define the elementary Laplacian $L_{(u,v)}$ to be the Laplacian of the graph containing just the edge of unit weight between vertices u and v. Then for an undirected graph $\mathbb{G}(t) = (\mathscr{V}, \mathscr{E}, \varepsilon(t))$ consisting of the vertex set $\mathscr{V}$, the edge set $\mathscr{E}$, and the weight function $\varepsilon : \mathscr{E} \to \mathbb{R}$, its Laplacian matrix has the following form:

$$L_{\mathbb{G}(t)} \triangleq \sum_{(u,v)\in\mathscr{E}} \varepsilon_{(u,v)}(t) \cdot L_{(u,v)}. \tag{10.6}$$

Moreover, we say graph $\mathbb{G}$ is unweighted if the weights $\varepsilon_{(u,v)} = 1$ for all $u \neq v$.

Now we introduce two graphical inequalities, which have been proved in Reference 17.

Lemma 10.2. *Let $c_1, \ldots, c_{n-1} > 0$. Then the following inequality is true:*

$$c\left(\sum_{i=1}^{n-1} c_i L_{(i,i+1)}\right) \succeq L_{(1,n)},$$

where $c = \sum_i \frac{1}{c_i}$.

If we take $c_1 = c_2 = \cdots = c_{n-1} = 1$, then Lemma 10.2 reduces to:

Lemma 10.3. *The following relationship is true for elementary Laplacians:*

$$(n-1)\left(\sum_{i=1}^{n-1} L_{(i,i+1)}\right) \succeq L_{(1,n)}.$$

10.4 Synchronization in an undirected network

10.4.1 Graph comparison with the complete graph

Theorem 10.3. *Suppose that graph $\mathbb{G}(t)$ is undirected and connected. Under Assumption 10.1, the synchronization manifold of system (10.1) is globally asymptotically stable if:*

$$\mathbb{G}(t) \succ \frac{a}{n}\mathbb{K}_n, \quad \textit{for all } t. \tag{10.7}$$

Proof. Let the graph $\mathbb{G}_0$ in Theorem 10.2 to be $\mathbb{K}_n$, then one has that the synchronization manifold of system (10.1) is globally asymptotically stable if $L_{\mathbb{K}_n} L_{\mathbb{G}(t)} \succ aL_{\mathbb{K}_n}$ for all t. Note that $L_{\mathbb{K}_n} = nI_n - J$ where J is the n-by-n all-one matrix. We know then $L_{\mathbb{K}_n} L_{\mathbb{G}(t)} = (nI_n - J)L_{\mathbb{G}(t)} = n L_{\mathbb{G}(t)} \succ aL_{\mathbb{K}_n}$. Thus, the synchronization manifold of system (10.1) is globally asymptotically stable if $L_{\mathbb{G}(t)} \succ \frac{a}{n} L_{\mathbb{K}_n}$. □

Theorem 10.3 implies that for any coupled oscillators whose couplings are described by a weighted undirected graph $\mathbb{G}(t)$, one can always examine whether $\mathbb{G}(t) \succ \frac{a}{n}\mathbb{K}_n$ holds by comparing $\mathbb{G}(t)$ to the complete graph with identical edge weight $\frac{a}{n}$. The inequality in Theorem 10.3 can be understood differently in terms of the condition on the second smallest eigenvalue of $L_{\mathbb{G}(t)}$, as illustrated in the next result.

Theorem 10.4. *For an undirected graph $\mathbb{G}(t)$, it holds that:*

$$\mathbb{G}(t) \succ \frac{a}{n}\mathbb{K}_n \Leftrightarrow \lambda_2(\mathbb{G}(t)) > a.$$

Its proof is omitted here. One can refer to Reference 12 for the details.

Remark 10.1. *Theorem 3 in Reference 8 provides a lower bound for $\lambda_2(\mathbb{G}(t))$ to guarantee network synchronization. In Theorem 10.4, we show the equivalence between the condition in the form of graph comparison and the condition bounding from below the second smallest eigenvalue of the network's Laplacian.*

Theorem 10.3 can be used to give graphical conditions for the synchronization of system (10.1). In the following, we present some sufficient conditions for synchronization using features of graph $\mathbb{G}(t)$. Consider a set of paths $\mathscr{P} = \{\mathscr{P}_{ij} | i, j = 1, \ldots, n, j > i\}$, one for each pair of distinct vertices i and j. We denote the length of path $\mathscr{P}_{ij}$ by $|\mathscr{P}_{ij}|$, which is the number of edges in $\mathscr{P}_{ij}$. We assume that there are altogether m edges in the edge set $\mathscr{E}$ of graph $\mathbb{G}(t)$. If we label the edges of $\mathbb{G}(t)$ by $1, \ldots, m$, then the lower bounds on the coupling strengths of all the edges can be constructed to guarantee that the inequality in Theorem 10.3 holds. We state a result from Reference 9 below.

Theorem 10.5. *Suppose that graph $\mathbb{G}(t)$ is undirected and connected. Under Assumption 10.1, the synchronization manifold of system (10.1) is globally asymptotically stable if:*

$$\varepsilon_k(t) > \frac{b_k}{n} a, \qquad \textit{for } k = 1, \ldots, m \textit{ and for all } t,$$

where $b_k = \sum_{j>i; k\in\mathscr{P}_{ij}} |\mathscr{P}_{ij}|$ is the sum of the lengths of all those paths $\mathscr{P}_{ij}$ in $\mathscr{P}$ that contain edge k.

We provide here a different and simpler proof of Theorem 10.5 using elementary Laplacians and graph comparison.

Proof. From the definition introduced by (10.6), we have:

$$\frac{a}{n} L_{\mathbb{K}_n} = \frac{a}{n} \sum_{i=1}^{n-1} \sum_{j>i} L_{(i,j)}.$$

For each pair of (i, j) where $j > i$, we choose one path $\mathscr{P}_{ij}$ in $\mathbb{G}$ that connects i and j. Then one can apply Lemma 10.3 by comparing the sum of all the Laplacian matrices L_k, $k \in \mathscr{P}_{ij}$, of all the edges along this chosen path and the Laplacian matrix $L_{(i,j)}$ of the single edge (i, j), which leads to:

$$|\mathscr{P}_{ij}| \sum_{k\in\mathscr{P}_{ij}} L_k \succeq L_{(i,j)}. \tag{10.8}$$

Choosing such paths in the topological graph $\mathbb{G}$ for all the pairs of i, j where $j > i$, one obtains:

$$\begin{aligned}
\frac{a}{n} L_{\mathbb{K}_n} &\preceq \frac{a}{n} \sum_{j>i} \left(|\mathscr{P}_{ij}| \sum_{k\in\mathscr{P}_{ij}} L_k \right) \\
&= \frac{a}{n} \sum_{k=1}^{m} \left(\sum_{\substack{j>i \\ k\in\mathscr{P}_{ij}}} |\mathscr{P}_{ij}| \right) L_k \\
&= \frac{a}{n} \sum_{k=1}^{m} b_k L_k \\
&\prec \sum_{k=1}^{m} \varepsilon_k(t) L_k = L_{\mathbb{G}(t)},
\end{aligned}$$

where $b_k = \sum_{\substack{j>i \\ k\in\mathscr{P}_{ij}}} |\mathscr{P}_{ij}|$ has been defined in Theorem 10.5. And the last inequality holds trivially when $\varepsilon_k(t) > \frac{a}{n} b_k$ for each edge k. Therefore, the constructed coupling strengths ε_k for $k = 1, \ldots, m$ guarantee that $\mathbb{G}(t) \succ \frac{a}{n}\mathbb{K}_n$ holds. Thus, we arrive at the conclusion. □

Remark 10.2. *Theorem 10.5 presents a synchronization condition for allocating coupling strengths for $\mathbb{G}(t)$. The same result has been obtained in Reference 9. Here, we give a different interpretation of the result and prove it by using combinatorial features of the topological graph $\mathbb{G}$, which leads to the construction of efficient algorithms for determining the coupling strengths as we will show later. In addition, our allocation method using graph comparison is much easier to implement in applications.*

So far, we have only compared graph $\mathbb{G}(t)$ with the complete graph $\mathbb{K}_n$. It is natural to ask what different synchronization criteria can be obtained if we compare $\mathbb{G}(t)$ with other graphs. We explore this in the next subsection.

10.4.2 Graph comparison with the star graph

We can compare any two undirected and connected graphs with the same vertex set, and obtain a graphical inequality as a result. One can find more details in References 17,18. With this property, we can compare the graph $\mathbb{G}(t)$ under study with a known synchronizable network graph $\mathbb{G}^*$ for self-dynamics $f(\cdot)$, and determine a set of coupling strengths for the edges in $\mathbb{G}(t)$ based upon combinatorial features of edges such that $\mathbb{G}(t) \succ \mathbb{G}^*$. Hence, the network coupled through the graph $\mathbb{G}(t)$ with the obtained weights also synchronizes. To show how to implement this idea, we give an example on comparing graph $\mathbb{G}(t)$ with a star graph. Similar results can be achieved when $\mathbb{G}(t)$ is compared with other graphs, such as ring graphs, path graphs, and any graphs with known second smallest eigenvalues.

Now consider an n-vertex star graph $\mathbb{S}_n$, in which without loss of generality we assume vertex 1 has $n-1$ neighbors. Then $L_{\mathbb{S}_n} = \sum_{i=2}^{n} L_{(1,i)}$. We consider two cases for all the edges $(1, i)$, $2 \leq i \leq n$, in $\mathbb{S}_n$:

1. Edge $(1, i)$ is not in the edge set $\mathscr{E}$ of $\mathbb{G}(t)$. Since $\mathbb{G}(t)$ is connected, there must exist some paths in $\mathbb{G}(t)$ connecting vertices 1 and i. We choose arbitrarily one of these paths, and denoted as $\mathscr{P}_{1,i}$. Then we have:

$$L_{(1,i)} \preceq |\mathscr{P}_{1,i}| \sum_{k \in \mathscr{P}_{1,i}} L_k. \tag{10.9}$$

2. Edge $(1, i)$ is in $\mathscr{E}$. There are two options: one is to use $(1, i)$ directly and the other is to choose arbitrarily another path between vertices 1 and i, if such a path exists. We set the probability of the first option to be $1 - \alpha_i$, and that for the second α_i where $0 \leq \alpha_i < 1$. If there are no paths between 1 and i other than the edge $(1, i)$, we will set $\alpha_i = 0$. Thus, we have:

$$L_{(1,i)} \preceq (1 - \alpha_i) L_{(1,i)} + \alpha_i \, |\mathscr{P}_{1,i}| \sum_{k \in \mathscr{P}_{1,i}} L_k. \tag{10.10}$$

Note that (10.9) is the special case of (10.10) when α_i is taken to be 1. Hence, we can use the inequality (10.10) with a proper choice of $\alpha_i \in [0, 1]$ for each $i \in \{2, \ldots, n\}$ leading to:

$$\begin{aligned}
L_{\mathbb{S}_n} &= \sum_{i=2}^{n} L_{(1,i)} \\
&\preceq \sum_{i=2}^{n} \left[(1-\alpha_i)L_{(1,i)} + \alpha_i \left|\mathscr{P}_{1,i}\right| \sum_{k \in \mathscr{P}_{1,i}} L_k \right] \\
&= \sum_{i=2}^{n} (1-\alpha_i)L_{(1,i)} + \sum_{i=2}^{n} \left[\alpha_i \left|\mathscr{P}_{1,i}\right| \sum_{k \in \mathscr{P}_{1,i}} L_k \right] \\
&= \sum_{k=1}^{m} \left[\sum_{\substack{i=2 \\ k \in \mathscr{P}_{1,i}}}^{n} \alpha_i \left|\mathscr{P}_{1,i}\right| \right] L_k + \sum_{i=2}^{n} (1-\alpha_i)L_{(1,i)} \\
&= \sum_{k=1}^{m} \left[\sum_{\substack{i=2 \\ k \in \mathscr{P}_{1,i}}}^{n} \alpha_i \left|\mathscr{P}_{1,i}\right| + \varphi(1-\alpha_i) \right] L_k,
\end{aligned}$$

where the real valued function $\varphi(1 - \alpha_i)$ satisfies:

$$\varphi(1-\alpha_i) = \begin{cases} 1-\alpha_i \neq 0, & \text{if } (1,i) \text{ is the edge } k, \\ 0, & \text{otherwise.} \end{cases}$$

Let:

$$b'_k = \sum_{\substack{i=2 \\ k \in \mathscr{P}_{1,i}}}^{n} \alpha_i \left|\mathscr{P}_{1,i}\right| + \varphi(1-\alpha_i). \tag{10.11}$$

Then we have $\mathbb{G}(t) \succ a\mathbb{S}_n$ if the weight of the edge k satisfies $\varepsilon_k(t) > a\, b'_k$ for $k = 1, \ldots, m$. From Lemma 10.1, one has $\lambda_2(\mathbb{G}(t)) > \lambda_2(a\mathbb{S}_n) = a\lambda_2(\mathbb{S}_n) = a$ if $\varepsilon_k(t) > a\, b'_k$ for all k. From Theorem 10.4, the synchronization manifold of the dynamical system (10.1) is globally asymptotically stable, if $\varepsilon_k(t) > a\, b'_k$ for $k = 1, \ldots, m$. Thus, we have proved the following theorem.

Theorem 10.6. *Suppose that graph $\mathbb{G}(t)$ is undirected and connected. Under Assumption 10.1, the synchronization manifold of system (10.1) is globally asymptotically stable if the coupling strength of edge k satisfies $\varepsilon_k(t) > a\, b'_k$ for all $k = 1, \ldots, m$ and for all t, where b'_k is given by (10.11).*

Remark 10.3. *Compared with the graphical condition in Theorem 10.5, the method in Theorem 10.6 greatly reduces the computational complexity. In the application of Theorem 10.6, we only need $n - 1$ candidate paths in the graph $\mathbb{G}$ and then utilize* *the combinational inequality in Lemma 10.3 for these paths. On the other hand, to apply Theorem 10.5, one needs to choose $n(n-1)/2$ paths in $\mathbb{G}$ and then takes the same operation using Lemma 10.3 for each of these paths.*

Remark 10.4. *Theorem 10.5 (respectively Theorem 10.6) is the special case when graph $\mathbb{G}$ is compared with the complete graph $\mathbb{K}_n$ (respectively the star graph $\mathbb{S}_n$). Similar results can be obtained when $\mathbb{G}$ is compared with other graphs, such as rings and paths. A proper choice of the graphs in comparison is helpful to obtain less conservative lower bounds for coupling strengths and reduce the computational complexity of the comparison at the same time.*

10.4.3 Illustrative examples

10.4.3.1 Benefits from comparing with star graphs

Now we give an example to demonstrate the advantages of using the synchronization condition in Theorem 10.6. To simplify our calculation, we consider a fractal tree with ten vertices, shown on the left of Figure 10.1. First, let the fractal graph $\mathbb{G}_{10}$ be compared with the star graph $\mathbb{S}_{10}$. Because of the fractal structure of graph $\mathbb{G}_{10}$, we only need to focus on the calculations of bounds for the edges $(1,2), (2,5), (2,6)$. And we have the comparison:

$$\begin{aligned} L_{\mathbb{S}_{10}} &= L_{(1,2)} + L_{(1,5)} + L_{(1,6)} + \cdots \\ &\preceq L_{(1,2)} + 2(L_{(1,2)} + L_{(2,5)}) + 2(L_{(1,2)} + L_{(2,6)}) + \cdots \\ &= 5L_{(1,2)} + 2L_{(2,5)} + 2L_{(2,6)} + \cdots . \end{aligned}$$

Thus, we have $b'_{(1,2)} = 5$, $b'_{(2,5)} = 2$, and $b'_{(2,6)} = 2$. From Theorem 10.6, we obtain the bounds for the edges $(1,2), (2,5), (2,6)$ in graph $\mathbb{G}_{10}$:

$$\varepsilon_{(1,2)} \geq 5a, \quad \varepsilon_{(2,5)} \geq 2a, \quad \varepsilon_{(2,6)} \geq 2a. \tag{10.12}$$

Next, we give another set of bounds for the edges in $\mathbb{G}_{10}$ using the method in Theorem 10.5. We implement graph comparison between graph $\mathbb{G}_{10}$ and the complete graph

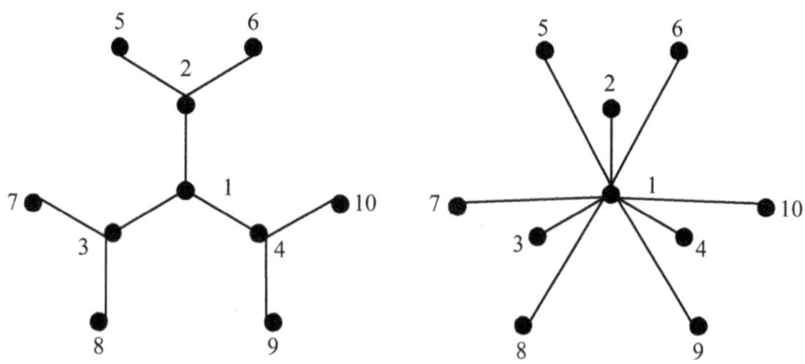

Figure 10.1 Comparison of a fractal tree with the star graph $\mathbb{S}_{10}$

$\mathbb{K}_{10}$. Thus, we need to consider the paths in $\mathbb{G}_{10}$ for every pair of vertices. The choice of the path in $\mathbb{G}_{10}$ for each pair of vertices is unique, because there is no cycle in $\mathbb{G}_{10}$. We only need to calculate the bounds for the edges (1, 2), (2, 5), (2, 6). To do so, we first list all the possible paths that pass through at least one of these edges, which are shown in Figure 10.2. Then, from $b_k = \sum_{j>i;k\in\mathscr{P}_{ij}} |\mathscr{P}_{ij}|$ in Theorem 10.5, we have:

$$\begin{aligned} b_{(1,2)} &= |\mathscr{P}_{1,2}| + |\mathscr{P}_{1,5}| + |\mathscr{P}_{1,6}| + |\mathscr{P}_{2,3}| + |\mathscr{P}_{2,4}| \\ &\quad + |\mathscr{P}_{2,7}| + |\mathscr{P}_{2,8}| + |\mathscr{P}_{2,9}| + |\mathscr{P}_{2,10}| \\ &\quad + |\mathscr{P}_{3,5}| + |\mathscr{P}_{3,6}| + |\mathscr{P}_{4,5}| + |\mathscr{P}_{4,6}| \\ &\quad + |\mathscr{P}_{5,7}| + |\mathscr{P}_{5,8}| + |\mathscr{P}_{5,9}| + |\mathscr{P}_{5,10}| \\ &\quad + |\mathscr{P}_{6,7}| + |\mathscr{P}_{6,8}| + |\mathscr{P}_{6,9}| + |\mathscr{P}_{6,10}| \\ &= 1 + 2 \times 4 + 3 \times 8 + 4 \times 8 = 65. \end{aligned}$$

Following the same reasoning, we have:

$$\begin{aligned} b_{(2,5)} &= |\mathscr{P}_{1,5}| + |\mathscr{P}_{2,5}| + |\mathscr{P}_{3,5}| + |\mathscr{P}_{4,5}| + |\mathscr{P}_{5,6}| \\ &\quad + |\mathscr{P}_{5,7}| + |\mathscr{P}_{5,8}| + |\mathscr{P}_{5,9}| + |\mathscr{P}_{5,10}| \\ &= 2 + 1 + 3 + 3 + 2 + 4 \times 4 = 27. \end{aligned}$$

And $b_{(2,6)} = 27$ can be calculated similarly.

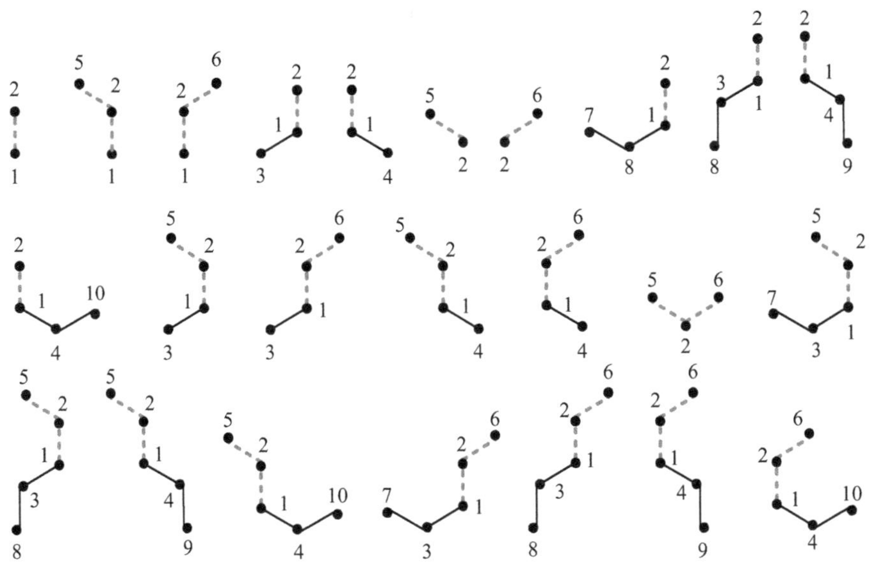

Figure 10.2 The paths in $\mathbb{G}_{10}$ that pass through one of the edges (1, 2), (2, 5), (2, 6)

According to $\varepsilon_k > \frac{b_k}{n} a$ in Theorem 10.5, we obtain the bounds for the coupling strengths of the edges $(1,2)$, $(2,5)$, and $(2,6)$ as:

$$\varepsilon_{(1,2)} \geq \frac{b_{(1,2)}}{10} a = 6.5a, \quad \varepsilon_{(2,5)} \geq \frac{b_{(2,5)}}{10} a = 2.7a, \quad \varepsilon_{(2,6)} \geq \frac{b_{(2,6)}}{10} a = 2.7a. \qquad (10.13)$$

The above calculations show that the computational complexity of graph comparisons is greatly reduced by using the method in Theorem 10.6, comparing with what obtained using Theorem 10.5. In addition, we have obtained another set of bounds for coupling strengths of $\mathbb{G}_{10}$, in which each bound is smaller. The proposed method is especially effective when networks are large and sparse.

10.4.3.2 Applications in dynamic networks with growing topologies

Now we apply Theorem 10.6 to allocate coupling strengths in dynamic networks with growing topologies where the vertices and edges are added to a graph in time. It is much easier to compare a growing graph with a growing star, with the same vertex set, than with a growing complete graph. If one more vertex is added to the star graph $\mathbb{S}_n$, only one more edge needs to be added. However, if one more vertex is added to the complete graph $\mathbb{K}_n$, n new edges need to get involved in calculation. Moreover, the second smallest eigenvalue of star graphs is the constant 1. These motivate us to apply Theorem 10.6 on dynamic networks with growing topologies. We can consider such graphs as layered graphs where at each time a new layer of vertices are added to the graph with edges connecting them to vertices in the previous layers.

In the following, we use complete binary trees to illustrate the application. The complete binary tree $\mathbb{T}_n$ with $n = 2^d - 1$ vertices is the graph with the edges of the form $(u, 2u)$ and $(u, 2u+1)$ for integer $u < n/2$ [17]. The depth of $\mathbb{T}_n$ is d. The complete binary tree $\mathbb{T}_n$ is shown on the left of Figure 10.3.

Proposition 10.1. *Under Assumption 10.1, for the complete binary tree $\mathbb{T}_n$ with $n = 2^d - 1$ vertices, the global synchronization of system (10.1) is guaranteed if the couplings between the $(k-1)$th layer and the kth layer satisfy:*

$$\varepsilon(k-1, k, d) \geq \sum_{j=k}^{d} (j-1) \times 2^{j-k} a, \qquad 2 \leq k \leq d.$$

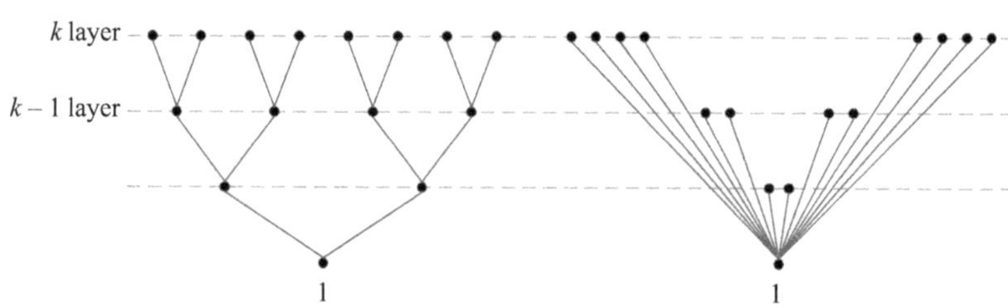

Figure 10.3 Comparison of the growing complete binary tree with the growing star graph

Proof. We start with the complete binary tree with two layers $d = 2$. Comparing it with the star $\mathbb{S}_3$, we have:

$$L_{\mathbb{S}_3} = L_{(1,2)} + L_{(1,3)}.$$

From Theorem 10.6, it is easy to obtain the bounds for the edges in $\mathbb{T}_3$: $\varepsilon_{(1,2)} \geq a$ and $\varepsilon_{(1,3)} \geq a$.

Compare the complete binary tree with $d = 3$ with $\mathbb{S}_7$:

$$\begin{aligned} L_{\mathbb{S}_7} &= L_{(1,2)} + L_{(1,3)} + L_{(1,4)} + L_{(1,5)} + L_{(1,6)} + L_{(1,7)} \\ &\preceq L_{(1,2)} + L_{(1,3)} + 2[L_{(1,2)} + L_{(2,4)}] + 2[L_{(1,2)} + L_{(2,5)}] \\ &\quad + 2[L_{(1,3)} + L_{(3,6)}] + 2[L_{(1,3)} + L_{(3,7)}] \\ &= (1 + 2 \times 2)L_{(1,2)} + (1 + 2 \times 2)L_{(1,3)} \\ &\quad + 2L_{(2,4)} + 2L_{(2,5)} + 2L_{(3,6)} + 2L_{(3,7)} \end{aligned}$$

and for the complete binary tree with $d = 4$:

$$\begin{aligned} L_{\mathbb{S}_{15}} &= L_{(1,2)} + L_{(1,3)} + L_{(1,4)} + \cdots + L_{(1,15)} \\ &\preceq (1 + 2 \times 2 + 3 \times 2^2)L_{(1,2)} \\ &\quad + (1 + 2 \times 2 + 3 \times 2^2)L_{(1,3)} \\ &\quad + (2 + 3 \times 2)L_{(2,4)} + (2 + 3 \times 2)L_{(2,5)} \\ &\quad + (2 + 3 \times 2)L_{(3,6)} + (2 + 3 \times 2)L_{(3,7)} \\ &\quad + 3L_{(4,8)} + 3L_{(4,9)} + 3L_{(5,10)} + 3L_{(5,11)} \\ &\quad + 3L_{(6,12)} + 3L_{(6,13)} + 3L_{(7,14)} + 3L_{(7,15)}. \end{aligned}$$

We use $b'(k-1, k, d)$ to denote the value of b' for the edges between the $(k-1)$th layer and the kth layer in the binary tree $\mathbb{T}_n$ with $n = 2^d - 1$, and $\varepsilon(k-1, k, d)$ to denote the couplings between the $(k-1)$th layer and the kth layer in $\mathbb{T}_n$. We obtain the weights for the edges in the complete binary tree $\mathbb{T}_n$ with $n = 2^d - 1$ by induction, and we postulate:

$$b'(k-1, k, d) = \sum_{j=k}^{d} (j-1) \times 2^{j-k} \tag{10.14}$$

for the edges of the form $(u, 2u)$ and $(u, 2u+1)$ with $u = 2^{k-2}, \ldots, 2^{k-1} - 1$ where $2 \leq k \leq d$ (i.e., the edges between the $(k-1)$th layer and the kth layer).

Now we use induction to prove our conjecture. Suppose that (10.14) holds for the complete binary tree $\mathbb{T}_n$ with $n = 2^d - 1$. Then we calculate $b'(k-1, k, d+1)$ for the binary tree with $n' = 2^{d+1} - 1$. The depth of $\mathbb{T}_{n'}$ is $d+1$. The binary tree $\mathbb{T}_{n'}$ has one more layer, and has 2^d more vertices which are labeled by $v_{2^d}, v_{2^d+1}, \ldots, v_{2^{d+1}-1}$. In order to assign couplings in $\mathbb{T}_{n'}$, we compare $\mathbb{T}_{n'}$ with the star graph $\mathbb{S}_{2^{d+1}-1}$, and calculate for the edges of the form $(u, 2u)$ and $(u, 2u+1)$ with $u = 2^{k-2}, \ldots, 2^{k-1}-1$ and $2 \leq k \leq d+1$ (i.e., the edges between the $(k-1)$th layer and the kth layer):

$$\begin{aligned} b'(k-1, k, d+1) &= b'(k-1, k, d) + d \times 2^{d-k+1} \\ &= \sum_{j=k}^{d} (j-1) \times 2^{j-k} + d \times 2^{d-k+1} \\ &= \sum_{j=k}^{d+1} (j-1) \times 2^{j-k}. \end{aligned}$$

This shows that (10.14) still holds for the binary tree $\mathbb{T}_{n'}$ with $d+1$ layers, and hence by induction we have proved that (10.14) is correct. Thus, for the binary tree $\mathbb{T}_n$ with $n = 2^d - 1$, the weights for the edges between the $(k-1)$th layer and the kth layer satisfy:

$$\varepsilon(k-1, k, d) \geq \sum_{j=k}^{d} (j-1) \times 2^{j-k} a,$$

where $2 \leq k \leq d$. □

10.5 Synchronization in a directed network

10.5.1 Graph comparison with the complete graph

First, we investigate graphical conditions for the synchronization in directed complex dynamical networks. In Theorem 10.2, we take graph $\mathbb{G}_0$ to be the complete graph $\mathbb{K}_n$. Therefore, the complete synchronization of system (10.1) is guaranteed if:

$$L_{\mathbb{K}_n} L_{\mathbb{G}(t)} \succeq a L_{\mathbb{K}_n}. \tag{10.15}$$

Note that $L_{\mathbb{K}_n} = nI_n - J$ where J is the $n \times n$ all-one matrix. From (10.15), one has:

$$nL_{\mathbb{G}(t)} - JL_{\mathbb{G}(t)} \succeq aL_{\mathbb{K}_n}. \tag{10.16}$$

In the case where $\mathbb{G}(t)$ is undirected, (10.16) can be further reduced to $\mathbb{G}(t) \succeq \frac{a}{n}\mathbb{K}_n$. However, in this section, we are looking at the more challenging scenario where the networks are directed. We will use the following property of directed graphs.

Lemma 10.4. *For a directed graph $\mathbb{G}(t)$, the condition $nL_{\mathbb{G}(t)} - JL_{\mathbb{G}(t)} \succeq aL_{\mathbb{K}_n}$ is equivalent to $\frac{n}{2}(L_{\mathbb{G}(t)} + L_{\mathbb{G}(t)}^{\top}) - \frac{1}{2}(JL_{\mathbb{G}(t)} + L_{\mathbb{G}(t)}^{\top}J) \succeq aL_{\mathbb{K}_n}$.*

Its proof is omitted here. One can refer to Reference 13 for the details.

With Lemma 10.4 at hand, we have a general graphical synchronization criterion for directed dynamical networks:

Theorem 10.7. *Suppose that Assumption 10.1 holds, and that graph* $\mathbb{G}(t)$ *contains a directed spanning tree. The synchronization manifold of system (10.1) is globally stable if:*

$$\frac{1}{2}(L_{\mathbb{G}(t)} + L_{\mathbb{G}(t)}^{\top}) - \frac{1}{2n}(JL_{\mathbb{G}(t)} + L_{\mathbb{G}(t)}^{\top}J) \succeq \frac{a}{n}\,L_{\mathbb{K}_n}, \quad \textit{for all } t. \tag{10.17}$$

In the following, we will show how to interpret (10.17) from the perspective of graph comparison. Here and in what follows, we use graph $\mathbb{G}$ as an abbreviation for $\mathbb{G}(t)$, and coupling ε_{ij} for $\varepsilon_{ij}(t)$ when it is convenient to do so. All conditions and criteria for the couplings are understood to hold for all times t, even if t is not explicitly specified. Let D_i^c denote the vertex unbalance [10] of vertex i, namely $D_i^c = \sum_{k=1}^{n} \varepsilon_{ki} = \sum_{k\neq i} \varepsilon_{ki} + \varepsilon_{ii} = \sum_{k\neq i} \varepsilon_{ki} - \sum_{k\neq i} \varepsilon_{ik}$, which is the difference between the out-degree and in-degree of vertex i. It is true that:

$$\begin{aligned} JL_{\mathbb{G}} &= \mathbf{1} \otimes \begin{bmatrix} -\sum_{k=1}^{n} \varepsilon_{k1} & -\sum_{k=1}^{n} \varepsilon_{k2} & \cdots & -\sum_{k=1}^{n} \varepsilon_{kn} \end{bmatrix} \\ &= -\mathbf{1} \otimes \begin{bmatrix} D_1^c & D_2^c & \cdots & D_n^c \end{bmatrix} \end{aligned}$$

and:

$$L_{\mathbb{G}}^{\top} J = -\mathbf{1}^{\top} \otimes \begin{bmatrix} D_1^c & D_2^c & \cdots & D_n^c \end{bmatrix}^{\top}.$$

It follows that the matrix $-(JL_{\mathbb{G}} + L_{\mathbb{G}}^{\top} J)$ is

$$\begin{bmatrix} 2D_1^c & D_1^c + D_2^c & \cdots & D_1^c + D_n^c \\ D_2^c + D_1^c & 2D_2^c & \cdots & D_2^c + D_n^c \\ \cdots & \cdots & \cdots & \cdots \\ D_n^c + D_1^c & D_n^c + D_2^c & \cdots & 2D_n^c \end{bmatrix},$$

where the (ij)th entry is $D_i^c + D_j^c$ for $i, j = 1, \ldots, n$. Since the sum of the out-degrees of all the vertices in $\mathbb{G}$ is equal to the sum of the in-degrees of all the vertices, we have $\sum_{i=1}^{n} D_i^c = 0$. The ith row-sum of the matrix $-(JL_{\mathbb{G}} + L_{\mathbb{G}}^{\top} J)$ is then $nD_i^c + \sum_{i=1}^{n} D_i^c = nD_i^c$ for $i = 1, \ldots, n$. Define the $n \times n$ matrix:

$$\Delta \triangleq \operatorname{diag}\{nD_1^c, nD_2^c, \ldots, nD_n^c\}.$$

Thus, the matrix $\frac{a}{n} L_{\mathbb{K}_n} + \frac{1}{2n}(JL_{\mathbb{G}} + L_{\mathbb{G}}^{\top} J) + \frac{1}{2n}\Delta$ is symmetric and has zero row sums. Since the ith row-sum of the matrix $L_{\mathbb{G}} + L_{\mathbb{G}}^{\top}$ is $-\sum_{k=1}^{n} \varepsilon_{ki} = -D_i^c$ for $i = 1, \ldots, n$, we know that the matrix $\frac{1}{2}(L_{\mathbb{G}} + L_{\mathbb{G}}^{\top}) + \frac{1}{2n}\Delta$ is symmetric and has zero row sums and non-positive off-diagonal entries. Now we are ready to compare the two symmetric matrices $\frac{a}{n} L_{\mathbb{K}_n} + \frac{1}{2n}(JL_{\mathbb{G}} + L_{\mathbb{G}}^{\top} J) + \frac{1}{2n}\Delta$ and $\frac{1}{2}(L_{\mathbb{G}} + L_{\mathbb{G}}^{\top}) + \frac{1}{2n}\Delta$. From (10.17), we have:

$$\frac{1}{2}\left(L_{\mathbb{G}(t)} + L_{\mathbb{G}(t)}^{\top}\right) + \frac{1}{2n}\Delta \succeq \frac{a}{n}\left(L_{\mathbb{K}_n} + \frac{1}{2a}\left(JL_{\mathbb{G}(t)} + L_{\mathbb{G}(t)}^{\top} J\right) + \frac{1}{2a}\Delta\right). \tag{10.18}$$

In order to better understand the graphical condition given by inequality (10.17), now we interpret Theorem 10.7 using the elementary Laplacians.

Theorem 10.8. *Suppose that Assumption 10.1 holds, and that graph $\mathbb{G}(t)$ contains a directed spanning tree. The synchronization manifold of system (10.1) is globally stable if:*

$$\sum_{j>i}\left(\frac{\varepsilon_{ij}+\varepsilon_{ji}}{2}-\frac{D_i^c+D_j^c}{2n}\right)L_{(i,j)}\succeq\frac{a}{n}\sum_{j>i}L_{(i,j)}; \tag{10.19}$$

or equivalently if:

$$\sum_{j>i}\frac{\varepsilon_{ij}+\varepsilon_{ji}}{2}L_{(i,j)}\succeq\frac{a}{n}\sum_{j>i}\left(1+\frac{D_i^c+D_j^c}{2a}\right)L_{(i,j)}; \tag{10.20}$$

for all t.

Proof. Using elementary Laplacians, the inequality (10.19) is a straightforward consequence of the graphical condition (10.17). Inequality (10.20) follows from the graphical criterion (10.18). □

Theorems 10.7 and 10.8 aim to symmetrize graph $\mathbb{G}$ first, and then provide graph comparison conditions based on the symmetrized graph. One can see that the left-hand sides of both (10.18) and (10.20) represent the symmetrized graph of $\mathbb{G}$. To symmetrize the graph $\mathbb{G}$, for any pair of unidirectionally coupled vertices i and j in $\mathbb{G}$, we replace the directed edge between them by an undirected edge with a weight $\varepsilon_{ij}/2$ that is half of the original coupling strength; for any bidirectionally coupled pair of vertices i and j, we replace the two edges between them by an undirected edge with the coupling strength $(\varepsilon_{ij}+\varepsilon_{ji})/2$. Let $\mathbb{G}^s$ be the obtained symmetrized graph from $\mathbb{G}$. One can then check that the Laplacian matrix of $\mathbb{G}^s$ is $L_{\mathbb{G}^s}=\frac{1}{2}(L_{\mathbb{G}}+L_{\mathbb{G}}^{\top})+\frac{1}{2n}\Delta$.

For the symmetrized graph $\mathbb{G}^s$, consider a set of paths $\mathscr{P}=\{\mathscr{P}_{ij}|i,\ j=1,\dots,n,\ j>i\}$, one for each pair of distinct vertices i and j. Next, we use Theorems 10.7 and 10.8 to construct graph theoretic conditions for the synchronization of network (10.1). We use $\mathscr{E}(\mathbb{G}^s)$ to denote the set of all the edges of $\mathbb{G}^s$ and assume that there are altogether m edges that are labeled by $1,\dots,m$. In the following theorem, we show that lower bounds on the coupling strengths $\varepsilon_k^{(s)}, k=1,\dots,m$, can be constructed to guarantee that inequality (10.18) holds.

Theorem 10.9. *Suppose that Assumption 10.1 holds, and graph $\mathbb{G}$ contains a directed spanning tree. The synchronization manifold of network (10.1) is globally stable if:*

$$\varepsilon_k^{(s)}>\frac{a}{n}b_k,\qquad \textit{for } k=1,\dots,m, \tag{10.21}$$

where $b_k=\sum_{j>i;k\in\mathscr{P}_{ij}}w(\mathscr{P}_{ij})$ is the sum of the weighted lengths $w(\mathscr{P}_{ij})$ of all the paths $\mathscr{P}_{ij}$ in $\mathscr{P}$ that contain the edge k that belongs to the symmetrized graph $\mathbb{G}^s$ and the weighted path length $w(\mathscr{P}_{ij})$ is defined by:

$$w(\mathscr{P}_{ij})\triangleq\begin{cases}|\mathscr{P}_{ij}|\,\chi\left(1+\frac{D_i^c+D_j^c}{2a}\right), & edge(i,j)\notin\mathscr{E}(\mathbb{G}^s),\\ 1+\frac{D_i^c+D_j^c}{2a}, & edge(i,j)\in\mathscr{E}(\mathbb{G}^s),\end{cases} \tag{10.22}$$

where for $z\in\mathbb{R}$, the function $\chi(z)=z$ if $z>0$, and $\chi(z)=0$ otherwise.

Proof. Since the two matrices $L_{\mathbb{G}^s}$ and $\frac{a}{n}\left(L_{\mathbb{K}_n} + \frac{1}{2a}(JL_{\mathbb{G}} + L_{\mathbb{G}}^{\top}J) + \frac{1}{2a}\Delta\right)$ are symmetric and have zero row and column sums, we can compare them as follows:

$$\begin{aligned}
&\frac{a}{n}\left(L_{\mathbb{K}_n} + \frac{1}{2a}(JL_{\mathbb{G}} + L_{\mathbb{G}}^{\top}J) + \frac{1}{2a}\Delta\right) \\
&\quad = \frac{a}{n}\sum_{j>i}\left(1 + \frac{D_i^c + D_j^c}{2a}\right)L_{(i,j)} \\
&\quad \preceq \frac{a}{n}\sum_{j>i;(i,j)\notin\mathscr{E}(\mathbb{G}^s)} \chi\left(1 + \frac{D_i^c + D_j^c}{2a}\right)L_{(i,j)} \\
&\qquad + \frac{a}{n}\sum_{j>i;(i,j)\in\mathscr{E}(\mathbb{G}^s)}\left(1 + \frac{D_i^c + D_j^c}{2a}\right)L_{(i,j)}.
\end{aligned} \tag{10.23}$$

In the last two terms, for edge $(i,j) \in \mathscr{E}(\mathbb{G}^s)$, we keep the term $\frac{a}{n}\left(1 + \frac{D_i^c + D_j^c}{2a}\right)L_{(i,j)}$; for edge $(i,j) \notin \mathscr{E}(\mathbb{G}^s)$, we choose a path $\mathscr{P}_{ij}$ in $\mathbb{G}^s$ that connects i and j. Then one can apply Lemma 10.3 by comparing the sum of all the Laplacian matrices $L_k, k \in \mathscr{P}_{ij}$ of all the edges along this chosen path and the Laplacian matrix $L_{(i,j)}$ of the single edge (i,j), which leads to $L_{(i,j)} \preceq |\mathscr{P}_{ij}| \sum_{k\in\mathscr{P}_{ij}} L_k$. Applying this inequality to each $L_{(i,j)}$ where $(i,j) \notin \mathscr{E}(\mathbb{G}^s)$, one obtains that the right-hand side of the inequality (10.23) is less than or equal to:

$$\begin{aligned}
&\frac{a}{n}\sum_{\substack{j>i;\\(i,j)\notin\mathscr{E}(\mathbb{G}^s)}}\left(\chi\left(1 + \frac{D_i^c + D_j^c}{2a}\right)\cdot|\mathscr{P}_{ij}|\sum_{k\in\mathscr{P}_{ij};k\in\mathscr{E}(\mathbb{G}^s)} L_k\right) \\
&\qquad\qquad + \frac{a}{n}\sum_{j>i;(i,j)\in\mathscr{E}(\mathbb{G}^s)}\left(1 + \frac{D_i^c + D_j^c}{2a}\right)L_{(i,j)}.
\end{aligned} \tag{10.24}$$

For the term in (10.24), we sum up all the weights for L_k where edge $k \in \mathscr{E}(\mathbb{G}^s)$ and we obtain:

$$\begin{aligned}
&\frac{a}{n}\sum_{k=1}^{m}\left(\sum_{j>i;k\in P_{ij}} w(\mathscr{P}_{ij})\right)L_k \\
&\quad = \frac{a}{n}\sum_{k=1}^{m} b_k L_k \\
&\quad \prec \sum_{k=1}^{m}\varepsilon_k^{(s)} L_k = L_{\mathbb{G}^s},
\end{aligned}$$

where $b_k = \sum_{j>i;k\in\mathscr{P}_{ij}} w(\mathscr{P}_{ij})$ has been defined in Theorem 10.9. The last inequality holds trivially when $\varepsilon_k^{(s)} > \frac{a}{n} b_k$ for each edge k. Therefore, the constructed coupling strengths $\varepsilon_k^{(s)}$ for $k = 1, \ldots, m$ guarantee that (10.18) holds. The proof is complete. □

Remark 10.5. *Theorem 10.9 is similar to Theorem 1 obtained by Belykh et al. in Reference 10, except for some minor differences in (10.21) and (10.22). Since we aim to propose several useful graphical comparison synchronization criteria and develop a new methodology to study network synchronization, we interpret Belykh et al.'s result [11] from a different viewpoint of spectral graph theory. Our approach utilizes more features of the graphs associated with the networks. In addition, using graph comparison, we have provided a simpler proof compared with those in References 10,11.*

Remark 10.6. *If $\mathbb{G}$ is directed and balanced, i.e., $D_i^c = 0$ for $i = 1, \ldots, n$, then the graphical synchronization criterion (10.17) becomes:*

$$\frac{1}{2}\left(L_{\mathbb{G}(t)} + L_{\mathbb{G}(t)}^{\mathrm{T}}\right) \succeq \frac{a}{n} L_{\mathbb{K}_n}.$$

And (10.20) accordingly becomes:

$$\sum_{j>i} \frac{\varepsilon_{ij} + \varepsilon_{ji}}{2} L_{(i,j)} \succeq \frac{a}{n} \sum_{j>i} L_{(i,j)}.$$

From Theorem 10.9, it follows that network (10.1) can be asymptotically synchronized if $\varepsilon_k^{(s)} > \frac{b_k}{n} a$ for $k = 1, \ldots, m$, where $b_k = \sum_{j>i;k\in\mathscr{P}_{ij}} |\mathscr{P}_{ij}|$. The result then becomes the same as Theorem 1 in Reference 11 in which the connection graph stability method on directed graphs with node balance is discussed.

10.5.2 Illustrative examples

Now we give several simple examples to show how to apply Theorem 10.9.

10.5.2.1 The directed ring graph

We consider one example as the directed ring graph with n vertices, shown in Figure 10.4(a). Suppose that the coupling strength for each edge is $\varepsilon(t)$. In the following, we use Theorem 10.7 to determine the lower bound for $\varepsilon(t)$ to synchronize the network. Note that the graph (a) is directed and balanced. And one has $D_i^c = 0$ for all $i = 1, \ldots, n$. Then (10.17) can be reduced to $\frac{1}{2}\left(L_{\mathbb{G}_{(a)}} + L_{\mathbb{G}_{(a)}}^T\right) \succeq \frac{a}{n} L_{\mathbb{K}_n}$. We symmetrize the directed graph by replacing each directed edge by an undirected edge with the coupling strength $\varepsilon(t)/2$, which is shown in Figure 10.4(b). Our problem reduces to determining $\varepsilon(t)$ such that $L_{\mathbb{G}_{(b)}} \succeq \frac{a}{n} \mathbb{K}_n$. One can use the technique of graph comparison to determine $\varepsilon(t)$. However, we choose to use an easier approach to solve this, since the undirected and unweighted ring graph $\mathbb{R}_n$ is a special graph and the eigenvalues of its Laplacian matrix can be explicitly written as $2 - 2\cos(2\pi k/n)$ for $0 \le k \le n/2$ [17]. One can check that the second smallest eigenvalue is $4\sin^2(\pi/n)$ when $k = 1$. From Theorem 3 in Reference 12, $L_{\mathbb{G}_{(b)}} \succeq \frac{a}{n} \mathbb{K}_n$ is equivalent to $\lambda_2(\mathbb{G}_{(b)}) \ge a$, i.e., $\frac{\varepsilon(t)}{2} \cdot 4\sin^2(\pi/n) \ge a$. It follows that $\varepsilon(t) \ge \frac{a}{2\sin^2(\pi/n)}$ for all times t.

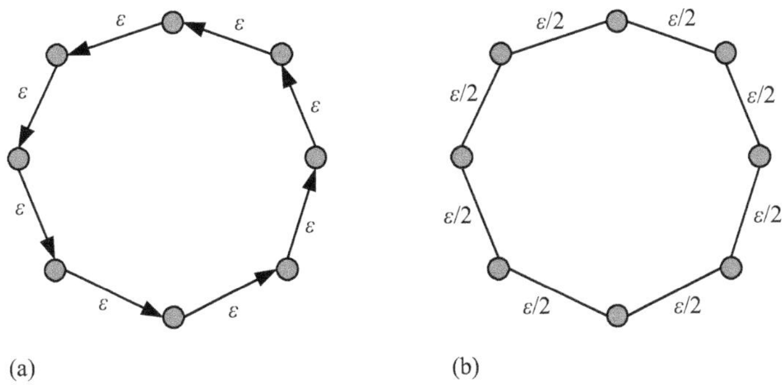

Figure 10.4 A directed ring graph with n vertices and its symmetrized graph. The directed edges in the graph on the panel (a) are replaced by the undirected edges with half weights in the graph on the panel (b)

10.5.2.2 Graph with three nodes

We consider the graph shown in Figure 10.5(a). Suppose the coupling strength for edge $(1, 2)$ is ε_1, and the couplings for $(2, 3)$ and $(3, 2)$ are ε_2. This example helps us understand how to apply graph comparison to obtain the bounds for coupling strengths in Theorem 10.9. Following the steps below, we aim to make the graphical inequality (10.20) holds for the graph (a).

Step 1. Calculate the node unbalance D_i^c for each node $i = 1, 2, 3$. We have $D_1^c = \varepsilon_1$, $D_2^c = -\varepsilon_1$, and $D_3^c = 0$.

Step 2. Symmetrize the graph by replacing each directed edge by the undirected edge with half weight. The symmetrized graph obtained is shown in Figure 10.5(b).

Step 3. According to inequality (10.20), compare the symmetrized graph with the modified complete graph $\mathbb{K}_3$ whose weights of edges $(i,\ j)$ are $\frac{a}{n} + \frac{D_i^c + D_j^c}{2n}$ for all $j > i$. The modified complete graph is shown in Figure 10.5(c).

Step 4. We use the technique of graph comparison to make $\mathbb{G}_{(b)} \succ \mathbb{G}_{(c)}$. For edge $(1, 3)$ that does not exist in graph (b), we choose the candidate path $(1, 2, 3)$ in the graph (b) for connecting vertices 1 and 3. Note that $L_{(1,3)} \preceq 2L_{(1,2)} + 2L_{(2,3)}$ from Lemma 10.3. Then we have:

$$
\begin{aligned}
L_{\mathbb{G}_{(c)}} &= \frac{a}{3} L_{(1,2)} + \left(\frac{a}{3} - \frac{\varepsilon_1}{6}\right) L_{(2,3)} + \left(\frac{a}{3} + \frac{\varepsilon_1}{6}\right) L_{(1,3)} \\
&\preceq \frac{a}{3} L_{(1,2)} + \left(\frac{a}{3} - \frac{\varepsilon_1}{6}\right) L_{(2,3)} \\
&\quad + \left(\frac{a}{3} + \frac{\varepsilon_1}{6}\right) \left(2L_{(1,2)} + 2L_{(2,3)}\right) \\
&= \left(a + \frac{\varepsilon_1}{3}\right) L_{(1,2)} + \left(a + \frac{\varepsilon_1}{6}\right) L_{(2,3)}.
\end{aligned}
$$

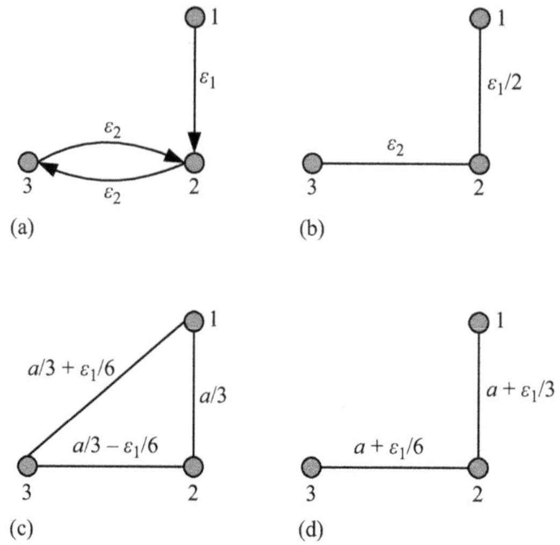

Figure 10.5 (a) A directed graph with three vertices. It has one directed edge (1, 2) and one bidirectional edge (2, 3), whose coupling strengths ε_1 and ε_2 are to be determined. (b) Its symmetrized graph. The directed edges in the graph (a) are replaced by the undirected edges with half weights in this graph. (c) The modified complete graph with three vertices whose edge weights are $\frac{a}{n} + \frac{D_i^c + D_j^c}{2n}$ for all $j > i$. (d) The graph has the same edge set as that of the graph (b), and satisfies $\mathbb{G}_{(d)} \succeq \mathbb{G}_{(c)}$

The terms in the last row only depend on the combinations of Laplacians of the edges in the graph (b), i.e., $L_{(1,2)}, L_{(2,3)}$. Graph (d) in Figure 10.5 shows the combinations. It follows that $\mathbb{G}_{(d)} \succeq \mathbb{G}_{(c)}$.

Step 5. In order to make $\mathbb{G}_{(b)} \succ \mathbb{G}_{(c)}$, we turn to make $\mathbb{G}_{(b)} \succ \mathbb{G}_{(d)}$. Thus, we obtain the following inequalities for edges $(1, 2)$ and $(2, 3)$:

$$\begin{aligned} \varepsilon_1/2 &> a + \varepsilon_1/3, \\ \varepsilon_2 &> a + \varepsilon_1/6. \end{aligned}$$

Therefore, one has $\varepsilon_1 > 6a$ and $\varepsilon_2 > a + \varepsilon_1/6$. These $\varepsilon_1, \varepsilon_2$ guarantee that $\mathbb{G}_{(b)} \succ \mathbb{G}_{(c)}$, i.e., inequality (10.20) holds. It returns the synchronization in the graph (a) in Figure 10.5.

In the above, we have used two simple examples to illustrate how to use graph comparison techniques to synchronize directed dynamical networks. For the complex networks with many nodes, we can implement Theorem 10.9 by algorithms and programming. Readers can refer to *Algorithm 1* in Reference 13 for the details.

A connected but *not* strongly connected directed graph can be decomposed into several locally strongly connected components that are unidirectedly connected in the original graph. (Note that strongly connected graphs cannot be decomposed any further.) We proposed *Algorithm 2* in Reference 13 in which one only needs to implement coupling strength allocations at local structures in a decomposable network. This algorithm can reduce the computational complexity for decomposable large-scale networks.

10.6 Conclusions

In this chapter, we have presented new ways to allocate coupling strengths using spectral graph theory in order to achieve synchronization in complex networks. The main idea is to bound the second-smallest eigenvalues of the Laplacian matrices associated with the given networks by comparing the corresponding network graphs to complete or other special graphs with the same vertex sets. The obtained results can simplify the computation and can be applied to growing networks. We have also presented algorithms to allocate coupling strengths to achieve synchronization in directed complex networks using graph comparison. By exploiting the symmetrization operation, we have dealt with the challenge that the Laplacian matrices associated with directed graphs are not guaranteed to be positive semidefinite anymore. The obtained algorithms can be applied to large but decomposable networks.

As the future work, we are interested in applying our results to practical engineered complex networks, such as the synchronization of generators in electric power grids and data fusion for signal processing in sensor networks.

Acknowledgments

The work was supported by the National Natural Science Foundation of China (No. 61403154), the Fundamental Research Funds for the Central Universities, China (HUST: 2014TS106), and the European Research Council Grant ERC-StG-2012-307207. Liu and Cao would also like to thank Prof. Zhiyun Lin at Zhejiang University, China, for his discussion and the corresponding support from the State Key Laboratory of Industrial Control Technology (Open Research Project ICT1510), Zhejiang University, China.

Appendix A: Comments on Assumption 10.1

Suppose that there are two oscillators i and j whose individual dynamics are $\dot{x}_i = f(x_i)$ and $\dot{x}_j = f(x_j)$, respectively. If the two oscillators are mutually coupled, and the coupling strength is ε_2, then their dynamics can be described by [20]:

$$\begin{aligned}\dot{x}_i &= f(x_i) - \varepsilon_2 P(x_i - x_j),\\ \dot{x}_j &= f(x_j) - \varepsilon_2 P(x_j - x_i).\end{aligned}$$

We use e to denote the difference of the states of the two oscillators, i.e., $e = x_i - x_j$. Then the dynamics of e satisfies:

$$\dot{e} = f(x_i) - f(x_j) - 2\varepsilon_2 P(x_i - x_j).$$

We construct a Lyapunov function $W(t) = (1/2)e^{\mathrm{T}} e$ in order to analyze the convergence of e, and obtain:

$$\dot{W} = (x_i - x_j)^{\mathrm{T}} \left(f(x_i) - f(x_j) - 2\varepsilon_2 P(x_i - x_j) \right). \tag{A.1}$$

Suppose that the two mutually coupled oscillators synchronize when their coupling strength is lower bounded by ε_2^*, i.e., the two oscillators can globally synchronize when $\varepsilon_2 > \varepsilon_2^*$. Through comparing the two inequalities (10.3) and (A.1), we have that the positive constant a in Assumption 10.1 is properly chosen when $a > 2\varepsilon_2^*$. Thus, one needs to first determine the bound ε_2^* when applying global synchronization criteria for complex networks proposed in this chapter.

The bound ε_2^* is determined by the dynamics of the individual oscillator and the inner coupling matrix P. In the following, we take Lorenz systems as an example to show how to determine the bound ε_2^*. Chen *et al.* [21] analyze theoretically the coupling bounds to synchronize two mutually coupled Lorenz systems under different inner coupling matrices P. The lemma below provides the bounds under two different P matrices.

Lemma 10.5 (Theorem 7 in Reference 21). *Lorenz system can be described by:*

$$\begin{cases} \dot{x} = \sigma(y - x), \\ \dot{y} = rx - y - xz, \\ \dot{z} = -bz + xy, \end{cases} \tag{A.2}$$

where σ, b, and r are system parameters. Consider two mutually coupled Lorenz systems with the inner coupling matrix P and the coupling strength ε_2.
(1) *Set* $P = \mathrm{diag}(1, 0, 0)$. *The two mutually coupled Lorenz systems with any initial conditions can synchronize as $t \to +\infty$, if the coupling strength satisfies $\varepsilon_2 > (1/2)(MR^2/4b - \sigma)$, where $M = \max\{1, b\}$ and:*

$$R^2 = \begin{cases} \dfrac{(\sigma + r)^2 b^2}{4(b-1)}, & \sigma \geq 1,\ b \geq 2, \\ (\sigma + r)^2, & \sigma > b/2,\ b < 2, \\ \dfrac{(\sigma + r)^2 b^2}{4\sigma(b-\sigma)}, & \sigma < 1,\ b \geq 2\sigma. \end{cases}$$

(2) *Set* $P = \mathrm{diag}(0, 1, 0)$. *The two mutually coupled Lorenz systems with any initial conditions can synchronize as $t \to +\infty$, if the coupling strength satisfies $\varepsilon_2 > (R^2/8\sigma) - 1/2$.*

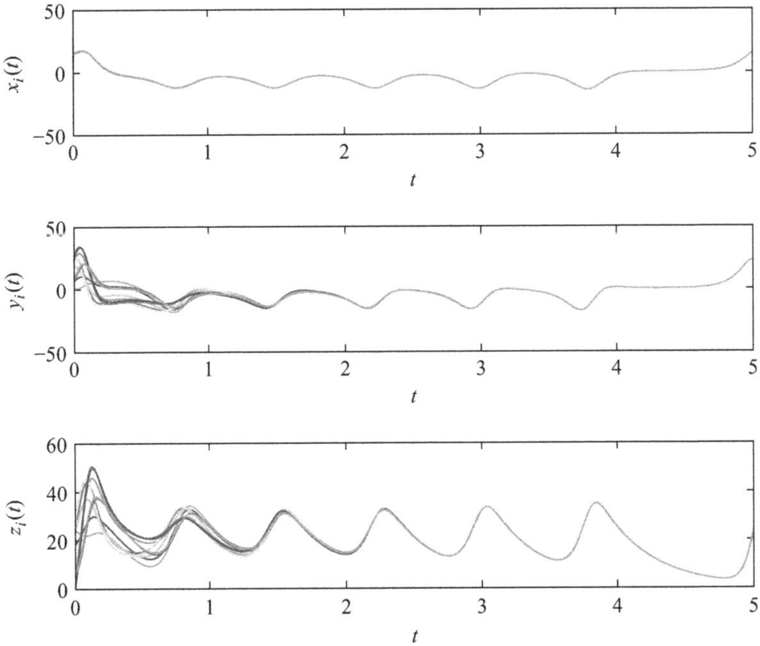

Figure 10.6 The states of the coupled Lorenz oscillators (B.1)

Appendix B: Numerical simulation

In this appendix, we provide a numerical example to validate Theorem 10.6. In this simulation, we consider the network in (10.1) consisting of n Lorenz systems coupled through the first components of their states. To be specific, the dynamics of the network are given by:

$$\begin{cases} \dot{x}_i = \sigma(y_i - x_i) + \sum_{j=1}^{n} \varepsilon_{ij}(t) x_j, \\ \dot{y}_i = r x_i - y_i - x_i z_i, \\ \dot{z}_i = -b z_i + x_i y_i, \end{cases} \tag{B.1}$$

and the inner coupling matrix is $P = \text{diag}\{1, 0, 0\}$. The parameters in (B.1) are set to be $\sigma = 10$, $r = 25$, $b = 8/3$. From Appendix A, the quantity $a > 2\varepsilon_2^* = 2 \times 158.33 = 316.66$*. So we take $a = 317$ in our simulation. We choose the fractal graph with $n = 10$ vertices on the left of Figure 10.1 to be the network topology used in the

*This value is less restrictive than the one in Reference [12, Section VI] using the result of Reference [9, Appendix A] in which $a > \frac{b(b+1)(r+\sigma)^2}{16(b-1)} - \sigma = 439.17$.

simulation. The bounds for coupling strengths have been calculated and given by (10.12), and we set the coupling strengths $\varepsilon_{(1,2)} = \varepsilon_{(1,3)} = \varepsilon_{(1,4)} = 5a$, and the coupling strengths for the other edges $2a$. The initial states are randomly chosen from $[0, 30]$. The three subfigures in Figure 10.6 show the state of the coupled network (B.1) in its x, y, z-dimension, respectively. From Figure 10.6, one can see that the coupled Lorenz oscillators asymptotically synchronize by adopting the coupling strength allocation (10.12) obtained according to Theorem 10.6. The simulation results illustrate the correctness of the theoretical analysis in Section 10.4.3.1.

References

[1] Pecora L M, Carroll T L. 'Master stability functions for synchronized coupled systems'. *Physical Review Letters*. 1998, 80: 2019–2112.

[2] Wu C W, Chua L O. 'Synchronization in an array of linearly coupled dynamical systems'. *IEEE Transactions on Circuits and Systems-I*, 1995, 42: 430–447.

[3] Strogatz S H. *SYNC: The Emerging Science of Spontaneous Order*. New York, NY: Hyperion, 2003.

[4] Newman M E J. *Networks: An Introduction*. Oxford, UK: Oxford University Press, 2010.

[5] Mei S, Zhang X, Cao M. *Power Grid Complexity*. Berlin: Springer-Verlag, 2011.

[6] Cao M, Morse A S, Anderson B D O. 'Reaching a consensus in a dynamically changing environment: A graphical approach'. *SIAM Journal on Control and Optimization*. 2008, 47: 575–600.

[7] Wu C W. *Synchronization in Complex Networks of Nonlinear Dynamical Systems*. Singapore: World Scientific, 2007.

[8] Wu C W. 'Perturbation of coupling matrices and its effect on the synchronizability in arrays of coupled chaotic systems'. *Physics Letters A*. 2003, 319: 495–503.

[9] Belykh V N, Belykh I V, Hasler M. 'Connection graph stability method for synchronized coupled chaotic systems'. *Physica D*. 2004, 195: 159–187.

[10] Belykh I V, Belykh V N, Hasler M. 'Generalized connection graph method for synchronization in asymmetrical networks'. *Physica D*. 2006, 224: 42–51.

[11] Belykh I V, Belykh V N, Hasler M. 'Synchronization in asymmetrically coupled networks with node balance'. *Chaos*. 2006, 16: 015102.

[12] Liu H, Cao M, Wu C W. 'Coupling strength allocation for synchronization in complex networks using spectral graph theory'. *IEEE Transactions on Circuits and Systems-I*. 2014, 61(5): 1520–1530.

[13] Liu H, Cao M, Wu C W, Lu J A, Tse C K. 'Synchronization in directed complex networks using graph comparison tools'. *IEEE Transactions on Circuits and Systems-I*. 2015, 62(4): 1185–1194.

[14] Guattery S, Leighton T, Miller G L. 'The path resistance method for bounding the smallest nontrivial eigenvalue of a Laplacian'. *Combinatorics, Probability and Computing*. 1999, 8: 441–460.

[15] Guattery S, Miller G L. 'Graph embeddings and Laplacian eigenvalues'. *SIAM Journal on Matrix Analysis and Applications*. 2000, 21(3): 703–723.

[16] Kahale N. 'A semidefinite bound for mixing rates of Markov chains'. *Random Structures and Algorithms*. 1998, 11: 299–313.

[17] Spielman D A. *Spectral graph theory and its applications* [online]. Lecture Notes. 2004. Available from http://www.cs.yale.edu/homes/spielman/eigs/ [Accessed Dec 2012].

[18] Spielman D A. 'Spectral graph theory'. In *Combinatorial Scientific Computing*. Naumann U, Schenk O. (eds.). 1st edn. Boca Raton, FL: CRC Press, 2012.

[19] Godsil C, Royle G. *Algebraic Graph Theory*. New York, NY: Springer-Verlag, 2001.

[20] Wu C W, Chua L O. 'A unified framework for synchronization and control of dynamical systems'. *International Journal of Bifurcation and Chaos*. 1994, 4(4): 979–998.

[21] Chen J, Lu J A, Wu X. 'Bidirectionally coupled synchronization of the generalized Lorenz systems'. *Journal of Systems Science and Complexity*. 2011, 24(3): 433–448.

Chapter 11

Experimental studies on reconfigurable networks of chaotic oscillators

Massimiliano de Magistris[1], *Carlo Petrarca*[1] *and Soudeh Yaghouti*[1]

This chapter recalls a collection of experimental results on reconfigurable topology networks of oscillators, with Chua's circuits as chaotic dynamical nodes. Collective behaviours arising from network node interactions, such as global synchronization, emergence of clusters, patterns and waves formation, have been observed in a setup expressly developed [1–7].

The motivation for considering such a dedicated setup comes first from the need of demonstrating the robustness of the appearance of some complex phenomena, expected from theory, to the experimental uncertainties and realization tolerances. Second, the simulation of large complex networks with chaotic nodes is time consuming and computationally costly, becoming often unaffordable when such networks have to be scanned in significant parameter's variation intervals.

A key issue in the considered realization is the implementation of a general structure link network, allowing to choose arbitrarily the coupling type and strength as well as the interconnecting topology among the nodes. Another important issue is the data acquisition and processing section, which allows high rate simultaneous recording of a large number of variables along with their real-time analysis. The entire setup is controlled with a user-friendly LABVIEW interface, allowing largely automated experimental campaigns. In this chapter, after summarizing the setup main features from a design and implementation perspective, and shortly resuming some notation and theoretical background, a collection of experimental results on a large set of network configurations, with various topologies from 4 up to 32 nodes, is presented.

11.1 Introduction

Networks of chaotic oscillators and their complex dynamics have been widely considered in the literature, e.g. References 8–10. In particular, networks with Chua's

[1]Department of Electrical Engineering and Information Technology, University of Naples FEDERICO II, Italy

circuits as nodes have been commonly used to study synchronization emerging in networks of dynamical systems [11,12], as well as to highlight the occurrence of fundamental collective nonlinear phenomena [13].

Most of the available results are of theoretical and numerical nature, whereas relatively few experimental studies have been presented [1,2,4–7,11,12,14–21]. This is mainly due to: (i) the large availability of reliable and affordable simulation tools in the circuits' area and (ii) difficulties in the prototypal realization and operation of networks with a large number of nodes. Indeed, most of the available experimental results have been obtained for networks of two or three nodes [11,12,19]. A larger scalable setup has been proposed in Reference 20 to study higher dimensional networks with a nearest neighbour structure in the context of cellular neural networks (CNNs) but little is available for networks with a general and reconfigurable structure. Most experimental setups (realized as electrical circuits) have been dedicated to investigate the synchronization phenomenon, possibly in the chaotic regime. Fewer examples are available for experimental efforts in the study of other interesting phenomena such as phase synchronization, clustering formation, the emergence of patterns and waves within the ensemble of oscillators. None of them, to our knowledge, has been realized in the form of a reconfigurable network, in terms of variable structure of connections (type, topology and strength) and setting of node's dynamics.

In this chapter, we describe the structure, implementation and main results of a recent developed setup, aimed at experimenting on complex networks of chaotic oscillators with general and reconfigurable structure [1,2,4,6]. Its peculiarity, with respect to other experimental arrangements, resides in: (i) the full and direct control of the network structure, the link type and coupling strength; (ii) the possibility of selecting each node main parameters and (iii) a modular high rate multichannel acquisition section for the simultaneous analysis of the high number of variables under observation. The setup allows to carry out general experiments on relatively large networks of nonlinear oscillators in reconfigurable structure, so realizing a general-purpose analogue simulator of such networks.

The chapter is organized as follows: after this introduction, Section 11.2 first resumes fundamentals on the reference models of nonlinear oscillators, with introducing the main variables and notations and giving a description of the experimental setup structure and implementation. Section 11.3 briefly gives fundamental analysis tools for complex networks of oscillators, enumerates several kinds of noticeable emerging dynamics and resumes the main theoretical tools for their identification and prediction. Section 11.4 reports a collection of experimental results on complete synchronization (both in presence of diffusive or dynamical links), examples of clustering, observation of pattern and waves in a wide set of topologies ranging from 4 to 32 nodes.

11.2 Realization of a network of nonlinear oscillators with linear coupling

As we just mentioned in the introduction, most of the available investigations on networks of nonlinear oscillators have been of numerical nature and just few experimental realizations are available, normally aimed at investigating specific topologies or validating some particular theoretical results. As well, it is of significant interest the possibility to experiment on networks of nonlinear oscillators like Chua's circuits with arbitrary connection topologies, allowing in this way the transition from regular topologies to arbitrary ones, possibly in a dynamical way, and with real-time system response. Such a complex network, with complexity arising from both the node dynamic and the connection topology, link type and strength, has to be modelled, designed and realized.

11.2.1 Designing a reconfigurable complex network of nonlinear oscillators

Without significant loss of generality, a general ensemble of N coupled nonlinear oscillators is satisfactorily modelled by a set of differential equations in the following form [9,10]:

$$\dot{\mathbf{x}}_i = \mathbf{f}_i(\mathbf{x}_i) + \sum_{j=1}^{N} g_{ij}\Gamma\mathbf{x}_j, \qquad i = 1, 2, \ldots, N, \tag{11.1}$$

where $\mathbf{x}_i \in \mathbb{R}^M$ is the state vector of node i, $\mathbf{f}_i(\mathbf{x}_i) : \mathbb{R}^M \rightarrow \mathbb{R}^M$ is the vector field describing the dynamics of the node i, $G = (g_{ij}) \in \mathbb{R}^{N\times N}$ is a coupling matrix (possibly a functional one for the case of dynamical links) which fully defines the effects of coupling between the nodes, whereas $\Gamma \in \mathbb{R}^{M\times M}$ is a constant diagonal binary matrix defining the choice of the coupled state variables.

Such model implicitly assumes linear (possibly diffusive) coupling among nodes. This choice, at the price of slightly restricting generality, allows a very useful description of the whole system in terms of sub-systems. In fact, in this case, the entire set of dynamical nonlinear nodes results to be interconnected by a linear N-pole system.

We pursued a circuit realization of the whole system described by (11.1) that allows setting (i) the parameters of the node's systems determining their own stand-alone dynamics and (ii) the control of the (linear) interconnection network. As system nodes, a classical robust implementation of Chua's circuits has been considered; they are universally recognized as paradigmatic systems in complex and chaotic dynamics. As for the interconnecting network, we have adopted a general N-pole realization, practically based on a switched set of impedances. A general schematic of the system, including the control and data acquisition section, is shown in Figure 11.1.

The setup consists of an ensemble of nominally identical Chua's nodes, with individually settable parameters, linked via adjustable linear impedances with a generic reconfigurable network structure. Links can be diffusive or dynamical, directed or undirected. From the topological point of view, the network can vary from a fully connected structure, to any possible combination of links, including regular lattices (as those typically considered for the CNN's case). The key design solution to achieve the required reconfigurability of the whole system, for the interconnection N-pole

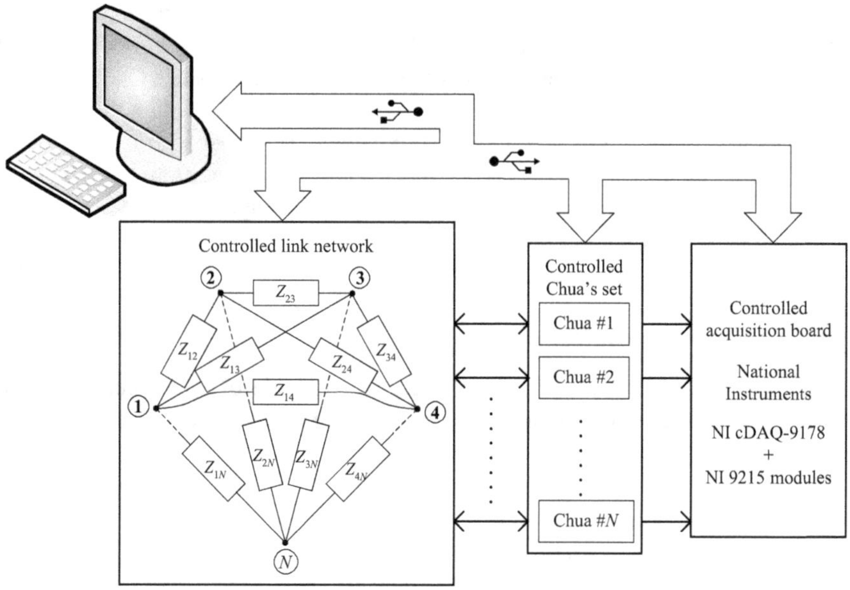

Figure 11.1 General schematic of the reconfigurable network of nonlinear oscillators

and the system's nodes, is the introduction of USB-controlled switches, allowing to set independently:

- the properties of each link, so letting to define in a very general manner the features of the interconnections among nodes (in terms of topology, nature of the links, link weights, etc.);
- the value of the parameters of each node (R in our Chua implementation), so that individual nodes can be made to exhibit a wide range of possible stand-alone dynamics (equilibria, limit cycles or chaos).

Finally, a modular USB multi-channel data acquisition system completes and integrates the experimental setup, allowing the full monitoring of the system state variables. We wish to emphasize that the resulting setup, as shown in Figure 11.1, is characterized by (i) simplicity and robustness of the implementation; (ii) complete reconfigurability of the interconnection topology; (iii) easy scalability to high number of nodes and (iv) possibility of running fully automated measurement campaigns.

In the following, we will describe the setup in terms of sub-systems and functionalities. Some more details on its actual implementation can be found in References 1,2,4,6.

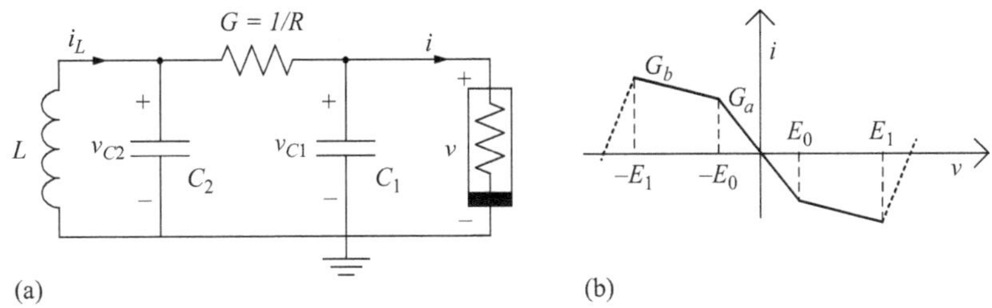

Figure 11.2 *(a) Schematics of the Chua's circuit with reference variables. (b) Piecewise linear characteristic of the "Chua's diode"*

11.2.2 Chua's circuits as system nodes

The Chua's circuit is universally recognized as simple and economical electronic implementation of a nonlinear oscillator, exhibiting bifurcations, complex dynamics and chaos [22], and has been extensively studied from a theoretical, numerical and experimental point of view (see Reference 23 and references therein). Moreover, many theoretical and numerical papers report studies on networks of Chua's circuits, demonstrating the large interest in the topic [24]. These reasons moved us towards its consideration for the realization of our reconfigurable network of nonlinear oscillators.

For the sake of clarity, we report here the classical schematic of the circuit, with symbols, variables and notations used throughout the chapter (Figure 11.2). In particular, in Figure 11.2(a), the electrical schematic of the ideal circuit is reported, whereas in Figure 11.2(b), the characteristic of the nonlinear piece-wise asymptotically passive one-port, normally referred as *Chua's diode* is reported. For the circuit equations, fundamental dynamics and further details, the reader is referred to Reference 23 and references therein.

11.2.3 A reconfigurable linear N-pole as interconnection network

A key point in the realization of our reconfigurable network of oscillators follows after the recognition that the linking term, as expressed in (11.1), can be viewed from a circuit implementation perspective in terms of a linear N-pole. If we consider, as first example, the case of bilateral diffusive coupling, then the coupling matrix G is a constant symmetric matrix. Assuming the linked variables as voltages (as for example the v_{C1} for the Chua's circuits), we immediately recognize the matrix $G = (G_{ij})$ as the admittance matrix of a general linear resistive N-pole, whose instances represent the realization of specific interconnection topologies. The properties of the coupling matrix G can then be easily inferred from those of linear N-poles, that for the case of symmetric (undirected) links are:

$$\sum_{j=1}^{N} G_{ij} = 0 \qquad \forall i \quad \Rightarrow \quad G_{ii} = -\sum_{\substack{i=1 \\ i \neq j}}^{N} G_{ij}. \tag{11.2}$$

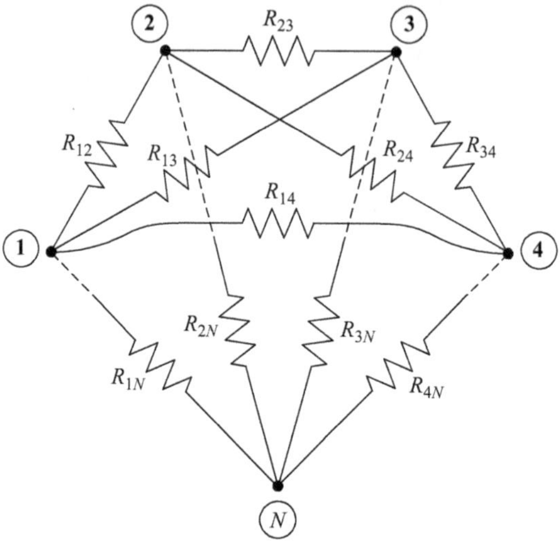

Figure 11.3 General N-nodes link network for the case of diffusive undirected links (for voltage variables)

In such symmetric case, the number of free entries in G is $N(N-1)/2$. The considered N-pole can be synthesized with the general scheme of Figure 11.3, such realization being general (although not unique), with the relationship between the link resistors values and the matrix elements G_{ij} given by:

$$G_{ij} = \frac{-1}{R_{ij}}, \qquad i \neq j. \tag{11.3}$$

If we restrict to unweighted networks, where by definition all the values of the linking terms are equal (so that $R_{ij} = R_{ji} = R_{link} \quad \forall\, i \neq j$), the structure of the matrix G can be further simplified and expressed in terms of a global coupling coefficient c and a simplified coupling matrix L as:

$$G_{ij} = cL_{ij}, \tag{11.4}$$

where it is easy to recognize $L = (l_{ij}) \in \mathbb{R}^{N\times N}$ as the *Laplacian* matrix that defines the coupling configuration and is associated with the wiring of the network graph [25].

The description of the link network in terms of the coupling matrix G, and its corresponding circuit implementation, is completely general and encompasses the many possible realizations of equivalent interconnection networks. In other terms, interconnection networks are different only if they differ in the corresponding admittance matrices regardless of equivalent physical realizations. This bypasses the problem of deriving alternative equivalent interconnection schemes (see, e.g. the "common point" link connection reported in Reference 20), as well as the problem of distinguishing locally or globally connected networks.

Finally, one has to note how such description and its corresponding realization can be straightforwardly extended to the case of directed links, affecting the symmetry of matrix G, by simply considering directed impedances (practically implemented by means of buffer configuration op-amps).

11.2.4 The actual system implementation

In this section, we give some practical detail on the physical realization of the experimental setup and its sub-systems, in order to allow the reader to be informed on the main implementation choices as well as on the actual ranges of the physical variables. More details are given in References 1,2,6.

11.2.4.1 Settable Chua's circuit implementation

For the realization of the Chua's oscillators as the network nodes, several well-known alternatives are possible, whose differences reside basically in the adopted components and technology. A well-known discrete robust realization is that introduced by Kennedy [26] with the further improvement of an inductor-less version as given by Aguirre *et al.* [27] (the inductor is equivalently realized by means of a capacitor and a gyrator with operational amplifiers). Such realization gives the following advantages: (i) tolerances on components and parasitic parameters are more easily controlled because of the absence of a real inductor; (ii) all state variables v_{C1}, v_{C2}, i_L are directly available in terms of some properly defined voltage, for coupling and measuring variables and (iii) it is possible to easily scale the dynamical parameters (by changing only resistors values) to low operating frequencies.

The setting of the single-node dynamics is pursued by digitally controlling parameters (at least one) in the single Chua. For the sake of simplicity, it has been implemented by means of a proper set of digitally controlled switched resistors, via a USB interface, in this way realizing an equivalent variable resistor R, within a prescribed range, for each Chua's circuit. In References 1,2, the realized circuit is fully described.

The possibility of setting the Chua's nodes dynamics (when uncoupled) is important from different perspectives. On the one hand, we can study the network properties referring to different basic regimes, as for example periodic or chaotic. Second, it can be used to experimentally analyse the collective behaviour emerging from ensembles of nodes characterized by different regimes when uncoupled, as for example their synchronization. Moreover, this system feature is useful when the analysis of networks of non-identical nodes is pursued.

11.2.4.2 Linear link network implementation

Based on the N-pole synthesis as given from (11.3), multi-valued discrete impedances (R, C) have been assembled with digitally controlled switches, enabling their insertion in a series–parallel scheme. This allows to set the values of such linking impedances within a prescribed range, according to the available number of steps. For example, for the case of resistive links, each link resistance R_{link} is realized as a series combination of eight resistance values; the desired value is obtained by setting the

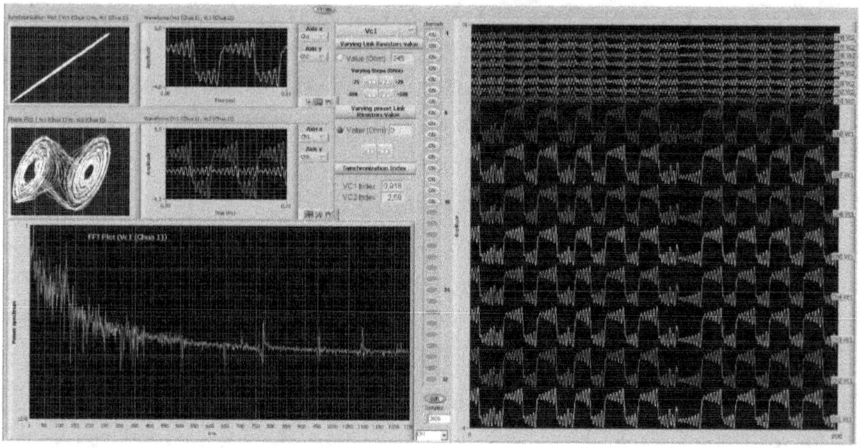

Figure 11.4 *Typical graphical output of the real-time acquisition system*

switches, which are controlled by a USB interface cascaded to a logic port and a decoder.

For the reported experiments, the resistor's module has been designed to scan the range 0–12 750 Ω in $2^8 = 256$ uniform steps of 50 Ω. Moreover, an "offset" precision resistor can be optionally placed in series, in order to widen the range of values assumed by the coupling resistors, while keeping the same accuracy.

Dynamic link coupling impedances have been implemented by capacitor sets C_{link} connected in parallel to resistors R_{link} [5]. The realization of the switched printed circuit board (PCB) is in principle quite similar to the one described above for resistive links, although series connections have to be replaced by parallel ones. For the reported experiments, each link capacitor value is obtained as a parallel of eight values. We considered the range 0–520 pF in steps of 19 pF. Each capacitor is placed in parallel to the corresponding link resistor in order to implement a *RC* dynamic link between the nodes of the network.

11.2.4.3 Data acquisition real-time analysis and control

The system has been designed, in terms of the isolated node's dynamics spectral range, in order to guarantee the possible simultaneous acquisition of a large number of variables and their processing in real time. A commercial National Instrument multi-channel USB Data Acquisition system (NI CompactDAQ with NI9215 modules) has been considered as natural choice for the developed setup, allowing a maximum of 64 channels per single USB unit. The experiment control and data acquisition is managed by a specific piece of software developed within the LABVIEW environment. Data batches are saved either automatically or at the operator request. In Figure 11.4, a typical graphical output of the real-time acquisition of the recorded signals is shown.

11.2.4.4 Tolerances, non-idealities and parasitics

The problem of admittable tolerances in the physical realization has been considered at the experiment design stage [2,6]. In particular, exhaustive numerical simulations were carried out to study effects on synchronization bounds and thresholds as a function of tolerances on components. More specifically, a preliminary Monte-Carlo analysis was carried out with SPICE simulator in some networks with a relatively small number of nodes in order to determine admittable tolerances on components. For an assigned link topology, emergence of synchronization in a small network of four nodes has been studied by simulation, as a function of the link resistance values, in the presence of passive components values which were randomly selected within assigned tolerance bounds. In the synchronization regime, after defining a proper bound in the difference of the corresponding variables, the admittable components tolerances have been estimated. After a conservative evaluation of such preliminary simulations, passive components for the physical implementation were carefully selected with tolerances below 1%, as reported in Reference 2.

11.3 Collective behaviours and relative analysis tools

A homogeneous complex network can be defined as an ensemble of N dynamical systems (assumed to be nominally identical), that are interconnected in some way. Complexity arises both from the structure and nature of the interconnection network, as well as from the nonlinear behaviour of the nodes [8,28]. As a result, the ensemble can exhibit rich complex dynamics and new emerging phenomena, when compared to the behaviour of each single dynamical unit detached from the rest.

11.3.1 Synchronization

Synchronization of chaotic oscillatory networks has been widely studied by several authors and from different perspectives [24,29–34]. Various concepts of synchronization have been introduced since it is revealed in many different ways [29]. In particular, a network of N oscillators is said to exhibit *complete synchronization* if their instantaneous states are exactly coincident ($\mathbf{x}_i(t) = \mathbf{x}_j(t)\ \forall i,j$). From a practical point of view (possibly non-identical oscillators), such concept is too restrictive, and the more realistic definition is *bounded synchronization* which is given as

$$\|\mathbf{x}_i(t) - \mathbf{x}_j(t)\| \leq \varepsilon \qquad \forall i,j(i \neq j), \qquad \forall t \geq t^*, \tag{11.5}$$

for some $\varepsilon > 0$, where $\mathbf{x}_k(t)$ indicates the state vectors of the generic kth node, $\|.\|$ is the Euclidean vector norm and t^* is the time after which any possible transient is considered as extinguished. *Phase synchronization* is defined as locking of the phases, while the amplitudes can remain chaotic and uncorrelated [35]; *lag synchronization* occurs when the state variables become identical after applying a specific time shift.

11.3.1.1 Measure of synchronization and cross-correlation

After these definitions, in real networks, it is crucial to define some figures measuring the degree of synchronization, and different possible approaches have been proposed. Any possible measure of the synchronization degree is basically based on the comparison between two time series but, when necessary, must be reformulated in order to take into account the coexistence of N nodes with M states each.

A quite general way of measuring the degree of complete/bounded synchronization between two corresponding states is based on the calculation of the cross-correlation index $I_{cc}(x_i, x_j)$ of the corresponding discrete time series $x_i(k)$ and $x_j(k)$, where $k = 1, 2, 3, \ldots, N_s$, being N_s the number of samples:

$$I_{cc}(x_i, x_j) = \frac{\sum_{k=1}^{N_s} [(x_i(k) - \overline{X_i})][(x_j(k) - \overline{X_j})]}{\sqrt{\sum_{k=1}^{N_s} [x_i(k) - \overline{X_i}]^2 [x_j(k) - \overline{X_j}]^2}}, \tag{11.6}$$

where $\overline{X_p} = \frac{1}{N_s} \sum_{k=1}^{N_s} x_p(k)$, $p = i, j$ is the mean value of the sequence $x_p(k)$. Such an index is bounded in $[-1, 1]$: it equals 1 when the two sequences are identical, it equals -1 if the signals are exactly opposite, and 0 when the sequences are uncorrelated.

In the case of N discrete signals x_p with $p = 1, \ldots, N$, the definition of the index can be extended by first calculating the averaged trajectory $\overline{x(k)}$ as:

$$\overline{x(k)} = \frac{1}{N} \sum_{p=1}^{N} \left(x_p(k) - \overline{X_p}\right), \tag{11.7}$$

then defining N correlation coefficients $I_{cc}(x_p, \overline{x})$ and calculating their mean value:

$$I_{cc} = \frac{1}{N} \sum_{p=1}^{N} I_{cc}(x_p, \overline{x}). \tag{11.8}$$

A first picture of such measure of the synchronization level is given in Figure 11.5(a), where we report results from accurate SPICE simulations for the case of a $N = 4$ network of Chua's circuits with chaotic double scroll trajectories, with all-to-all link topology. In particular, more expressive quantity $(1 - I_{cc})$ is reported in a log–log plot as a function of the link resistance R_{link}, for the case of identical nodes and non-identical nodes with 1% tolerances on randomly chosen passive component values. Both the curves share a common sharp transition from the bounded synchronization to uncorrelated node's dynamics, so defining a threshold coupling strength for complete (bounded) synchronization. Such threshold results intimately related to the stability of synchronous solution in the presence of transverse perturbations, as it will be shown later. It is interesting to note that for identical circuits, the synchronization level increases almost indefinitely with increasing coupling strength (decreasing link resistance), leading to real complete synchronization. This does not happen when

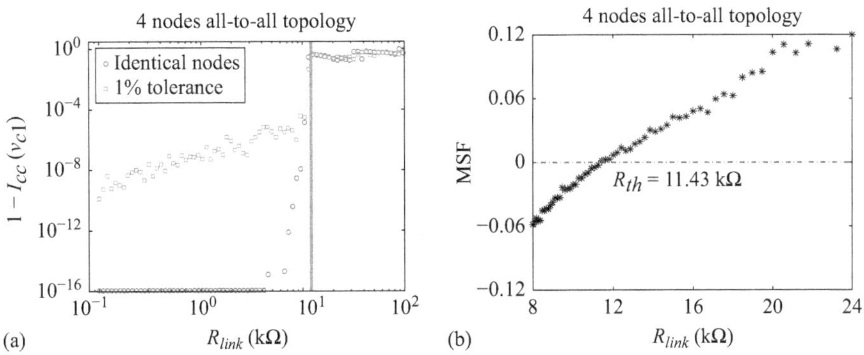

Figure 11.5 Four nodes all-to-all topology: (a) accurate simulation of $(1 - I_{cc})$ *index as a function of coupling resistance for the two cases of identical nodes and 1% tolerance, with the MSF threshold as solid vertical line; (b) MSF as a function of the coupling resistance* R_{link} *(note that the behaviour is reciprocal with respect to coupling strength c)*

non-identical circuits with realistic component tolerances are considered, and only bounded synchronization is achieved.

11.3.1.2 Synchronization thresholds and master stability function

The problem of studying stability of the synchronous solution in a network, described by (11.1), as a function of the coupling strength and the network topology, has been addressed with different approaches [10]. The most known among them is by far the so-called master stability function (MSF) [32], valid for the case where nodes are identical and coupled through an arbitrary network topology. Following such approach, the threshold value of the coupling strength for the stability of synchronization is estimated as a function of the eigenvalues λ_i, which are the spectrum of the Laplacian matrix L and the node dynamics. It can be briefly resumed as follows: it is first assumed the existence of an invariant synchronous solution $\mathbf{x}_s = \mathbf{x}_1(t) = \mathbf{x}_2(t) = \cdots = \mathbf{x}_N$, which is the solution of an isolated node, i.e. $\dot{\mathbf{x}}_s = \mathbf{f}(\mathbf{x}_s)$; then, an infinitesimal perturbation $\delta\mathbf{x}_i = \mathbf{x}_i - \mathbf{x}_s$ is considered, and its stability is studied by linearizing perturbation dynamics around the synchronous solution. A further diagonalization allows to obtain a set of uncoupled equations given by:

$$\delta\dot{\mathbf{y}}_i = [Jf(\mathbf{x}_s) - c\lambda_i\Gamma]\delta\mathbf{y}, \quad i = \{1, \ldots, N\}, \tag{11.9}$$

where $0 = \lambda_1 < \lambda_2 < \cdots < \lambda_N$ are the eigenvalues of coupling matrix L with $\delta\mathbf{y}_i$, $i = \{2, \ldots, N\}$ representing the perturbations transverse to the synchronous manifold. The sign of maximum Lyapunov exponent of the equation determines stability of the i_{th} mode corresponding to λ_i. Note that for $\lambda_1 = 0$ the Lyapunov exponents correspond to the ones of an isolated node. The threshold coupling strength c_{th} can be obtained by considering a fixed interval $k \in [k_{min}, k_{max}]$ where $k = c\lambda_i$ for $i = \{2, \ldots, N\}$ and calculating the corresponding Lyapunov exponents. In fact the sign of the largest

Lyapunov exponent as a function of k defines in general the regions where the synchronous trajectory is stable or unstable. Although not restrictive for the discussed setup, we consider in the following experiments the coupling of Chua's oscillators exclusively through the v_{C1} state variable. For such specific coupling, the MSF results to be monotonically decreasing as a function of k, yielding a single synchronization threshold at the crossing point $\Lambda_{max}(k_{th}) \leq 0$ given by $c_{th} = k_{th}/\lambda_2$ [24].

We report here as an example, for the case of four nodes with all-to-all connections, the evaluation of the MSF as a function of the coupling strength (Figure 11.5(b)). Such value is also reported in Figure 11.5(a) as a solid vertical line, to be compared with the results of the simulations, showing the excellent capability of the MSF approach to get the real stability threshold for synchronization also for non-identical node cases.

In Section 11.4, the synchronization thresholds calculated with MSF for the reported network configurations will be compared directly with experimental results.

11.3.2 Clustering

Global synchronization in a network of chaotic oscillators, although impressive, is neither the only nor the most significant collective behaviour. Important and intriguing phenomena typically emerge within the region ranging from the complete synchronization to the uncorrelated dynamics, which are rich of information and application potential. Therefore, necessarily, characterization of synchronization transition and understanding the condition under which nodes in the network correlate in different patterns is an important challenge which is still an open question. According to the recent findings in complex network research, this transition is strongly depending on the network structure [36–38]. A particular emergent dynamic is cluster synchronization, when nodes in the network partition in clusters, such that nodes belong to the same cluster are highly correlated and weakly or not correlated with nodes from other clusters [39].

A quite interesting feature to investigate in this area is the clustering driven by topological symmetries in the network, where the dynamics of nodes which are paired by some structural symmetry show capability to synchronize. Following these findings, different approaches have been proposed to study clusters and their stability. However, the problem has not been addressed sufficiently by means of experimental studies. In Section 11.4.3, we will show results of some experiment on topological clustering, as obtained with the considered setup in a simple network of five nodes. In this way, we start addressing this relatively new research area from experimental point of view, demonstrating at the same time the large experimental potential of the discussed implementation.

11.4 Experimental results and validation of theoretical predictions

To illustrate the flexibility and viability of the setup, we report next a set of representative experiments showing a collection of different complex phenomena. Its unique feature of allowing to change the structure of interconnections among the nodes has been used for implementing a wide set of topologies, as for example those shown in Figure 11.6 [6], investigating their effects on the node's collective behaviour.

11.4.1 Experiments on synchronization with diffusive links

As first issue, we report an extensive evaluation of the onset of complete synchronization in the case of nominally identical chaotic nodes and bilateral (undirected) diffusive coupling.

A typical example of synchronized dynamics onto a double-scroll chaotic attractor, as directly observable in the setup, is shown in Figure 11.7, for the case of 16 nodes in a "ring" topology with $R_{link} = 159\ \Omega$. Figure 11.7(a) shows synchronized waveforms of the state variable v_{C1}; Figure 11.7(b) shows the $v_{C1} - v_{C1}$ state plane (double scroll attractors) for each node; Figure 11.7(c) shows the synchronization plots for v_{C11} vs. v_{C1k}, $k = 1, \ldots, N = 16$; Figure 11.7(d) shows the Poincaré sections calculated at $t^* : v_{C2k}(t^*) = \text{mean}[v_{C2k}(t)]$.

The problem of studying synchronization thresholds and bounds, as a function of parasitic effects and parameters mismatch in practical implementations has been treated in References 2,40, and it is still an open research problem from a general point of view. We have numerically investigated the effects of nodes mismatch in the simple case of bilateral, diffusive (resistive) coupling via v_{C1} in a four nodes all-to-all connected network, evaluating the degree of synchronization as a function of the coupling resistances for a fixed (and realistic) tolerance level. In Figure 11.8(a), the comparison of the $(1 - I_{cc})$ index between the simulation with 1% tolerance and the experimental values is shown. It is evident how both the threshold value and the corresponding jump in cross correlation are highly related in the experimental and simulated networks, as well as the behaviour before and after the threshold value.

For the same network (four nodes all-to-all coupling), we report in Figure 11.8(b) the experimentally recorded values of the index $(1 - I_{cc})$ as functions of the coupling resistance R_{link} for each state variable, v_{C1}, v_{C2}, i_L. As expected, strong synchronization is observed at low resistance values; in this range, also, the synchronization of variables which are not directly coupled as v_{C2}, i_L is still very good, although with less cross correlation (index $(1 - I_{cc})$ never goes above 10^{-4}). Such value of cross correlation can then be considered the maximum reachable from synchronized variables which are not directly connected by a link.

An extensive study of the transition to synchronous regimes has been carried out for all the topologies shown in Figure 11.6 [6]. The results are collected in Figure 11.8(c) and (d) for networks of four and eight nodes. On the same graphs, the theoretical thresholds as predicted by the MSF model for each case are also reported (vertical lines). This experimentally confirms the prediction ability of the MSF approach on a larger number of networks with different structures and size than is typically considered in the existing literature. Also some robustness is demonstrated for the MSF approach that, despite being derived under the assumption of all nodes being identical, still gives good predictions at least when they are within a small tolerance range (1%).

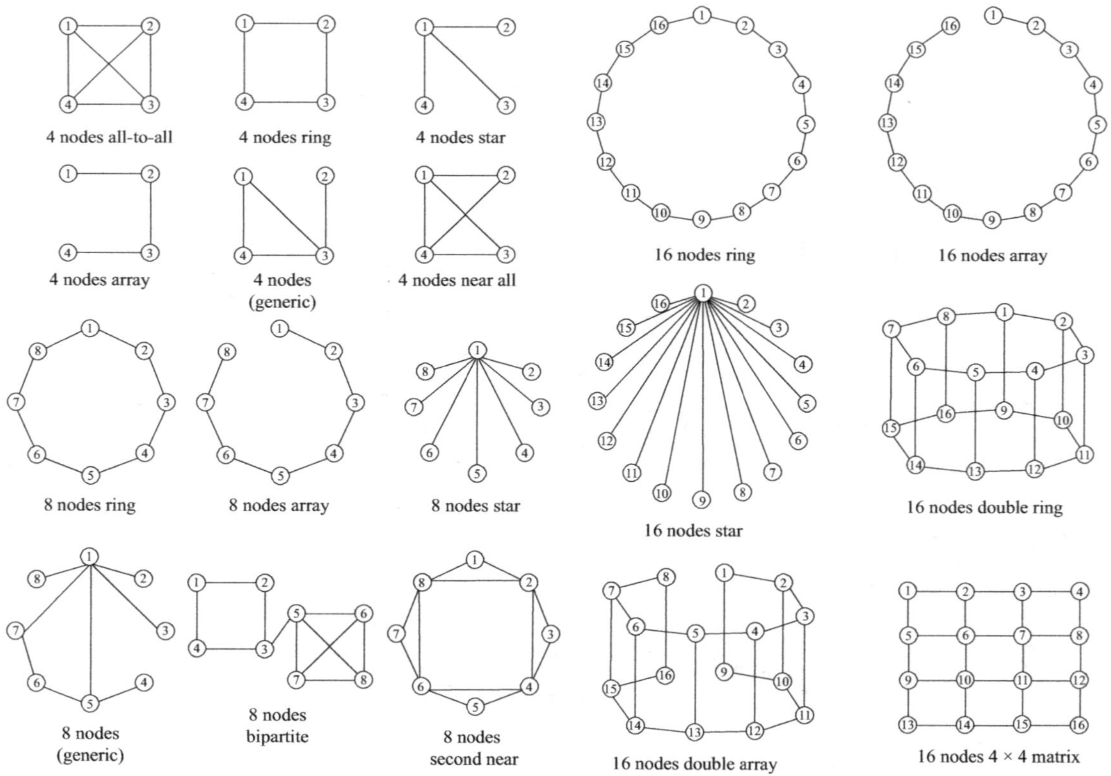

Figure 11.6 A collection of topologies up to 16 nodes, as implemented experimentally with the described setup. Figure reprinted with permission from de Magistris, M., di Bernardo, M., Manfredi, S., Petrarca, C. and Yaghouti, S., 'Modular experimental setup for real-time analysis of emergent behavior in networks of Chua's circuits', International Journal of Circuit Theory and Applications, DOI: 10.1002/cta.2179, Wiley Online Libray, 2015, License n. 3876950081617

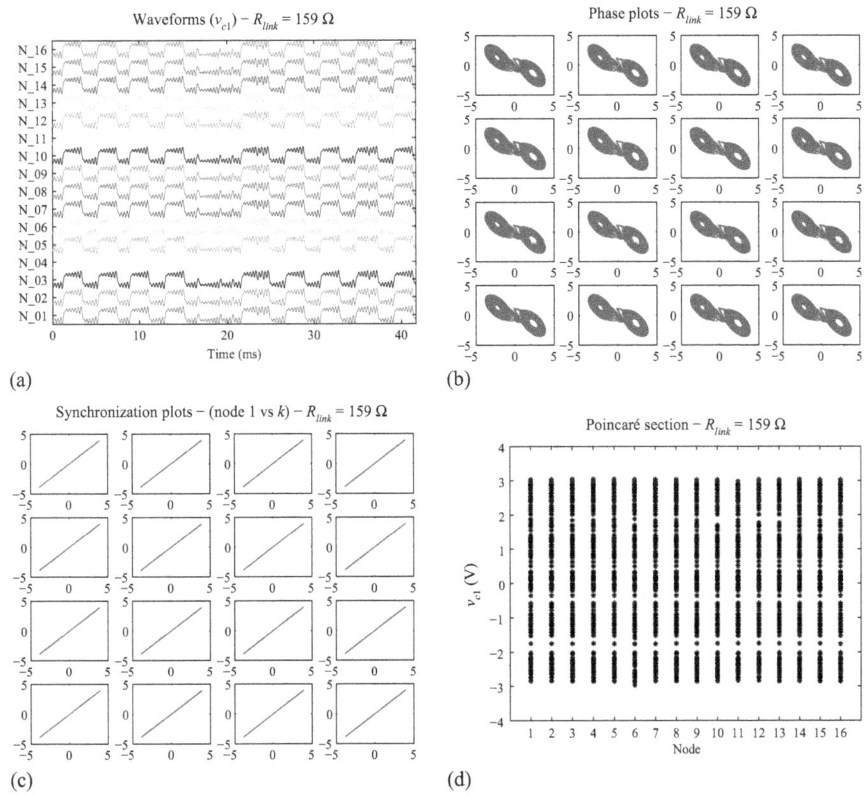

Figure 11.7 An example of experimental synchronized chaotic dynamics for the case of 16 nodes, ring topology: (a) v_{C1} traces; (b) state plane $v_{C1} - v_{C2}$ for each node; (c) synchronization diagrams $v_{C11} - v_{C1k}$; (d) Poincaré sections v_{C1k} calculated at t^* : $v_{C2k}(t^*) = mean[v_{C2k}(t)]$

11.4.2 Experiments on synchronization with dynamic links

A quite important extension of observable phenomena in the synchronization of nonlinear oscillators with linear coupling is given when dynamical links are considered. Also for this research area, the large effort in theoretical studies has not been complemented by an equivalent experimental effort, with few laboratory tests described in the literature [41]. In this section, we illustrate the main results of the described setup for the case of dynamic links [5]. Few simple modifications to the interconnection network allow, in fact, to create dynamical link (R, C) impedances. For each link, a set of eight capacitors of different value has been added, that can be connected in parallel to the corresponding set of eight link resistors via digitally controlled switches.

As an example, in the following, we present results on experiments carried out on a network of four Chua's oscillators coupled on the state variable v_{C1} via dynamic

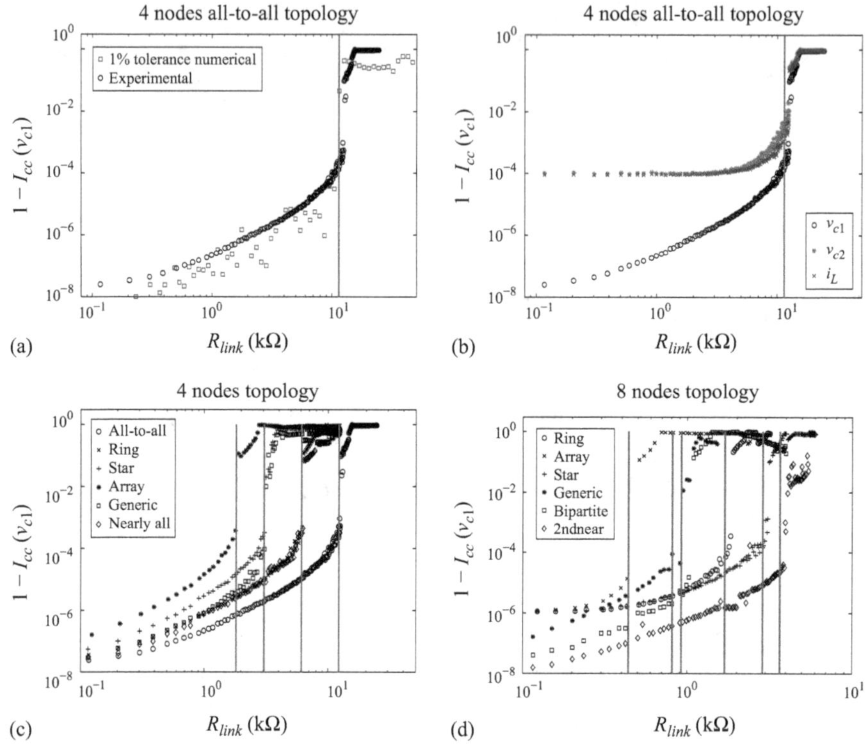

Figure 11.8 (a) Cross correlation for 1% tolerance nodes evaluated in SPICE simulations, in comparison with experimental data; (b) typical experimental behaviour of the cross correlation for the three state variables when coupling is given on v_{C1}; (c) and (d) measured cross correlation for some of the topologies considered in Figure 11.6, as a function of link resistance R_{link}; solid vertical lines correspond to theoretical thresholds calculated by MSF

R–*C* links, with the four nodes connected in an "all-to-all" configuration. The link capacitance C_{link} spans the range 0–520 pF in steps of 19 pF, whereas the link resistance R_{link} the range 0–74 kΩ in steps of 100 Ω. A total of more than 18 000 coupling configurations have been analysed in this way. For each configuration, the recorded waveforms are analysed in term of cross correlation. In Figure 11.9, the index $(1 - I_{cc})$, calculated on v_{C1} is reported, as a function of the link resistance, with the coupling capacitance C_{link} as a parameter. For low values of link capacitance ($C_{link} = 33$ pF), we can observe that synchronization is definitely lost for R_{link} values above the threshold $R_{th} = 22.8\ k\Omega$. When using a larger coupling capacitance (i.e. $C_{link} = 300$ pF), we notice that the transition between different dynamics shows an interesting feature, which was not observed with resistive links: the threshold slightly increases

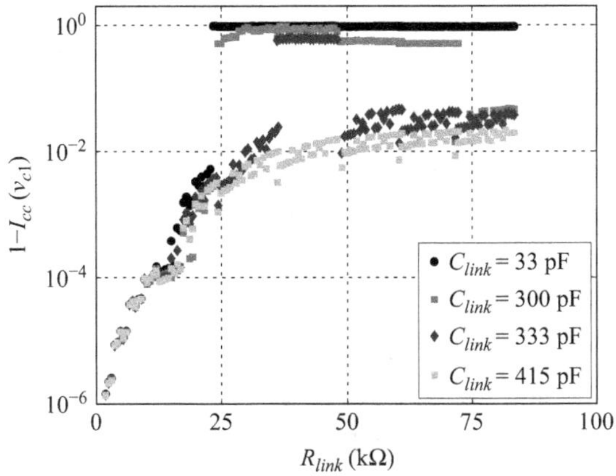

Figure 11.9 Experimental synchronization index of v_{C1} as a function of R_{link}, using C_{link} as parameter

(R_{th} = 24 kΩ) and the unsynchronized state is maintained only in a defined range 24 kΩ $\leq R_{link} \leq$ 72 kΩ values, above which the synchronized state is recovered in the network. If the link capacitance is further increased (C_{link} = 333 pF), the threshold becomes higher (R_{th} = 36 kΩ) and the resistance range in which the network is unsynchronized is sensibly reduced. In fact for $R_{link} >$ 48 kΩ, the state variables v_{C1} tend towards the synchronized state. By increasing the value of the coupling capacitance, we find a threshold $C_{th} >$ 380 pF above which no transition occurs: the value of the index $(1 - I_{cc})$ increases very slowly with R_{link}, but synchronization is never completely lost. The results put in evidence the role played by the capacitance C_{link} in affecting the coupling strength of each link: for higher capacitance values, the equivalent impedance of each link becomes sufficiently small as to leave the waveforms in a synchronized state even for high values of the link resistance R_{link}.

All results from the 18 000 experimentally tested configurations are well summarized in Figure 11.10, where synchronization regions are described as functions of the link conductance G_{link} and the link capacitance C_{link}: the colour scale from white to black indicates the regions with good synchronization (white) up to no synchronization (black) areas as functions of the link parameters. Such experimental results confirm the theoretical findings in Reference 42 where the Harmonic Balance method together with an MSF approach have been used for the analysis of networks of Chua's with dynamic coupling.

11.4.3 Experiments on clustering

In this section, we resume some results of experiments on clustering as obtained with the considered setup [7], aimed to better investigate the parameters region intermediate between the global synchronization and the unsynchronized state. In particular, the

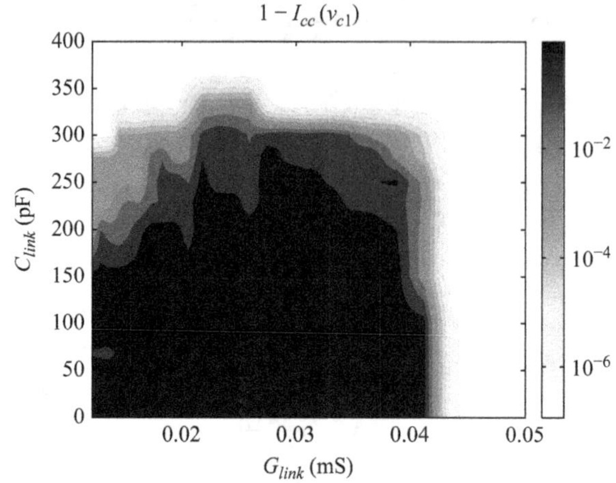

Figure 11.10 Experimental synchronization index as a function of coupling conductance G_{link} and capacitance C_{link}

appearance of clusters is evaluated in reason of the topological symmetries and as a function of individual links coupling values.

As reference network we consider, for the sake of simplicity, a five-node network as shown in Figure 11.11(a), where some possible symmetry is graphically represented by a dotted line. The nodes of the network are nominally identical and evolve onto a double-scroll chaotic regime when uncoupled; they are linked diffusively via the state variable v_{C1}. The network collective dynamic has been examined in three different cases: (i) variation of the resistance R_{link}, kept equal in each link; (ii) removal of one link [Figure 11.11(c) and (d)] and consequent change of the symmetry and (iii) variation of the coupling strength of only one link (i.e. $R_{link2-5}$) (Figure 11.11(b)).

As a first result we show in Figure 11.12 the index $(1 - I_{cc})$ of the reference network in Figure 11.11(a) as a function of the coupling resistance R_{link}, spanning the range (100 Ω–12 kΩ).

Synchronization is lost at $R_{link} = 9\,460\ \Omega$ corresponding to an abrupt jump of two orders of magnitude of the index; the vertical line in the same figure indicates the theoretical synchronization threshold $R_{th} = 8\,843\ \Omega$ calculated with the MSF approach. Behaviour of index $(1 - I_{cc})$ above the complete synchronization threshold shows, after the first jump, a quite slow continuous increase to higher values, where the correlation becomes negligible. It suggests that, although not completely synchronized, the waveforms still preserve some correlation. In fact it is possible to observe (Figure 11.13) that, although the time behaviour of v_{C1} at node 1 sensibly differs from all other waveforms, v_{C1} at nodes 2, 3, 4 and 5 show lag synchronization.

The effects generated in the network by a change in topology are then investigated with a new experiment: network of Figure 11.11(a) has been reconfigured twice, each time by removing one link. By removing the link between nodes 1 and 4, we get the

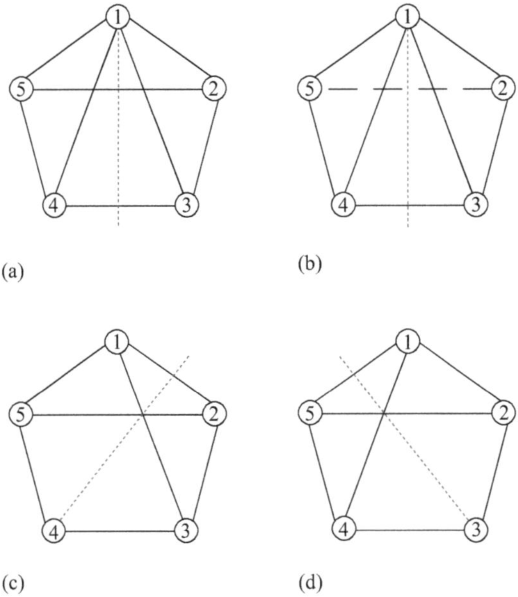

Figure 11.11 *Implemented topologies: (a) five-node network with symmetry; (b) five-node network with symmetry and only link 2-5 varies; (c) links 1–4 removed from network (a); (d) links 1–3 removed from network (a); the dashed line indicates some topological symmetries*

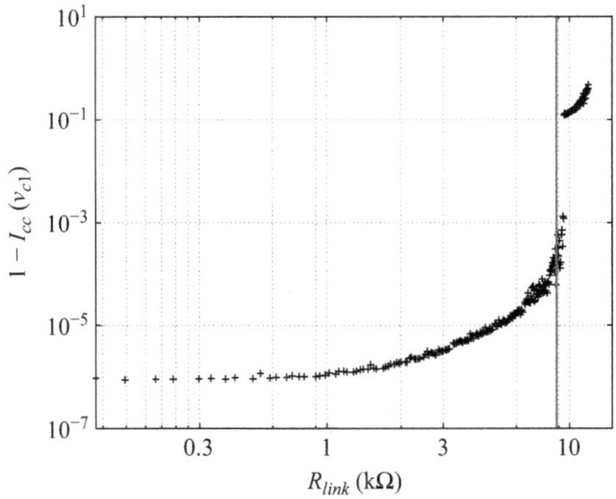

Figure 11.12 *Measured index ($1 - I_{cc}$) vs. R_{link} for topology 11.11(a); solid vertical line represents the MSF theoretical threshold for complete synchronization*

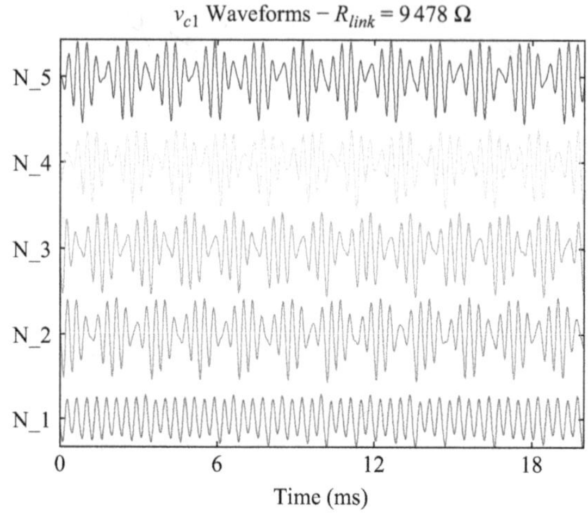

Figure 11.13 Example of experimentally observed lag synchronization among v_{C1} waveforms of nodes 2, 3, 4 and 5 for topology 11.11(a)

network of Figure 11.11(c); by removing the links 1–3, we obtain the network of Figure 11.11(d). In both cases, the symmetry of the network changes as compared to the original configuration, and the transition from the synchronized state towards the unsynchronized state assumes different characteristics, as shown in Figure.11.14. In fact, the index $(1 - I_{cc})$ reveals a much smoother transition, starting approximately at $R_{link} = 6\,620\ \Omega$, until $R_{link} = 8\,170\ \Omega$. At $R_{link} = 8\,170\ \Omega$, the five waveforms are no longer synchronized, but the network still keeps some coherence between the nodes by creating clustered waveforms, formed by two couples of nodes (1–2) and (3–5) which reflect the symmetry in the topology. In fact, in Figure 11.15, the waveforms at nodes (1–2) are overlapped, as well as nodes (3–5), whereas waveform at node 4 has its own different time evolution. This clustered synchronization is finally lost after a second jump in the index at $R_{link} = 1\,1306\ \Omega$. Note that the behaviour of the index calculated for the topology of Figure 11.11(d) is similar, as expected due to the symmetry of the network and is reported in Figure 11.14(b). According to the new symmetry, now clusters are formed between the couples of nodes (1–5) and (2–4).

The behaviour of clustering in the two considered cases can be brought to evidence by the calculation of the specific cross-correlation index $I_{cc}(x_i, x_j)$ which is shown in Figure 11.16 both for v_{C1} waveforms of nodes 3,5 and for the waveforms of nodes 1,2. It is plotted as a function of R_{link} and the transition region ($6\,620\ \Omega < R_{link} < 8\,170\ \Omega$) in which the global synchronization lost is highlighted by two vertical dashed lines. In that region, although the global cross-correlation index tends towards unity (Figure 11.14), the cluster between the couple of nodes 1,2 and 3,5 are maintained; cluster 1,2 tends to disappear for values above $R_{link} = 9\,000\ \Omega$ and cluster 3,5 vanishes at $R_{link} = 11\,200\ \Omega$.

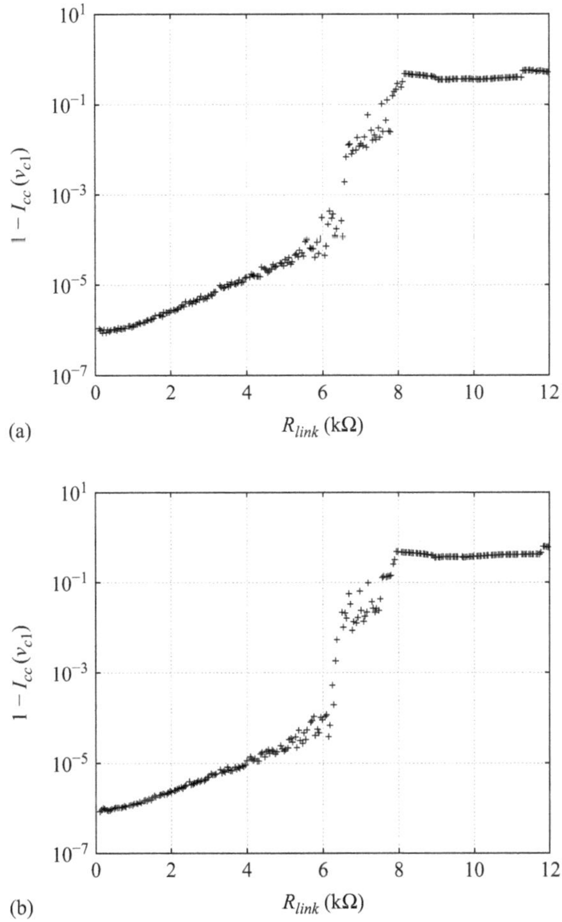

Figure 11.14 *(a) Measured index $(1 - I_{cc}(v_{c1}))$ vs. R_{link} for topology of figure 11.11(c) and (b) for topology of figure 11.11(d)*

Finally, an experiment has been carried out for the case of Figure 11.11(b), where leaving the symmetry unaffected, the weight (i.e. the value of the link resistance) between the nodes 2 and 5 spans the range 100 Ω–12 kΩ, keeping all other links fixed at $R_{link} = 10$ kΩ. Note that this is the only case where individual link has been settled at values different from other links. First of all it is observed that even for very small value of the link resistance $R_{link(2-5)}$, the global synchronization is never achieved, since the cross-correlation index is never over the value of 0.9 [Figure 11.17(a)]. Interestingly, in Figure 11.17(b), we can notice that for low values of the link resistance, clusters between the couples of nodes 2–5 and 3–4 are formed. With increasing the link resistance, these clusters disappear at $R_{link(2-5)} = 7\,856$ Ω

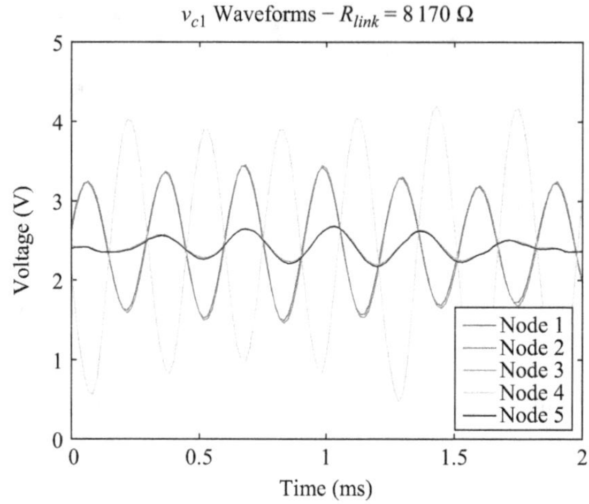

Figure 11.15 Example of experimentally observed formation of clusters for topology 11.11(c)

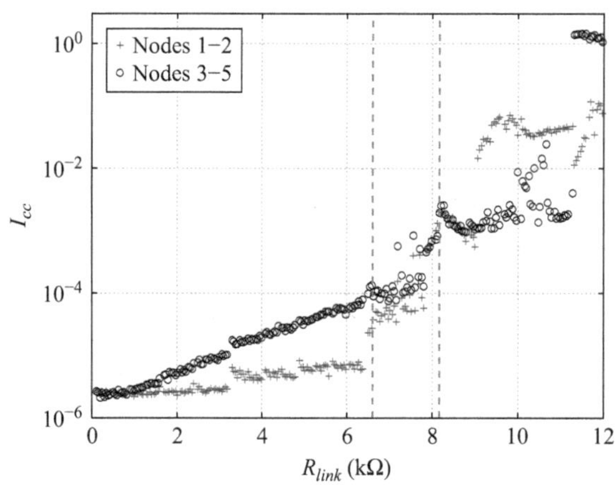

Figure 11.16 Experimental relative cross-correlation index $I_{cc}(i,j)$ vs. R_{link} in topology 11.11(c)

and suddenly they are substituted by different clusters formed between the new couple of nodes 2–3 and 4–5 in the whole interval [7 900–10 800] Ω. At higher values of the link resistance above 10 800 Ω, again clusters (2–5) and (3–4) re-appear where clusters (2–3) and (4–5) disappear, as shown in Figure 11.17(b), where the formation and disappearance of clusters is visually distinguishable; when cluster (2–5) is synchronized, cluster (2–3) is not synchronized and vice versa.

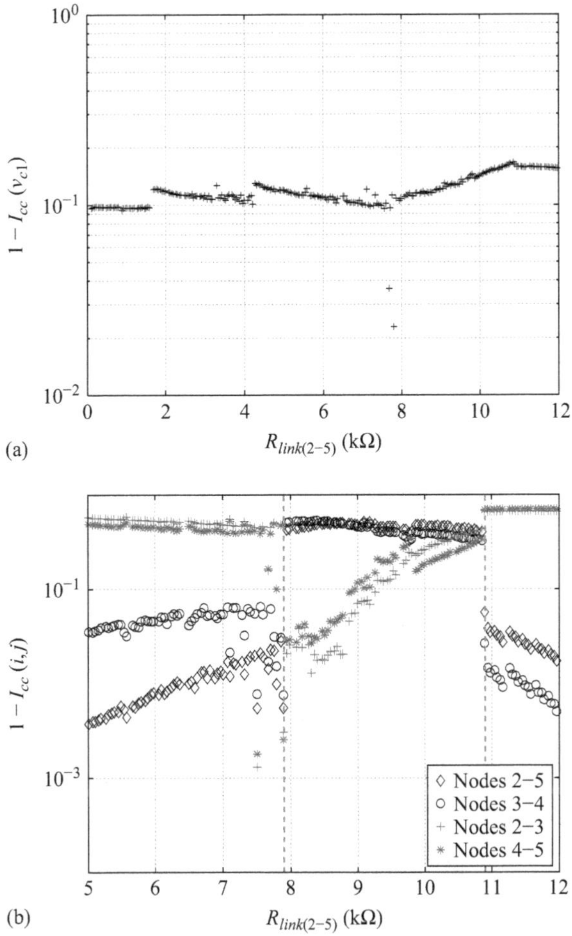

Figure 11.17 *(a) Measured index $(1 - I_{cc}(v_{c1}))$ and (b) relative cross-correlation index $I_{cc}(i,j)$ vs. R_{link} for topology 11.11(b), as function of link resistance of link (2–5)*

11.4.4 Observation of patterns and waves

In previous sections, the noticeable regimes of global synchronization and the emergence of clusters have been experimentally reported. In some sense, they represent the simplest collective behaviours emerging in complex networks of nonlinear oscillators. It is known as well that such networks can exhibit different types of spatio-temporal structures and patterns such as coexistence of chaotic and periodic oscillations, nonlinear waves and solitons, Turing patterns, etc. (see References 43–48 and references therein).

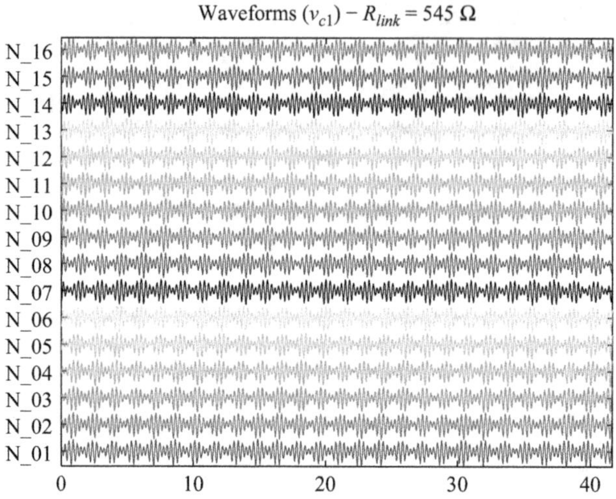

(a)

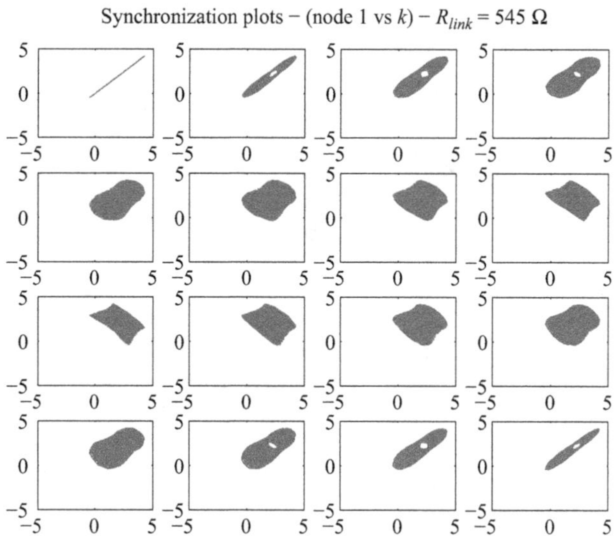

(b)

Figure 11.18 Example of experimentally observed lag synchronization of chaotic dynamics for the case of 16 nodes, ring topology: (a) v_{C1} traces; (b) synchronization diagrams v_{C1k} vs. v_{C11}

In this section, we recall some topical examples of complex behaviour, different from synchronization or clustering, in this way showing some other behaviours expected from theory or simulation in networks of chaotic oscillators. This, besides confirming the possibility of real experimental observation of such phenomena, demonstrates once more the actual versatility of the considered network realization.

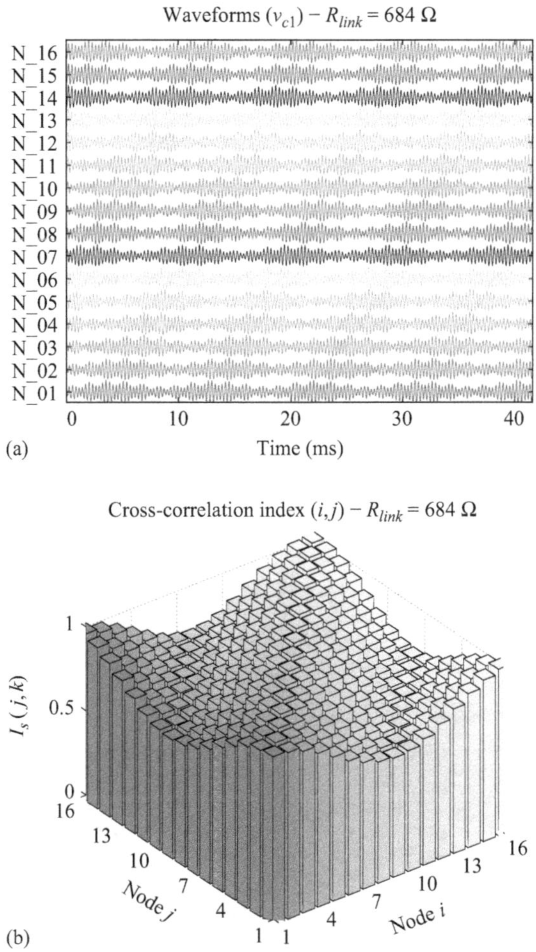

Figure 11.19 Example of experimentally observed travelling (rotating) waves in a ring of 16 nodes, (a) v_{C1} traces; (b) cross-correlation index

All the following cases, although in different configurations and number of nodes, share the same setting of the isolated node dynamics as double-scroll chaotic regime, and the diffusive equally weighted links on v_{C1}.

As a first example, we show the case of a "lag synchronization" regime [49], for the case of 16 nodes in a ring topology. As shown in Figure 11.18(a), the waveforms follow each other with a time shift of $2\pi k/N$ (with k being the node index and N the total number of nodes); the absence of complete synchronization is more evident from the synchronization plots reported in Figure 11.18(b).

For the same topology (ring with 16 nodes), the emergence of travelling waves can be observed at higher values of the link resistance. The periodic structure of the waves is clearly visible in Figure 11.19(a) and can be further highlighted by plotting

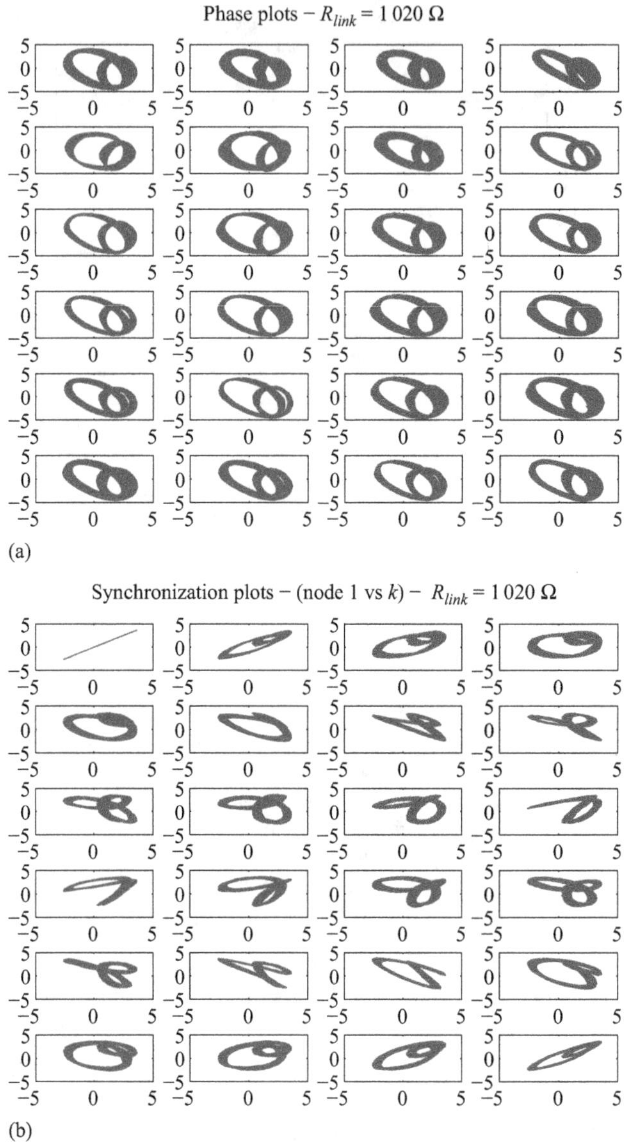

Figure 11.20 Some experimental synchronization patterns in weakly chaotic regime for a 24 nodes ring network configuration: (a) phase plots; (b) synchronization plots

the cross-correlation matrix $I_{cc}(i,j)$, as shown in Figure 11.19(b). In particular, the correlation between trajectories is observed to decrease as the distance between nodes increases, with minimum correlation between nodes located at the largest distance from each other along the ring (i.e. nodes (1–9), nodes (2–10), etc.).

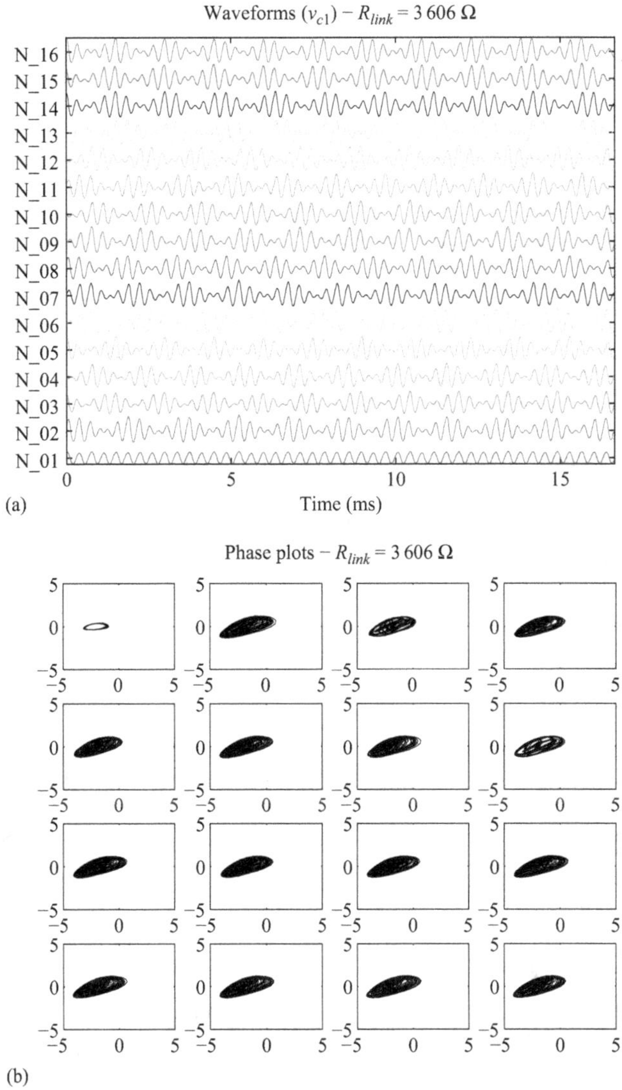

Figure 11.21 Experimentally observed chaotic time shift in a 16-node star configuration ($N = 1$ is the common node): (a) waveforms and (b) phase plots

In Figure 11.20, we show, for a network of 24 nodes with ring topology, some synchronization patterns in a weakly chaotic regime. In particular, in Figure 11.20(b), the almost periodic structure of the synchronization plots is evidenced.

A typical behaviour of a "star" configuration, for the case of 16 nodes is reported in Figure 11.21, where the central node is labelled as #1. We observe that after synchronization is lost, for high enough values of the coupling resistance, the central node dynamic shrinks onto a small amplitude periodic solution, while the other nodes start oscillating with a phase shift of $k2\pi/(N-1)$. Interestingly, every time the experiment is repeated, each node exhibits a different time shift picked at random from the same sequence.

As for the emergence of waves in chaotic dynamics regime, we show observations in a network of 24 nodes with a ring topology. In Figure 11.22, travelling waves are clearly recognizable, both in the clockwise direction (Figure 11.22(a), $R_{link} = 309\,\Omega$) and counterclockwise direction (Figure 11.22(b), $R_{link} = 334\,\Omega$).

The same 24-ring topology, with slightly greater values of the link resistance ($R_{link} = 345\,\Omega$), shows waveforms that can be viewed as the superposition of two counter rotating waveforms, Figure 11.23(a). Periodicity is evidenced from the Poincaré section diagram, Figure 11.23(b).

As last examples we report the case of a ring with nodes $N = 32$ (Figure 11.24) and an array topology with $N = 32$ nodes (Figure 11.25), where, respectively, chaotic waves and increasing wavelength oscillations are observed. For such case, a complete scan in the coupling resistance parameter range has been carried out [6], evidencing dynamical behaviours agreeing to theoretical and numerical predictions as those given in Reference 50.

11.5 Concluding remarks

In this chapter, based on a general structure reconfigurable experimental setup which has been recently proposed, we have recalled and described a quite wide spectrum of experimental results on the complex dynamics emerging in networks of nonlinear and chaotic oscillators. The novelty of the presented approach, as demonstrated by the quite diverse observed phenomena, resides in its general-purpose structure, as well as its modularity. In fact, differently from other experimental works in this area, investigations of different topics have been realized with the same setup, and many other are still possible. In particular, we have resumed: (i) investigations on complete synchronization in chaotic regime of different topology networks, comparing the real cross correlation in the presence of node's mismatch and the corresponding stability thresholds; (ii) the synchronization analysis in presence of dynamical links, with a very detailed investigation in term of the parameters space, as practically allowed by the setup; (iii) the experimental study (although preliminary) of the emergence of clusters at the edge of complete synchronization, evidencing the dependence on topological symmetries as well as single-link coupling strength; and (iv) a resume of phenomena other than synchronization and clustering, such as travelling waves and pattern formation in chaotic regime, arising in networks of 24 or 32 nodes.

Such collection of diverse experimental results is unique, at least from the perspective of being observed within the same experimental environment, revealing its flexibility as well as the easiness of its realization and adaptation. We wish to emphasize that the structure of the discussed setup could be possibly easily adapted to study networks of chaotic oscillators different from Chua's, when needed.

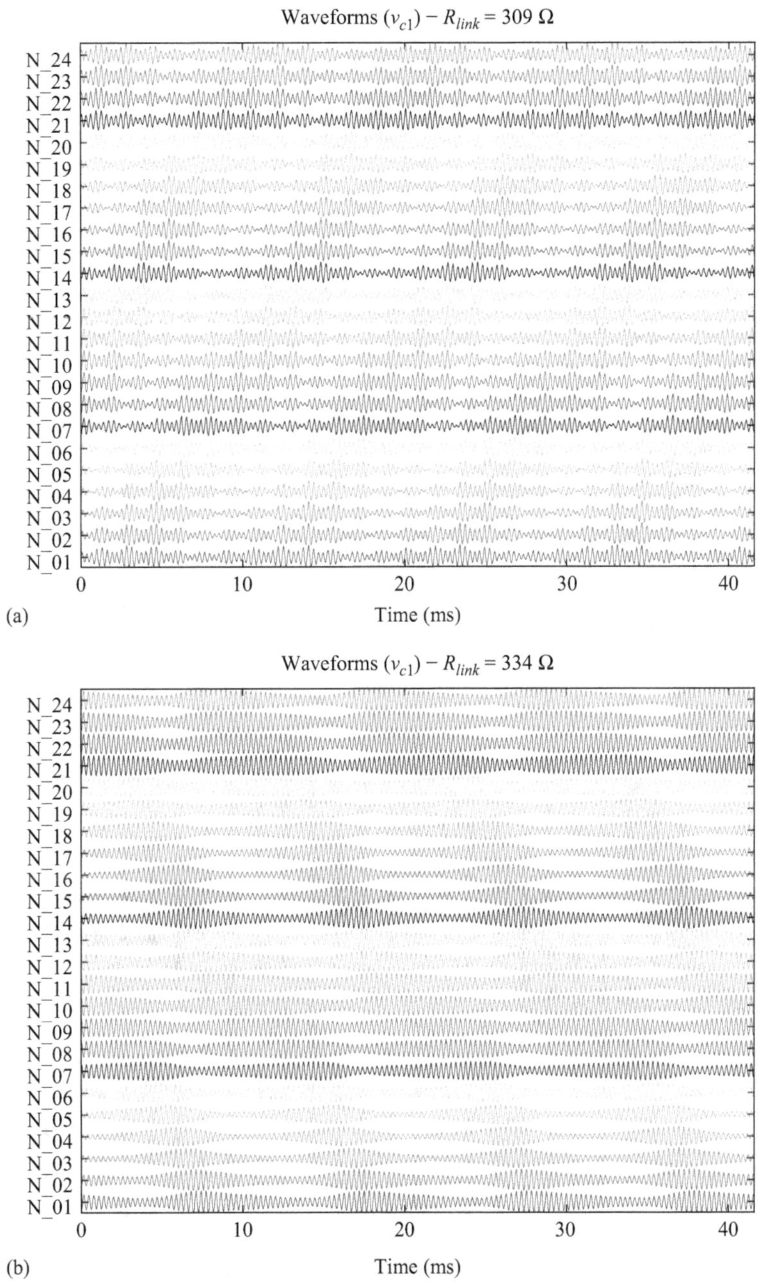

Figure 11.22 *Experimentally observed space periodic ($N = 24$) rotating waves: (a) "clock-wise" and (b) "counter-clock-wise". Figure reprinted with permission from de Magistris, M., di Bernardo, M., Manfredi, S., Petrarca, C. and Yaghouti, S., 'Modular experimental setup for real-time analysis of emergent behavior in networks of Chua's circuits', International Journal of Circuit Theory and Applications, DOI: 10.1002/cta.2179, Wiley Online Libray, 2015, License n. 3876950081617*

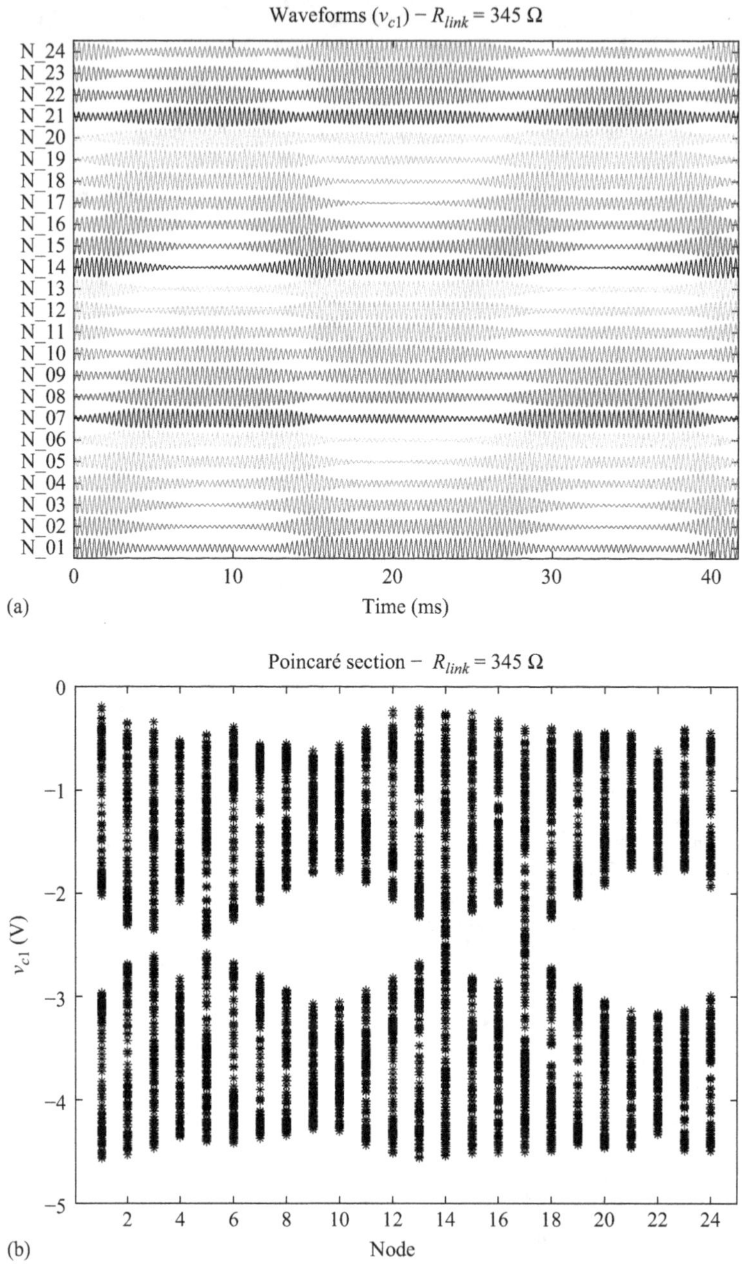

Figure 11.23 Experimentally observed travelling waves in a ring network of 24 Chua's circuits: (a) waveforms and (b) Poincaré sections (spatial periodicity)

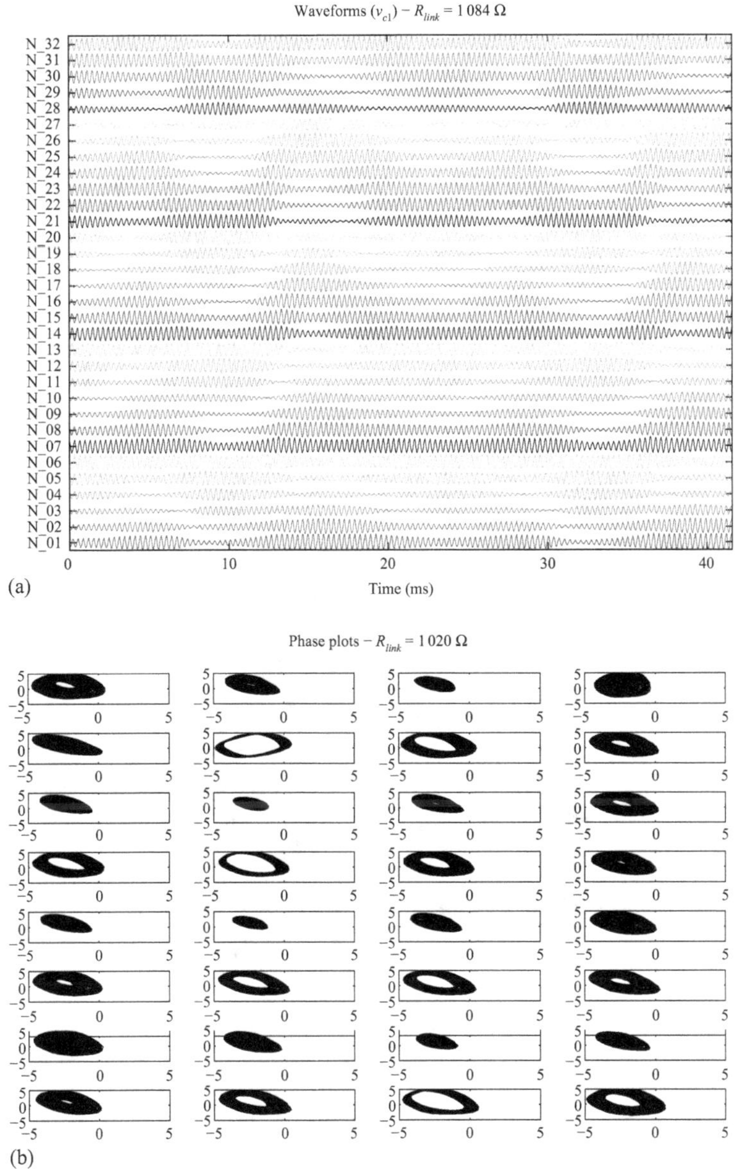

Figure 11.24 Experimentally observed chaotic waves in a ring network of 32 Chua's circuits: (a) waveforms; (b) phase plots

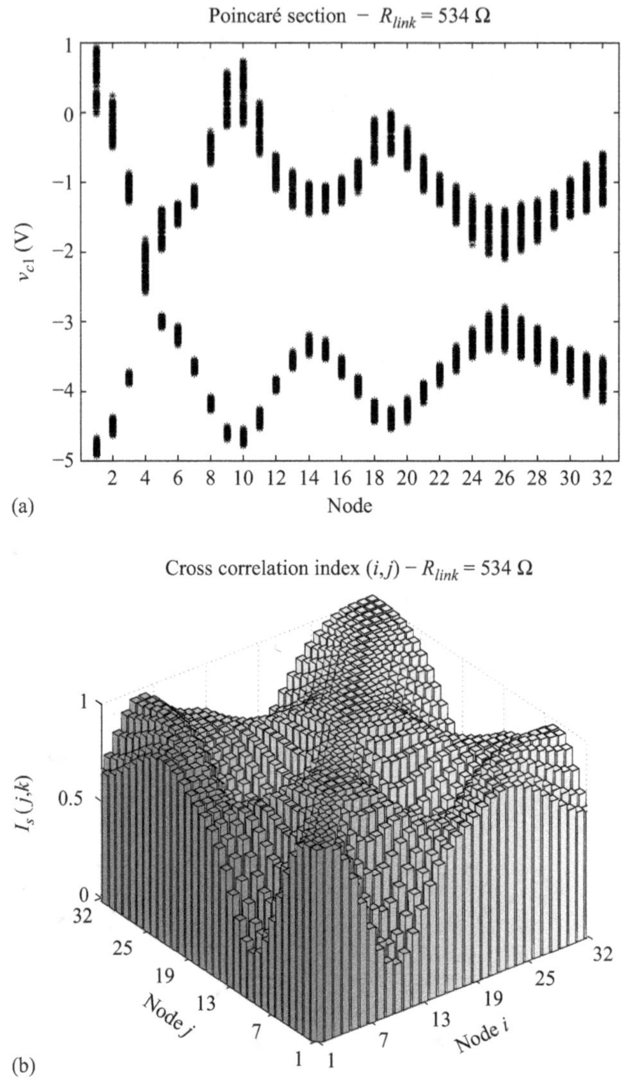

Figure 11.25 Experimentally observed asymmetric waves and non-uniform wavelength in an array of 32 nodes: (a) Poincaré sections; (b) distance matrix

There are still different areas of experimental work possibly to explore within the same experimental framework, which range from the analysis of more complicated structures of the interconnection network (where link type and strength are settled individually), to the possibility of investigating dynamical evolution of the network's topology according to some desired emerging behaviour. All that will hopefully constitute object of future work.

References

[1] Marco Colandrea, Massimiliano de Magistris, Mario di Bernardo, and Sabato Manfredi. A fully reconfigurable experimental setup to study complex networks of Chua's circuits. In *Nonlinear Dynamics of Electronic Systems, Proceedings of NDES 2012*, pages 1–4. VDE, Wolfenbüttel, Germany, 2012.

[2] Massimiliano de Magistris, Mario Di Bernardo, Edmondo Di Tucci, and Sabato Manfredi. Synchronization of networks of non-identical Chua's circuits: Analysis and experiments. *Circuits and Systems I: Regular Papers, IEEE Transactions on*, 59(5):1029–1041, 2012.

[3] M Colandrea, M de Magistris, C Petrarca, M Di Bernardo, and S Manfredi. Realization of a fully configurable complex network of non-linear Chua's oscillators. In *Cellular Nanoscale Networks and Their Applications (CNNA), 2012 13th International Workshop on*, pages 1–2. IEEE, Piscataway, NJ, 2012.

[4] Massimiliano de Magistris, Mario di Bernardo, and Carlo Petrarca. Experiments on synchronization in networks of nonlinear oscillators with dynamic links. *Nonlinear Theory and Its Applications, IEICE*, 4(4):462–472, 2013.

[5] Carlo Petrarca, Soudeh Yaghouti, and Massimiliano de Magistris. Experimental dynamics observed in a configurable complex network of chaotic oscillators. In *Nonlinear Dynamics of Electronic Systems*, pages 203–210. Springer, Berlin, 2014.

[6] Massimiliano Magistris, Mario Bernardo, Sabato Manfredi, Carlo Petrarca, and Soudeh Yaghouti. Modular experimental setup for real-time analysis of emergent behavior in networks of Chua's circuits. *International Journal of Circuit Theory and Applications*, 44 (8): 1551–1571, 2015.

[7] Soudeh Yaghouti, Carlo Petrarca, and Massimiliano Magistris. Experiments on clustering and synchronous patterns in a configurable network of chaotic oscillators. In *Nonlinear Dynamics of Electronic Systems, Proceedings of NDES 2015*, Italy, 2015.

[8] Steven H Strogatz. Exploring complex networks. *Nature*, 410(6825):268–276, 2001.

[9] Stefano Boccaletti, Vito Latora, Yamir Moreno, Martin Chavez, and D-U Hwang. Complex networks: Structure and dynamics. *Physics Reports*, 424(4):175–308, 2006.

[10] Yang Tang, Feng Qian, Huijun Gao, and Jürgen Kurths. Synchronization in complex networks and its application – A survey of recent advances and challenges. *Annual Reviews in Control*, 38(2):184–198, 2014.

[11] Leon O Chua, Ljupco Kocarev, Kevin Eckert, and Makoto Itoh. Experimental chaos synchronization in Chua's circuit. *International Journal of Bifurcation and Chaos*, 2(3):705–708, 1992.

[12] Leon O Chua, Makoto Itoh, Ljupco Kocarev, and Kevin Eckert. Chaos synchronization in Chua's circuit. *Journal of Circuits, Systems, and Computers*, 3(1):93–108, 1993.

[13] Guo-Qun Zhong, Chai Wah Wu, and Leon O Chua. Torus-doubling bifurcations in four mutually coupled Chua's circuits. *Circuits and Systems I: Fundamental Theory and Applications, IEEE Transactions on*, 45(2):186–193, 1998.

[14] Tomasz Kapitaniak, Leon O Chua, and Guo-Qun Zhong. Experimental hyperchaos in coupled Chua's circuits. *Circuits and Systems I: Fundamental Theory and Applications, IEEE Transactions on*, 41(7):499–503, 1994.

[15] Yoshifumi Nishio and Akio Ushida. Spatio-temporal chaos in simple coupled chaotic circuits. *Circuits and Systems I: Fundamental Theory and Applications, IEEE Transactions on*, 42(10):678–686, 1995.

[16] Yoshifumi Nishio and Akio Ushida. Quasi-synchronization phenomena in chaotic circuits coupled by one resistor. *Circuits and Systems I: Fundamental Theory and Applications, IEEE Transactions on*, 43(6):491–496, 1996.

[17] M Gomez-Gesteira, M DeCastro, V Perez-Villar, and LO Chua. Experimental Chua's circuit arrays as an autowave simulator. *IEEE Transactions on Circuits and Systems. 1, Fundamental Theory and Applications*, 46(4):495–499, 1999.

[18] Yoshifumi Nishio and Akio Ushida. Phase synchronization in a ring of chaotic circuits. In *Proceedings of NOLTA'01*, Miyagi, Japan, pages 163–166, 2001.

[19] C Posadas-Castillo, C Cruz-Hernández, and RM López-Gutiérrez. Experimental realization of synchronization in complex networks with Chua's circuits like nodes. *Chaos, Solitons & Fractals*, 40(4):1963–1975, 2009.

[20] Ákos Tar, Gaurav Gandhi, and György Cserey. Hardware implementation of CNN architecture-based test bed for studying synchronization phenomenon in oscillatory and chaotic networks. *International Journal of Circuit Theory and Applications*, 37(4):529–542, 2009.

[21] Yoko Uwate and Yoshifumi Nishio. Synchronization in several types of coupled polygonal oscillatory networks. *IEEE Transactions on Circuits and Systems I: Regular Papers*, 5(59):1042–1050, 2012.

[22] Rabinder N Madan. *Chua's Circuit: A Paradigm for Chaos*, volume 1. World Scientific, Singapore, 1993.

[23] Luigi Fortuna, Mattia Frasca, and Maria Gabriella Xibilia. *Chua's Circuit Implementations: Yesterday, Today and Tomorrow*. World Scientific, Singapore, 2009.

[24] Liang Huang, Qingfei Chen, Ying-Cheng Lai, and Louis M Pecora. Generic behavior of master-stability functions in coupled nonlinear dynamical systems. *Physical Review E*, 80(3):036204, 2009.

[25] Russell Merris. Laplacian matrices of graphs: A survey. *Linear Algebra and Its Applications*, 197:143–176, 1994.

[26] Michael Peter Kennedy. Robust op amp realization of Chua's circuit. *Frequenz*, 46(3–4):66–80, 1992.

[27] Leonardo A. B. Torres, Luis A. Aguirre. Inductorless Chua's circuit. *Electronics Letters*, 36(23):1915–1916, 2000.

[28] Leon O Chua. Local activity is the origin of complexity. *International Journal of Bifurcation and Chaos*, 15(11):3435–3456, 2005.

[29] MG Rosenblum, A Pikovsky, and J Kurths. *Synchronization – A Universal Concept in Nonlinear Sciences*. Cambridge University Press, Cambridge, 2001.

[30] VN Belykh, NN Verichev, LJ Kocarev, and Leon O Chua. On chaotic synchronization in a linear array of Chua's circuits. *Journal of Circuits, Systems, and Computers*, 3(2):579–589, 1993.

[31] Chai Wah Wu and Leon O Chua. Application of graph theory to the synchronization in an array of coupled nonlinear oscillators. *Circuits and Systems I: Fundamental Theory and Applications, IEEE Transactions on*, 42(8):494–497, 1995.

[32] Louis M Pecora and Thomas L Carroll. Master stability functions for synchronized coupled systems. *Physical Review Letters*, 80(10):2109, 1998.

[33] Mauricio Barahona and Louis M Pecora. Synchronization in small-world systems. *Physical Review Letters*, 89(5):054101, 2002.

[34] Mahdi Jalili, Ali Ajdari Rad, and Martin Hasler. Enhancing synchronizability of dynamical networks using the connection graph stability method. *International Journal of Circuit Theory and Applications*, 35(5–6):611–622, 2007.

[35] Michael G Rosenblum, Arkady S Pikovsky, and Jürgen Kurths. Phase synchronization of chaotic oscillators. *Physical Review Letters*, 76(11):1804–1807, 1996.

[36] Changsong Zhou and Jürgen Kurths. Hierarchical synchronization in complex networks with heterogeneous degrees. *Chaos: An Interdisciplinary Journal of Nonlinear Science*, 16(1):015104, 2006.

[37] Jesús Gómez-Gardenes, Yamir Moreno, and Alex Arenas. Paths to synchronization on complex networks. *Physical Review Letters*, 98(3):034101, 2007.

[38] Chenbo Fu, Weijie Lin, Liang Huang, and Xingang Wang. Synchronization transition in networked chaotic oscillators: The viewpoint from partial synchronization. *Physical Review E*, 89:052908, May 2014.

[39] Louis M Pecora, Francesco Sorrentino, Aaron M Hagerstrom, Thomas E Murphy, and Rajarshi Roy. Cluster synchronization and isolated desynchronization in complex networks with symmetries. *Nature Communications*, 5: 1–8, 2014.

[40] Igor Mishkovski, Miroslav Mirchev, Fernardo Corinto, and Mario Biey. Synchronization analysis of networks of identical and nearly identical Chua's oscillators. In *Circuits and Systems (ISCAS), 2012 IEEE International Symposium on*, pages 2115–2118. IEEE, Piscataway, NJ, 2012.

[41] Ayana Shimada, Yoko Uwate, Yoshifumi Nishio, and Jingmin Xin. Synchronization phenomena of two simple RC chaotic circuits coupled by a capacitor. *Journal of Signal Processing*, 17(4):99–102, 2013.

[42] Marco Righero, Fernando Corinto, and Mario Biey. Master stability function for networks of Chua's circuits with static and dynamic couplings. In *2011 IEEE International Symposium of Circuits and Systems (ISCAS)*, Rio de Janeiro, Brazil, 2011.

[43] A Shabunin, V Astakhov, and V Anishchenko. Developing chaos on base of traveling waves in a chain of coupled oscillators with period-doubling: Synchronization and hierarchy of multistability formation. *International Journal of Bifurcation and Chaos*, 12(8):1895–1907, 2002.

[44] V Perez-Munuzuri, AP Munuzuri, M Gómez-Gesteira, V Pérez-Villar, L Pivka, and LO Chua. Nonlinear waves, patterns and spatio-temporal chaos in cellular neural networks. *Philosophical Transactions of the Royal Society of London A: Mathematical, Physical and Engineering Sciences*, 353(1701):101–113, 1995.

[45] Kunihiko Kaneko. Clustering, coding, switching, hierarchical ordering, and control in a network of chaotic elements. *Physica D: Nonlinear Phenomena*, 41(2):137–172, 1990.

[46] Kunihiko Kaneko. Pattern dynamics in spatiotemporal chaos: Pattern selection, diffusion of defect and pattern competition intermittency. *Physica D: Nonlinear Phenomena*, 34(1):1–41, 1989.

[47] Masayuki Yamauchi, Masahiro Wada, Yoshifumi Nishio, and Akio Ushida. Wave propagation phenomena of phase states in oscillators coupled by inductors as a ladder. *IEICE Transactions on Fundamentals of Electronics, Communications and Computer Sciences*, 82(11):2592–2598, 1999.

[48] Valentina Lanza, Fernando Corinto, and Marco Gilli. Localized oscillations in diffusively coupled cyclic negative feedback systems. *International Journal of Bifurcation and Chaos*, 22(1):1250014, 2012.

[49] Michael G Rosenblum, Arkady S Pikovsky, and Jürgen Kurths. From phase to lag synchronization in coupled chaotic oscillators. *Physical Review Letters*, 78(22):4193, 1997.

[50] Maciej J Ogorzalek, Zbigniew Galias, Andrzej M Dąbrowski, and Wladyslaw R Dąbrowski. Chaotic waves and spatio-temporal patterns in large arrays of doubly-coupled Chua's circuits. *Circuits and Systems I: Fundamental Theory and Applications, IEEE Transactions on*, 42(10):706–714, 1995.

Chapter 12
Fundamental operation and design of high-frequency high-efficiency tuned power oscillator

Hiroo Sekiya[1]

Tuned power oscillators are important power-electronics circuits for achieving high power density. In addition, it is possible to apply the soft-switching techniques and high power-conversion efficiency can be achieved at high frequencies. This chapter presents fundamental operation and design of high-frequency high-efficiency tuned power oscillator. Concretely, the class-E oscillator family is introduced along with operation principle and design example.

12.1 Introduction

The importance of power electronics is rapidly increasing due to the raise in the awareness of environmental issues. The main function of the power-electronics circuits is to transform the voltage and current for our easy-use. For example, dc–ac inverter, ac–dc rectifier, dc–dc converter, and ac–ac converter are typical circuits of the power electronics, which provide proper shape of waveform and power level. They are used in most of electric systems from high power to low power. Additionally, the electric powers are used via many power-electronics converters in series from a power source to end points. Therefore, the power-conversion efficiency of each converter should be as high as possible. The power-conversion efficiency improvement of the power converters is traditional but still important research topic in electrical systems. It can be stated that the big impact can be obtained to the power save and environmental issue if performance of power-electronics circuits is improved. This is because huge amount of power converters are used in the world.

The increase in the power density is a major purpose in the power-electronics research field. For enhancing the power density, the circuit scale should be smaller. The magnetic components such as inductor and transformer are dominant factor

[1]Graduate School of Advanced Integration Science, Chiba University, Chiba, Japan

for determining the circuit volume. Therefore, it can be stated that high-frequency operation is effective to reduce the circuit volume. The switching power-loss reduction becomes an important problem at high-frequency converters because switching number is in proportion to the switching frequency. The cooling and/or heat dissipation systems for switching power losses increase the circuit volume. Therefore, the soft switching techniques such as zero-voltage switching (ZVS) and zero-current switching (ZCS) are mandatory techniques for switching-loss reduction.

In high-frequency power converters, high-speed switching devices are necessary. Recently, next-generation semiconductor devices, e.g. gallium nitride (GaN) and silicon carbide (SiC), appear, which provide new evolvement of power converters. These devices contribute to the increase in the power density. Especially, GaN devices have a high potential to apply to high-frequency converters. By the way, switching-device driver design is a covert problem in high-speed switching devices. The square-driving voltage is strained due to parasitic capacitances and resistances of the switching device. Because a precise switching pattern is required for achieving the soft-switching conditions at high frequencies, the driver-circuit designs, taking into account the parasitic components, are the technical barrier for high-frequency power-electronics circuit.

The tuned power oscillators, which are autonomous circuit without driver, have a potential to solve the problems of high-frequency power-electronics circuit. The tuned power oscillator is driven by the feedback signal from the sinusoidal output voltage. In addition, it is possible to apply the soft-switching techniques and high power-conversion efficiency can be achieved. Namely, the tuned power oscillators are suitable to high-power density converters. Actually, there are wide-area applications of tuned power oscillators, e.g. dc–ac inverter part of the dc–dc converters and transmitters of the wireless power transfer systems and wireless communications. Tuned power oscillators, however, have problems of design difficulty and frequency instability. It is useful and effective for tuned power oscillator usages to understand the operation mechanism and design strategies of high-frequency high-efficiency tuned power oscillators.

This chapter covers fundamental operation and design of high-frequency high-efficiency tuned power oscillator. Concretely, the class-E oscillator family is introduced. The class-E oscillator satisfies the class-E zero-voltage switching and zero-derivative switching (ZVS/ZDS) conditions. Therefore, high power-conversion efficiency can be achieved at high frequencies. The organizations of this chapter are: in Section 12.2 the tuned power amplifiers, the class-D and class-E amplifiers, are introduced and their problems are pointed out. The soft switching techniques and their necessity are understood in this section. In Section 12.3, the class-E tuned power oscillator is shown. The class-E oscillator is regarded that just a feedback network from the output load to the gate of the Metal-Oxide-Semiconductor Field-Effect Transistor (MOSFET) is added to the class-E amplifier instead of switching-device driver. Therefore, high-efficiency operation, which is a feature of the class-E power amplifier, can be kept in the tuned power oscillator. In addition, several extended versions of the class-E oscillator are introduced. Section 12.4 explains the design strategy and numerical design method of the class-E oscillator. For designs of the class-E oscillator, it is necessary to satisfy the class-E ZVS/ZDS conditions and proper phase shift

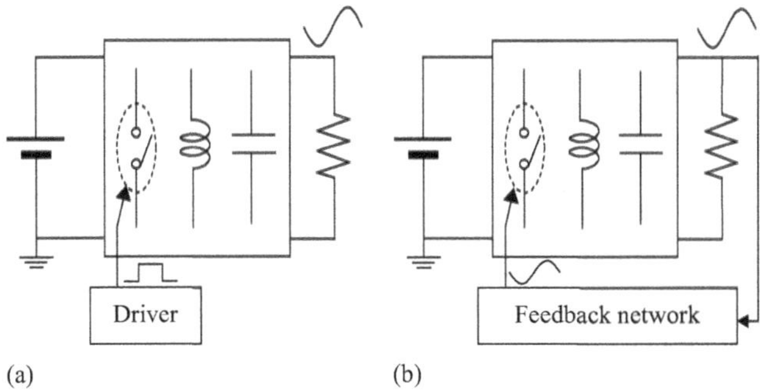

Figure 12.1 Configuration diagram of tuned power amplifier and oscillator. (a) Amplifier configuration. (b) Oscillator configuration

and amplitude gain of the feedback network. The optimization technique is necessary for satisfying them simultaneously. Section 12.5 summarizes this chapter.

12.2 Power amplifiers

Figure 12.1(a) shows a fundamental configuration diagram of power amplifiers. Power amplifiers [1–4] usually consist of dc-input voltage, switching devices, memory components, and load resistance and have a function to transform dc voltage to ac one. In general, switching operations are adopted for generating the frequency component of ac voltage. Therefore, switching devices, such as Metal-Oxide-Semiconductor (MOS), GaN, and SiC devices, are included in power amplifiers. Especially, the power amplifiers using switching devices are called switching amplifiers. In addition, capacitances and inductances are also included in power amplifiers, which construct output network, i.e. resonant filter.

12.2.1 Class-D amplifier

Figure 12.2(a) shows a circuit topology of the class-D amplifier, which is one of the typical switching amplifiers [5–8]. The class-D amplifier is composed of input voltage source V_I, two switching devices S_1 and S_2, series resonant filter $L - C$ whose resonant frequency is the same as the operating frequency, and load resistance R. Figure 12.2(b) shows example waveforms of the class-D amplifier. In this figure, $\theta = \omega t = 2\pi ft$ represents the angular time, where f is the operating frequency. The MOSFETs S_1 and S_2 turn ON and OFF alternatively. Namely, the duty ratios of the both switching devices are $D = 0.5$. Therefore, the square-waveform voltage v_{S1}, whose amplitude is V_{DD}, appears across S_1. Through the resonant filter, fundamental frequency component of v_{S1} is obtained and the sinusoidal output waveform appears at the load resistance as shown in Figure 12.2(b). In the class-D amplifier operation, the switch voltage and current never appear simultaneously because of ideal switching operations. Therefore, the power losses at the switching device are regarded as zero

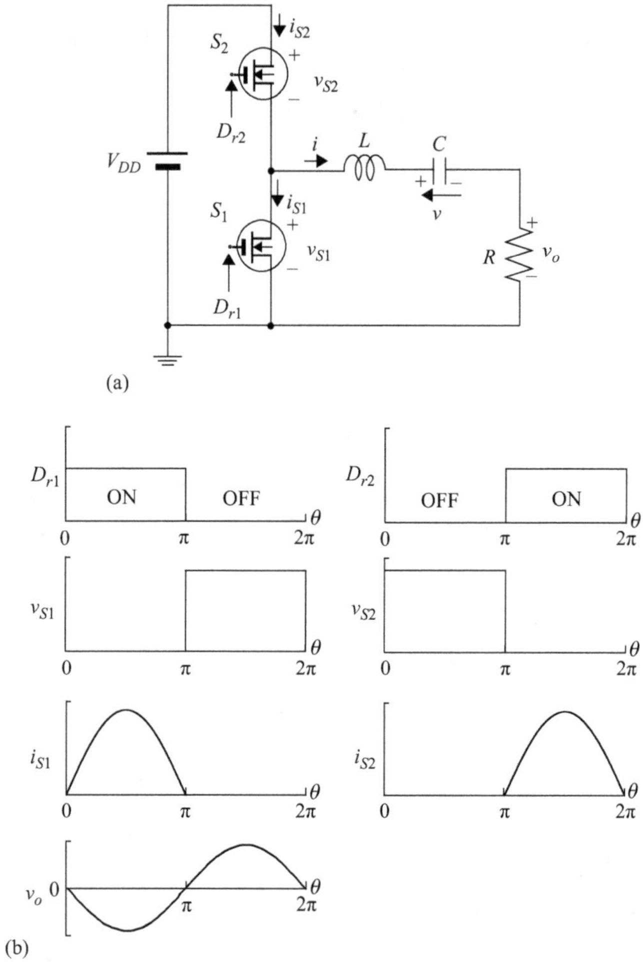

Figure 12.2 Class-D amplifier. (a) Circuit topology. (b) Example waveforms

and 100% power-conversion efficiency can be achieved, where the power-conversion efficiency is defined as

$$\eta = \frac{\text{Output power}}{\text{Input power}} = \frac{\text{Output power}}{\text{Output power} + \text{Power losses}}, \tag{12.1}$$

in this section.

In the class-D amplifier, however, there are two power loss factors. Therefore, real circuits cannot achieve 100% power-conversion efficiency.

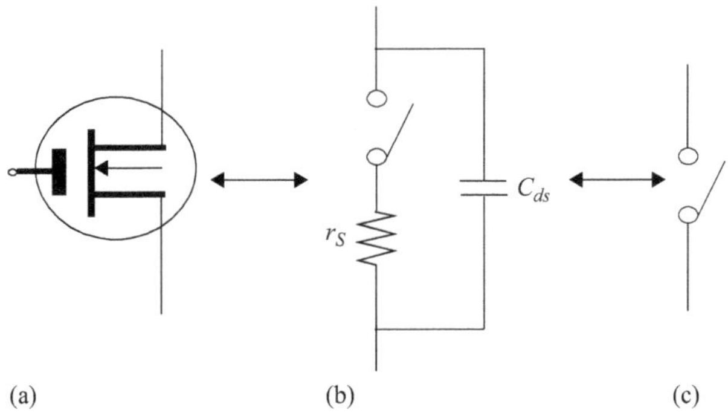

Figure 12.3 Equivalent model of the switching device. (a) Device expression. (b) Model with switch ON-resistance and parasitic capacitance. (c) Ideal switch model

(1) *Power losses due to the parasitic resistances*
Each circuit component has parasitic resistances, which generates power loss due to the current flow. Especially, the power losses at the ON-resistance of the switching devices and at parasitic resistance on the resonant inductor are dominant factors. Now the equivalent model of the switching device is shown in Figure 12.3. It is possible to express the switching devices as an ideal switch as shown in Figure 12.3(c). There is, however, an ON-resistance r_S in actual devices as shown in Figure 12.3(b). For example, if we use a MOSFET, ON-resistance is 0.1–1.0 Ω.

(2) *Switching losses*
Though they do not appear on the circuit topology in Figure 12.2, the switching devices have parasitic capacitance between drain and source as shown in Figure 12.3(b). When a switching device is in OFF state, the switch voltage drain to source is the same as input voltage, namely $v_S = V_{DD}$. Because the class-D amplifier has two switches, the switching losses of the class-D amplifier are expressed as

$$p_{Cds} = 2 \times \frac{1}{2} C_{ds} v_S^2 f = C_{ds} V_I^2 f, \tag{12.2}$$

where C_{ds} is the drain-to-source capacitance of the switching device. Note that the switching loss occurs at every turn-ON switching. Usually, the value of C_{ds} is small and the switching losses can be ignored at low frequencies. At high frequencies, however, the switching losses should be considered because of the increase in the operating frequency f. When the operating frequency is in the megahertz order, the power-conversion efficiency degradation due to switching losses is main problem of the class-D amplifiers. Additionally, the switch voltage varies rapidly at the turn-ON instant. Namely, switching noise is also a problem of the high-frequency class-D amplifier, which is the main factor of ElectroMagnetic Interference (EMI) occurrence. For the above reasons, we have to consider the switching loss reduction from a circuit-design viewpoint.

12.2.2 Class-E amplifier

The class-E amplifier was proposed by Sokal, which is known as a high-frequency high-efficiency amplifier [9–15]. By satisfying the switching conditions, which are called the class-E ZVS/ZDS conditions, the switching losses are reduced to be zero. Therefore, the class-E amplifier achieves high power-conversion efficiency at high frequencies. Figure 12.4(a) shows a circuit topology of the class-E amplifier. The class-E amplifier consists of dc-supply voltage V_{DD}, input inductance L_C, switching device S with shunt capacitance C_S, resonant filter $L - C$, and load resistance R. The most important component in the class-E amplifier is a shunt capacitance C_S. The shunt capacitance includes the drain-to-source parasitic capacitance of the switching device. By showing the shunt capacitance in the circuit topology explicitly, it is possible to discuss the circuit operation with parasitic capacitance effects.

Figure 12.4(c) shows example waveforms of the class-E amplifier at the nominal operation. When radio-frequency (RF) choke is used as the input inductance, the combination of the dc-supply voltage V_{DD} and inductance L_C is regarded as input current source. Similar to the class-D amplifier, the class-E amplifier includes the switching device S, which works as an ideal switch. Therefore, the equivalent circuit

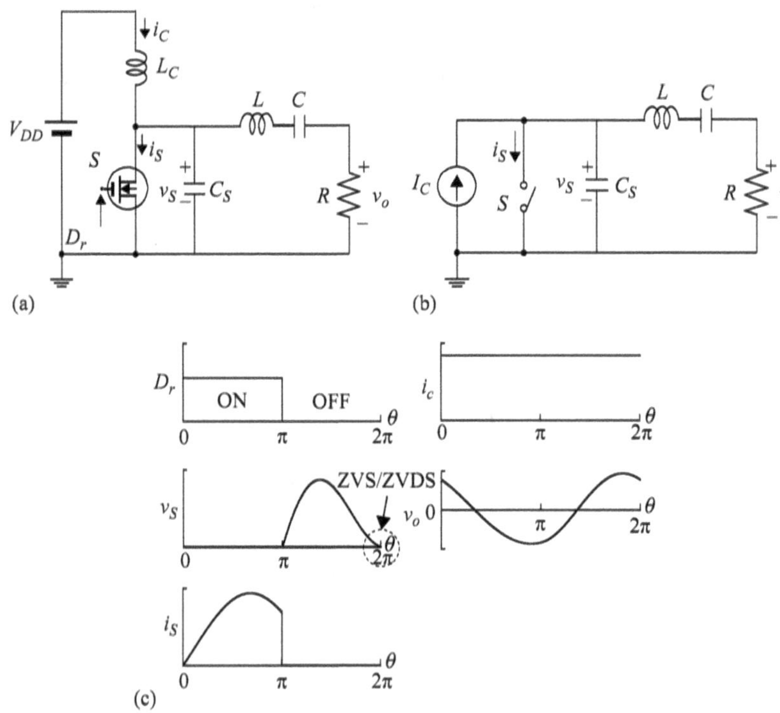

Figure 12.4 Class-E amplifier. (a) Circuit topology. (b) Equivalent circuit model. (c) Nominal waveforms

of the class-E amplifier is modeled as shown in Figure 12.4(b). When the switch is in ON state, the current flows through the switch and the voltages across the switch and the shunt capacitance are zero. Conversely, the current flows through the shunt capacitance when the switch is in OFF state. The slope of the current through the shunt capacitance $i_{CS} = \omega C_S dv_S/d\theta$ changes from positive to negative in switch-OFF state. Therefore, the pulse shape switch voltage appears as shown in Figure 12.4(c). The switch turns ON when the switch voltage is zero and the derivative of the switch voltage is also zero. Namely, the switch voltage satisfies

$$v_S(2\pi) = 0, \tag{12.3}$$

and

$$\left.\frac{dv_S}{d\theta}\right|_{\theta=2\pi} = 0, \tag{12.4}$$

simultaneously. The condition in (12.3) is called the ZVS condition. It is seen from (12.2) that the switching losses are zero when the ZVS condition is satisfied. When only the ZVS condition is considered, however, there is possibility that large switching loss appears when the switching timing is shifted slightly. The switching timing shift occurs due to the component tolerances. For suppressing this problem, the class-E amplifier applies the ZDS condition as given in (12.4), which is the feature of the class-E amplifier operation. By applying the ZDS condition, the switch current increases smoothly from zero posterior to turn-ON instant. Namely, smooth voltage and current waveforms can be realized in vicinity of the switch turn-ON. The power losses are a product of the switch voltage and current. Therefore, the class-E amplifier can be kept low switching losses even if the switching timing is shifted slightly from the nominal operation by applying the class-E ZVS/ZDS conditions. As a result, it is possible to operate the class-E amplifier at high frequencies. Because of smooth voltage and current waveforms, quite low noise yields in the class-E amplifier. We have to note, however, that the class-E ZVS/ZDS conditions are not achieved automatically. It is necessary to adjust and choose the component values for satisfying the class-E ZVS/ZDS conditions. The difficulty of the component-value decisions is one of the major problems of the class-E amplifier. Actually, many design strategies have been proposed since the class-E amplifier was proposed.

Generally, the input inductance is sufficiently high to generate direct current. Additionally, the resonant inductance is also high for achieving high-loaded quality factor. Namely, the adjustable components are the capacitances C_S and C. These component values should be determined properly for satisfying two switching conditions, namely the class-E ZVS/ZDS conditions. When the input voltage and the load resistance are given as design specifications, the component values are fixed uniquely.

12.2.3 Driver circuit

The switching devices of amplifiers are driven by the driving signal from a driver circuit. When a commercial driver is used, the drive waveform is usually square waveform, which is no problem at low frequencies. It is, however, a problem that the square waveform strains at high frequencies. The strain occurs due to the gate-to-source parasitic capacitance and provides difference of switch-ON duty ratio. The duty-ratio error affects the output power and the switching conditions. Therefore, we suffer from the power-conversion efficiency degradation because of driver operations. Additionally, high driving power is necessary for high-frequency and/or high-output power, which also increases the difficulty of the driver designs. The driver design is also a major problem of switching amplifiers for high-frequency and high-output power, in particular.

12.3 Tuned power oscillator

A fundamental configuration diagram of tuned power oscillators is shown in Figure 12.1(b). The constructional components of tuned power oscillators are almost the same as those of power amplifiers. The switching devices of tuned power oscillators are driven the voltage through the feedback network from the output voltage, which is different from the power amplifier. By applying the feedback voltage as the driving signal, the circuit works with self-oscillation and designers can be relieved from the implementation difficulty of driver circuit. The driving signal is not a square waveform but a sinusoidal waveform in the tuned power oscillator because the output voltage is regarded as a sinusoidal waveform. Namely, the feedback network should have roles to adjust phase shift between output voltage and gate-source voltage and amplitude of the gate voltage, which should be less than the permissive value.

12.3.1 Free-running class-E oscillator

Figure 12.5(a) shows a circuit topology of the free-running class-E oscillator [16–26]. The class-E oscillator consists of the class-E amplifier and feedback network C_1, C_2, and L_f. R_{d1} and R_{d2} give the bias voltage, which is the same as the threshold voltage V_{th}, for the gate of the switching device. Figure 12.6 shows example waveforms of the class-E oscillator with nominal conditions. The switching device of the class-E oscillator is driven by the feedback voltage v_f, which is from the output voltage v_o. The feedback voltage is a sinusoidal waveform because the feedback current flows through the resonant filter, which consists of L_f, gate-source parasitic capacitance C_g, and gate-source parasitic resistance r_g. Because C_g and r_g are fixed, which depend on the MOSFET type, the feedback network can be designed by choosing the component values of C_1, C_2, and L_f. By adjusting them, the amplitude and phase shift between the output voltage and the gate voltage are adjusted. The fundamental operation is the same as the class-E amplifier. Namely, the switch voltage achieves the class-E ZVS/ZVDS conditions as given in (12.3) and (12.4). Because of the class-E ZVS/ZDS conditions, the class-E oscillator achieves high power-conversion efficiency at high frequencies.

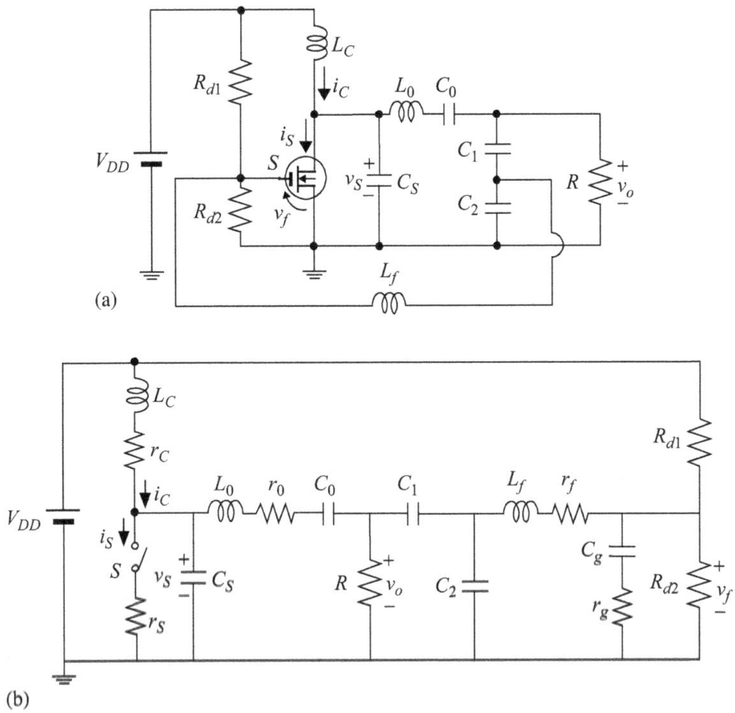

Figure 12.5 Class-E free-running oscillator. (a) Circuit topology. (b) Equivalent model

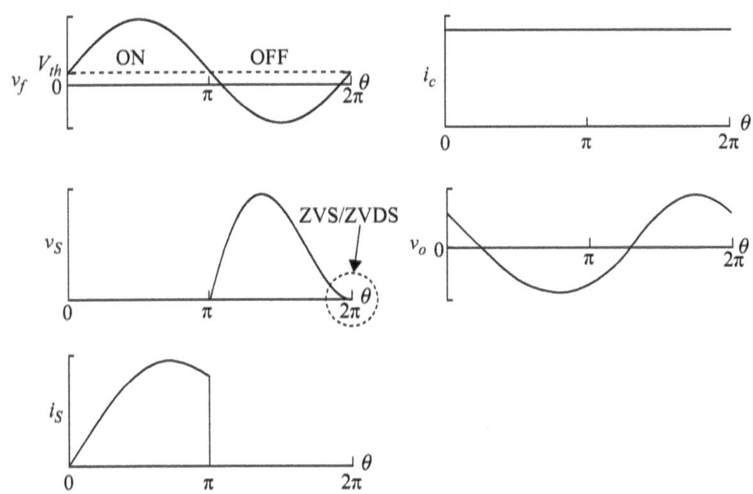

Figure 12.6 Nominal waveforms of the class-E free-running oscillator

It is a general problem of free-running oscillators that the output frequency has an error from the target frequency. This is because the free-running frequency is determined by circuit component values, which have tolerances in real components. Generally, accurate fixed frequency is necessary for most of the applications of tuned power oscillators. For obtaining the output frequency accurately and stably, injection-locking technique is effective.

In the class-E amplifier and oscillator, the switch-voltage waveform is smooth because of the class-E ZVS/ZDS conditions as shown in Figures 12.4(c) and 12.6. The switch-current waveform, however, includes jumps at every turn-off instant. When current fall time cannot be negligible, these jumps cause the power dissipations. For avoiding this problem, a fast-speed MOSFET or high driving power is required. The fast-speed MOSFET suffers from the implementation cost. The high-driving power deteriorates the power-added efficiency.

12.3.2 Injection-locked class-E oscillator

Figure 12.7 shows an example topology of the injection-locked class-E oscillator [27,28]. The small-power signal v_{inj} is injected to the gate terminal as shown in Figure 12.7. Because the injection-signal power is low, it is possible to obtain the injection-locked oscillator by just adding the injection signal to the original free-running oscillator. If the feedback voltage of the class-E free-running oscillator is synchronized with the injection-signal v_{inj}, the oscillation frequency is locked with the injection-signal frequency f_{inj}, which means the frequency of the output voltage is fixed with f_{inj}. It is easy to achieve synchronization as the injection-signal power increases. However, high power injection affects the waveforms of the feedback voltage and switch-ON duty ratio, which yields the design complexity. It is necessary to conduct the total design of the free-running oscillator and injection circuits for large perturbation. Additionally, low injection-signal power is good from a power-added efficiency perspective.

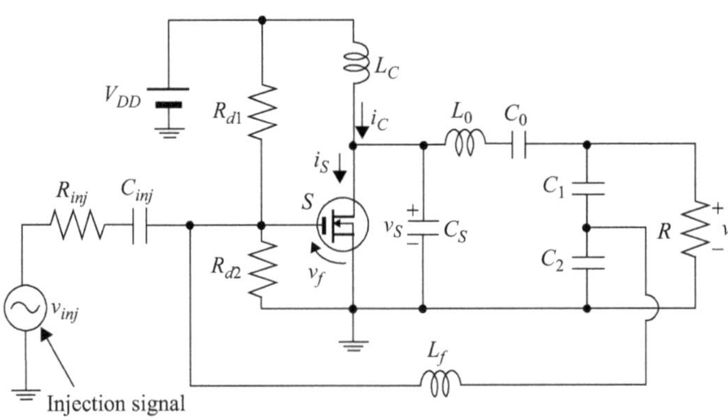

Figure 12.7 Circuit topology of the injection-locked class-E oscillator

12.3.3 Class-E$_M$ oscillator with second harmonic injection

Figure 12.8(a) shows a circuit topology of the class-E$_M$ oscillator with second harmonic injection [29], which is composed of the main circuit and the injection circuit. The injection circuit is usually operated as the class-E frequency doubler [30,31]. The

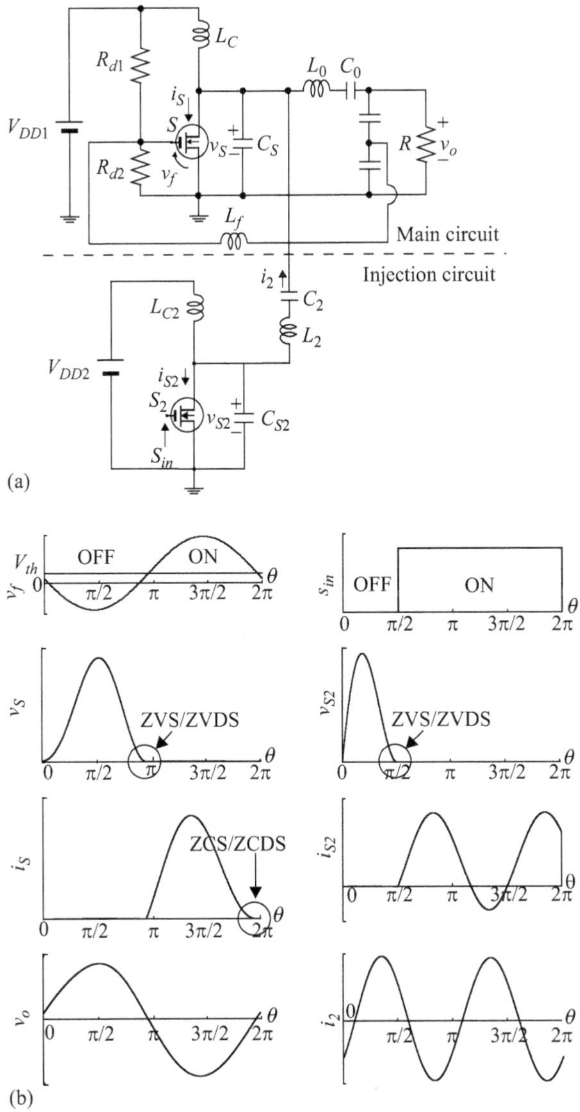

Figure 12.8 Class-E$_M$ oscillator with second harmonic injection. (a) Circuit topology. (b) Nominal waveforms

nominal waveforms of the oscillator are shown in Figure 12.8(b). The main circuit is driven by the feedback voltage v_f from the output voltage. The switch voltage v_{S1} satisfies the class-E ZVS/ZVDS conditions at transistor turn-on instant. Additionally, the switch current i_{S1} achieves the ZCS and zero-current-derivative switching (ZCDS) conditions simultaneously at the transistor turn-off instant. Because of the ZCS/ZCDS conditions, the waveforms of both the switch voltage and current at the transistor turn-off are also smooth. Because of these switching conditions, which are called the class-E$_{\mathrm{M}}$ ZVS/ZVDS/ZCS/ZCDS conditions, there are no jumps on the switch-voltage and switch-current waveforms in the main circuit. Therefore, the class-E$_{\mathrm{M}}$ amplifier enhances high power-conversion efficiency even if the main-circuit transistor has long turn-off-switching time and suppresses the implementation cost. For achieving the ZCS/ZCDS conditions in the class-E$_{\mathrm{M}}$ amplifier, the injection circuit is mandatory [32–34]. The injection circuit should provide the second-harmonic current i_2 with the proper phase shift and the proper amplitude for achieving the ZCS/ZCDS conditions in the main-circuit switch. The switch voltage v_{S1} is transformed into the sinusoidal output voltage v_o through the resonant filter $L_1 - C_1$. The injection circuit is driven by the input signal s_{in} whose fundamental frequency is the same as the output voltage. In other words, the output frequency is locked with the input frequency. In this sense, the proposed oscillator is regarded as one of the injection-locked oscillators.

From the above explanations, it can be stated that the injection circuit has multiple roles in the class-E$_{\mathrm{M}}$ oscillator. First, it offers the class-E$_{\mathrm{M}}$ ZVS/ZVDS/ZCS/ZCDS conditions, which enhance the power-conversion efficiency and allow to use a slow switching device. It is possible to reduce the circuit-implementation cost, especially, the main-circuit-MOSFET cost. Second, the output-voltage frequency is locked with the input-signal frequency, which is half as high as the injection-current frequency. Finally, the output power becomes high by adding the injection circuit, which is useful for high-power applications.

12.4 Design of free-running class-E oscillator

Roughly, there are two strategies for designing the class-E oscillator. One is the designs using analytical expressions and the other is those with numerical calculations.

12.4.1 Design strategies

The analysis-based design procedure for the free-running class-E oscillator in References 16,17 is divided into two parts, which are the design of the class-E amplifier [9–11] and that of the feedback network. In the feedback network design, the component values are determined by taking into account the phase shift between output voltage and gate-to-source voltage. The output voltage of the oscillator, namely the input voltage of the feedback network, can be assumed as a sinusoidal waveform. Therefore, fundamental-component analysis can be applied for expressing the phase shift relationship analytically. As a result, output Q should be high and it is impossible to design the class-E oscillator at low Q by the analysis-based design procedure.

Low Q oscillator is, however, useful for high power output for suppressing the voltage across the resonant filter. Additionally, parasitic resistances of the memory components and switch ON-resistance of the switching devices affect the power-conversion efficiency. Namely, it is preferred that the parasitic and ON-resistance of the switching device are included in the circuit model for comprehending the limitation of power-conversion efficiency. The numerical design procedure satisfies the above requirements perfectly. This section explains the numerical design algorithm in detail.

12.4.2 Numerical design procedure

This section presents a numerical design procedure of the class-E free-running oscillator. The fundamental consideration is the same as the analysis-based procedure. Namely, the feedback network is regarded as additional part to the class-E amplifier, which is designed for achieving that the phase of gate-to-source voltage agrees with the turn-on/off timing of the class-E amplifier.

12.4.2.1 Assumptions and parameters

First, the following parameters of the circuit are defined.

1. $\omega_0 = 2\pi f_0 = 1/\sqrt{L_0 C_0}$: The resonant angular frequency in the amplifier.
2. $\omega_f = 1/\sqrt{L_f C_g}$: The resonant angular frequency in the feedback network.
3. $A = (\omega_0/\omega)^2 = (f_0/f)^2$: The square value of the ratio of oscillator resonant frequency to the free-running frequency.
4. $B = C_0/C_S$: The ratio of resonant capacitance to shunt capacitance.
5. $H = L_0/L_C$: The ratio of oscillator resonant inductance to input inductance.
6. $J = C_0/C_1$: The ratio of oscillator resonant capacitance to a feedback network capacitance.
7. $K = C_1/C_2$: The ratio of output-voltage division capacitances.
8. $M = (\omega_f/\omega)^2 = (f_f/f)^2$: The square value of the ratio of feedback network resonant frequency to free-running frequency.
9. $Q = \omega L_0/R$: The loaded quality factor of the oscillator resonant circuit.
10. $Q_f = \omega L_f/r_g = 1/\omega M C_g r_g$: The loaded quality factor of the feedback network $L_f - C_g - r_g$.
11. D: The ON-duty ratio of the switching device.

In addition, the oscillator design given below is based on the following assumptions.

(a) The switching device S has zero switching times, infinite off resistance and ON-resistance r_S.
(b) The inductances have equivalent series resistance (ESR).
(c) All passive components including switch ON-resistance and ESRs work as linear components.
(d) The shunt capacitance includes parasitic capacitances of the MOSFET.
(e) High Q_f is assumed, namely $Q_f > 5$. High Q_f means that the feedback waveform v_f is sinusoidal.
(f) The MOSFET turns on at $\theta = 0$ and turns off at $\theta = 2\pi D$.

12.4.2.2 Circuit equation

Figure 12.5(b) shows an equivalent circuit model of the free-running class-E oscillator. The circuit operation in the interval $0 \leq \theta \leq 2\pi$ is considered. From the equivalent circuit model, the circuit equations are formulated as follow:

$$\begin{cases} \dfrac{di_C}{d\theta} = \dfrac{H}{QR}(V_D - v_S - r_C i_C), \\ \dfrac{dv_S}{d\theta} = ABQR\left(i_C - \dfrac{v_S}{R_S} - i\right), \\ \dfrac{dv}{d\theta} = AQRi, \\ \dfrac{di}{d\theta} = \dfrac{1}{QR}(v_S - v - v_1 - v_2 - r_0 i), \\ \dfrac{dv_1}{d\theta} = AJQR\left(i - \dfrac{v_1 + v_2}{R}\right), \\ \dfrac{dv_2}{d\theta} = AJKQR\left(i - \dfrac{v_1 + v_2}{R} - i_f\right), \\ \dfrac{di_f}{d\theta} = \omega MC_g\left(v_2 - v_g - \omega r_g C_g \dfrac{dv_g}{d\theta} - r_f i_f\right), \\ \dfrac{dv_g}{d\theta} = \dfrac{1}{\omega C_g}\left(i_f + \dfrac{V_D - v_f}{R_{d1}} - \dfrac{v_f}{R_{d2}}\right), \end{cases} \tag{12.5}$$

where $v_f = v_g + \omega r_g C_g \dfrac{dv_g}{d\theta}$. Additionally, R_S is the resistance of the switch S. From the assumption (f), we have

$$R_S = \begin{cases} r_S (0 \leq \theta \leq 2\pi D), \\ \infty \ (2\pi D \leq \theta \leq 2\pi). \end{cases} \tag{12.6}$$

In the numerical design procedure, a forced oscillation with the timing of switching in (12.6) is assumed. Additionally, the component values of the feedback network are determined for matching the phase of the feedback voltage v_f with the timing of switching.

When variation vector $\boldsymbol{x}(\theta) = [x_1, x_2, \ldots, x_8]^T = [i_C, v_S, v, i, v_1, v_2, i_f, v_g]^T \in \boldsymbol{R}^8$ and parameter vector $\boldsymbol{\lambda} = [A, B, H, Q, J, K, M, C_g, \omega, V_D, r_S, r_g, r_C, r_0, r_f, R, R_{d1}, R_{d2}]^T \in \boldsymbol{R}^{18}$ are prepared, (12.5) can be written as

$$\frac{d\boldsymbol{x}}{d\theta} = f(\theta, \boldsymbol{x}, \boldsymbol{\lambda}). \tag{12.7}$$

12.4.2.3 Conditions for the design

It is assumed that (12.5) has a solution $\boldsymbol{x}(\theta) = \boldsymbol{\varphi}(\theta, \boldsymbol{x}_0, \boldsymbol{\lambda}) = [\varphi_1, \varphi_2, \ldots, \varphi_8]^T$ is defined on $-\infty < \theta < \infty$ with every initial condition $\boldsymbol{x}_0$ and every $\boldsymbol{\lambda} : \boldsymbol{x}(0) = \boldsymbol{\varphi}(0, \boldsymbol{x}_0, \boldsymbol{\lambda})$. If the oscillator is in the steady state,

$$\boldsymbol{\varphi}(2\pi, \boldsymbol{x}_0, \boldsymbol{\lambda}) - \boldsymbol{\varphi}(0, \boldsymbol{x}_0, \boldsymbol{\lambda}) = 0 \in \boldsymbol{R}^8 \tag{12.8}$$

is satisfied, which is the boundary conditions between $\theta = 0$ and $\theta = 2\pi$ for ensuring the steady-state operation.

For design of the class-E free-running oscillator, the class-E ZVS/ZDS conditions should be considered for achieving high power-conversion efficiency at high frequencies. The class-E switching conditions are rewritten as

$$\varphi_2(2\pi, \boldsymbol{x}_0, \boldsymbol{\lambda}) = 0, \tag{12.9}$$

and

$$\left.\frac{d\varphi_2(\theta, \boldsymbol{x}_0, \boldsymbol{\lambda})}{d\theta}\right|_{\theta=2\pi} = ABQR(\varphi_1(2\pi, \boldsymbol{x}_0, \boldsymbol{\lambda}) - \varphi_4(2\pi, \boldsymbol{x}_0, \boldsymbol{\lambda})) = 0 \tag{12.10}$$

are given. Additionally, the phase matching of the feedback voltage should be also achieved for free-running oscillation. The phase matching condition is that the feedback voltage v_f is equal to V_{th} when switch S turns on. Therefore, we have

$$\varphi_8(0, \boldsymbol{x}_0, \boldsymbol{\lambda}) + \omega r_g C_g \frac{\varphi_8(0, \boldsymbol{x}_0, \boldsymbol{\lambda})}{d\theta} = V_{th} \tag{12.11}$$

and

$$\varphi_8(2\pi D, \boldsymbol{x}_0, \boldsymbol{\lambda}) + \omega r_g C_g \frac{\varphi_8(2\pi D, \boldsymbol{x}_0, \boldsymbol{\lambda})}{d\theta} = V_{th}. \tag{12.12}$$

Because of the assumption (e), the feedback voltage is expressed by the sum of the pure sinusoidal waveform and the dc-bias voltage. Namely, we can consider either (12.11) or (12.12). The duty ratio is adjusted by the dc-bias voltage, which is generated by R_{d1} and R_{d2}. The dc-bias voltage is expressed as

$$V_{bi} = V_{th} - V_m \sin\left[(\pi - 2\pi D)/2\right]. \tag{12.13}$$

From V_{bi}, the relationship between R_{d1} and R_{d2} is

$$\frac{V_{bi}}{V_{DD}} - \frac{R_{d1}}{R_{d1} + R_{d2}} = 0. \tag{12.14}$$

The component-value decisions of R_{d1} and R_{d2} are independent of other conditions. When R_{d1} and R_{d2} are sufficiently large, the power losses on them can be ignored.

From the above considerations, it is recognized that the design problem of the class-E free-running oscillator narrows to the solution derivations of the algebraic equations (12.8)–(12.11). From the above conditions, we have 11 algebraic equations and 8 unknown initial values, namely $\boldsymbol{x}_0 \in \boldsymbol{R}^8$. Therefore, three parameters can be set as design parameters from $\boldsymbol{\lambda} \in \boldsymbol{R}^{18}$. In this design example, A, B, and M are set as unknown parameters. The other parameters are given as the design specifications. As a result, we have the design equations as

$$F(\boldsymbol{x}_0, A, B, M) = \begin{bmatrix} \boldsymbol{\varphi}(2\pi, \boldsymbol{x}_0, A, B, M) - \boldsymbol{\varphi}(0, \boldsymbol{x}_0, A, B, M) \\ \varphi_2(2\pi, \boldsymbol{x}_0, A, B, M) \\ \varphi_1(2\pi, \boldsymbol{x}_0, A, B, M) - \varphi_4(2\pi, \boldsymbol{x}_0, A, B, M) \\ \varphi_8(2\pi D, \boldsymbol{x}_0, A, B, M) + \omega r_g C_g \dfrac{\varphi_8(2\pi D, \boldsymbol{x}_0, A, B, M)}{d\theta} - V_{th} \end{bmatrix} = \boldsymbol{O}. \tag{12.15}$$

12.4.2.4 Numerical calculation algorithm

Equation (12.15) is solved by using Newton's method that is the general algorithm to solve the algebraic equations. When the unknown values are expressed as $\boldsymbol{u} = [\boldsymbol{x}_0^T, A, B, M]^T$, the computations

$$\boldsymbol{u}^{k+1} = \boldsymbol{u}^k - F(\boldsymbol{u}^k) F'^{-1}(\boldsymbol{u}^k) \tag{12.16}$$

are iterated for $||\boldsymbol{u}^{k+1} - \boldsymbol{u}^k|| < \delta$ in order to find the design parameters, where $F'(\boldsymbol{u}^k)$ is an Jacobian matrix of $F(\boldsymbol{u}^k)$, k is an iteration number, and $\delta \ll 1$. When $\boldsymbol{u}^k$ in this algorithm is converged, $\boldsymbol{u}^k$ is a solution of (12.15). For computations of (12.16), $\boldsymbol{\varphi}(2\pi, \boldsymbol{u}^k)$ can be derived by applying Runge–Kutta method to (12.5). Additionally, the elements in Jacobian matrix $F'(\boldsymbol{u}^k)$ of (12.16) can be obtained by solving the first-order variational equations [14] or by calculating approximate partial differentials [15].

For solving the algebraic equations in (12.15) by using Newton's method, it is important to set the proper initial values for iterative calculations. First of all, by using certain values of the variations $\boldsymbol{u}^0$, the values of $F(\boldsymbol{u}^0)$ are obtained as $F(\boldsymbol{u}^0) = \boldsymbol{a}$, which can be obtained from long-duration Runge–Kutta method to (12.5). Second, the algebraic equations $F(\boldsymbol{u}^1) = (1 - \varepsilon)\boldsymbol{a}$ are solved by using the Newton's method with $\boldsymbol{u}^0$ as initial values, where $\varepsilon \leq 1/K \ll 1$ and K is an integer value. The algebraic equations $F(\boldsymbol{u}^k) = (1 - k\varepsilon)\boldsymbol{a}$ are solved by using the Newton's method with the initial values of $\boldsymbol{u}^{k-1}$. After K iterations, we can obtain the solution of the algebraic equations $F(\boldsymbol{u}^K) = (1 - K\varepsilon)\boldsymbol{a} = \boldsymbol{0}$.

From the above computations, the unknown parameters $\boldsymbol{u}$ can be found, and the design values, i.e. A, B, and M are determined.

12.4.3 Design examples and experimental measurements

In this chapter, design examples of three parameter specifications and their laboratory experimental results are shown. The validity and accuracy of the design procedure can be confirmed from these experimental results.

The basic design specifications were given as: operating frequency $f = 1$ MHz, switch-ON duty ratio $D = 0.5$, dc-supply voltage $V_{DD} = 12$ V, and load resistance $R = 25\ \Omega$. Under these specifications, three types of the class-E free-running oscillators were designed, namely for high Q and high input inductance, low

Table 12.1 The parameters of IRF530 MOSFET, which were applied to laboratory experiments

V_{th}	r_S	C_g	r_g
3.0 V	0.16 Ω	1.78 nF	2.17 Ω

Q and high input inductance, and high Q and low input inductance. The IRF530 MOSFET was selected as switching devices for all the oscillators, whose parameters are given in Table 12.1. In Table 12.1, V_{th} and r_S are obtained from the Field-Effect Transistor (FET) datasheet. r_g and C_g were measured by impedance analyzer. Because of $v_{th} = 3$ V and $V_{DD} = 12$ V, $R_{d1} = 750$ kΩ and $R_{d2} = 250$ kΩ were obtained from (12.14).

12.4.3.1 High Q and high input inductance

For high Q and high input inductance design, $Q = 10$ and $H = 0.01$ were also given as design specifications. From $f = 1$ MHz, $R = 25$ Ω, and $Q = 10$, the resonant inductance are fixed as $L_0 = 39.8$ μH. From this L_0 value and $H = 0.01$, the input inductance is $L_c = 39.8$ mH. Therefore, input and resonant inductors were made prior to numerical optimizations. From the implemented inductors, ESRs of L_C and L_0 could be measured by impedance meter as $r_{LC} = 0.119$ Ω and $r_{L0} = 0.78$ Ω. Additionally, $r_f = 0.3$ Ω was assumed from the relationship between L_0 and r_0.

By applying the above parameters to design algorithm, the unknown design parameters were obtained as $A = 0.931$, $B = 0.441$, and $M = 0.988$. From the obtained parameter values, the component values of the free-running class-E oscillator were derived as given in Table 12.2. Figure 12.9(a) shows the experimental and numerical waveforms. It is seen from Figure 12.9(a) that the input current i_C and the output voltage V_o were dc-current waveform and sinusoidal one, respectively, because of low H and high Q. It is also seen from Figure 12.9 that the switch-voltage waveform satisfied the class-E ZVS/ZDS conditions. Additionally, the phase shift between the output voltage and the driving signal in the circuit experiment agreed with the numerical prediction well. It can be stated from Figure 12.9(a) that the deign procedure with numerical algorithm provides accurate design values.

12.4.3.2 Other design examples

Generally, analytical expressions of the class-E amplifier and oscillator are based on the assumptions of dc input current and pure sinusoidal output voltage. Therefore, it is impossible to design the class-E oscillator for low Q or low input inductance by the theoretical approach. The numerical design strategy explained in this chapter, however, requires only the circuit equations and constraint conditions. Following the design procedure in this chapter, it is possible to design the class-E free-running oscillator at any parameters.

Table 12.2 Theoretical design values of free-running class-E oscillators

	$Q = 10$ and $H = 0.01$	$Q = 3$ and $H = 0.01$	$Q = 10$ and $H = 1$
f_{free}	1 MHz	1 MHz	1 MHz
D	0.5	0.5	0.5
V_{DD}	12.0 V	12.0 V	12.0 V
R	25 Ω	25 Ω	25 Ω
L_C	3.98 mH	1.20 mH	39.8 μH
L_0	39.8 μH	12.0 μH	39.8 μH
L_f	14.6 μH	14.5 μH	14.7 μH
C_S	1.67 nF	1.96 nF	1.84 nF
C_0	735 pF	3.97 nF	729 pF
C_1	4.17 nF	4.36 nF	3.34 nF
C_2	155 nF	166 nF	112 nF
R_{d1}	750 kΩ	750 kΩ	750 kΩ
R_{d2}	250 kΩ	250 kΩ	250 kΩ
r_{LC}	0.119 Ω	0.042 Ω	0.780 Ω
r_{L0}	0.780 Ω	0.257 Ω	0.823 Ω
r_{Lf}	0.3 Ω	0.3 Ω	0.3 Ω
I_c	0.415 A	0.352 A	0.364 A
V_o	10.4 V	9.54 V	9.64 V

Second design example gave specifications of $Q = 3$ and $H = 0.01$, which means low Q and high input inductance. In this case, non-sinusoidal output voltage waveform may appear. Following the procedures of the first design example, the component values were obtained as given in Table 12.2. Figure 12.9(b) shows experimental and numerical waveforms. Compared with Figure 12.9(a), it is seen that the output current included high-frequency component. Nevertheless, the switch voltage achieved the class-E ZVS/ZDS conditions perfectly and the phase shift is also agreed with numerical prediction.

The specifications of third design example are $Q = 10$ and $H = 1$, which is an example of high Q and low input inductance. The input current ripple appears in these specifications. Table 12.2 also gives the numerical component values and Figure 12.9(c) shows experimental and numerical waveforms. It is seen from Figure 12.9(c) that the input current includes ac component. The constraint conditions, namely the class-E ZVS/ZDS conditions and the phase shift condition, were achieved in the experimental waveforms, which showed the validity, accuracy, and effectiveness of the numerical design procedure.

It can be stated from three design examples with experimental results that the experimental waveform achieved the constrained conditions regardless of design specifications. By replacing the feedback voltage as driving signal, namely by considering the phase shift and amplitude characteristics of the feedback network, it is possible to design the class-E free-running oscillator with constraint conditions.

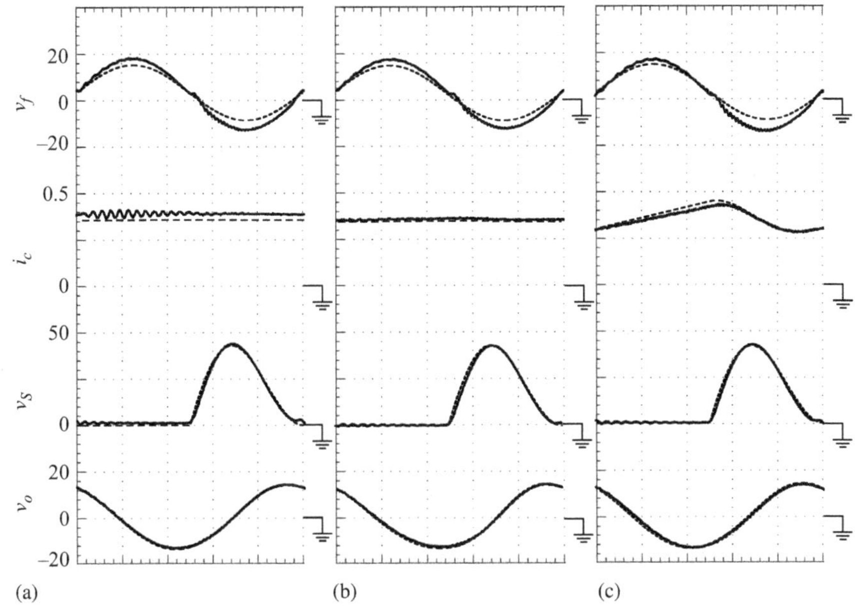

Figure 12.9 Waveforms of the class-E free-running oscillator for (a) $Q = 10$ and $H = 0.01$, (b) $Q = 3$ and $H = 0.01$, and (c) $Q = 10$ and $H = 1$. Solid line: experimental waveforms and dotted line: numerical waveforms. Vertical: v_f: 20 V/div, i_C: 0.25 A/div, v_S: 25 V/div, and v_o: 20 V/div. Horizontal: 0.5 μs/div

This design strategy can be applied to designs of other power oscillators with feedback network.

12.5 Conclusion

This chapter has presented fundamental operation and design of the class-E tuned power oscillator. The sinusoidal feedback voltage from output one is used as the driving signal in tuned power oscillator. Because of the class-E ZVS/ZDS conditions, the class-E oscillator achieves high power-conversion efficiency at high frequencies. It is an effective design strategy that a forced oscillation is assumed and the component values of the feedback network are determined for matching the phase of the feedback voltage v_f with the timing of switching. Additionally, it is necessary to consider the switching conditions. For carrying out the optimization, the Newton's method is applied in this chapter. The accuracy and effectiveness of the design procedure were confirmed from the experimental results, which agreed with numerical predictions well. It can be seen from the experimental results that the tuned power oscillator, which satisfy the class-E ZVS/ZDS conditions, achieved high power-conversion efficiency at high frequencies.

References

[1] A. Grebennikov and N. O. Sokal, *Switchmode RF Power Amplifiers*, Newnes, Burlington, MA, 2007.

[2] M. K. Kazimierczuk, *RF Power Amplifier,* John Wiley & Sons, West Sussex, 2008.

[3] M. Albulet, *RF Power Amplifier,* Noble, Atlanta, GA, 2001.

[4] S. C. Cripps, *RF Power Amplifier for Wireless Communications*, 2nd edn., Artech House, London, 2006.

[5] P. J. Vaxandall, "Transistor sinewave oscillators," *Proc. IEE*, vol. 106, no. 16, pp. 748–758, May 1959.

[6] O. Lucia, J. M. Burdio, I. Millan, J. Acero, and L. A. Barragan, "Efficiency-oriented design of ZVS half-bridge series resonant inverter with variable frequency duty cycle control," *IEEE Trans. Power Electron.*, vol. 25, no. 7, pp. 1671–1674, Jul. 2010.

[7] M. K. Kazimierczuk, "Class-D zero-voltage-switching inverter with only one shunt capacitor," *IEE Proc. B*, vol. 139, no. 5, pp. 449–456, Sep. 1992.

[8] X. Wei, H. Sekiya, T. Nagashima, M. K. Kazimierczuk, and T. Suetsugu, "Steady-state analysis and design of class-D ZVS inverter at any duty ratio," *IEEE Trans. Power Electron.*, vol. 31, no. 1, pp. 394–405, Jan. 2016.

[9] N. O. Sokal and A. D. Sokal, "Class E – A new class of high-efficiency tuned single-ended switching power amplifiers," *IEEE J. Solid-State Circuits*, vol. 10, no. 3, pp. 168–176, Jun. 1975.

[10] F. H. Raab, "Idealized operation of the class E tuned power amplifier," *IEEE Trans. Circuits Syst.*, vol. 24, no. 12, pp. 725–735, Dec. 1977.

[11] M. K. Kazimierczuk and K. Puczko, "Exact analysis of class E tuned power amplifier at any Q and switch duty cycle," *IEEE Trans. Circuits Syst.*, vol. CAS-34, no. 2, pp. 149–159, Feb. 1987.

[12] X. Wei, H. Sekiya, S. Kuroiwa, T. Suetsugu, and M. K. Kazimierczuk, "Design of class-E amplifier with MOSFET linear gate-to-drain and nonlinear drain-to-source capacitances," *IEEE Trans. Circuits Syst. I*, vol. 58, no. 10, pp. 2556–2565, Oct. 2011.

[13] T. Nagashima, X. Wei, T. Suetsugu, M. K. Kazimierczuk, and H. Sekiya, "Waveform equations, output power, and power conversion efficiency for class-E inverter outside nominal operation," *IEEE Trans. Ind. Electron.*, vol. 61, no. 4, pp. 1799–1810, 2014.

[14] H. Sekiya, I. Sasase, and S. Mori, "Computation of design values for class E amplifier without using waveform equations," *IEEE Trans. Circuits Syst. I*, vol. 49, no. 7, pp. 966–978, Jul. 2002.

[15] H. Sekiya, T. Ezawa, and Y. Tanji, "Design procedure for class E switching circuits allowing implicit circuit equations," *IEEE Trans. Circuits Syst. I*, vol. 55, no. 11, pp. 3688–3696, Dec. 2008.

[16] J. Ebert and M. Kazimierczuk, "Class E high-efficiency tuned power oscillator," *IEEE J. Solid-State Circuits*, vol. 16, no. 2, pp. 62–66, Apr. 1981.

[17] D. V. Chernov, M. K. Kazimierczuk, and V. G. Krizhanovski, "Class-E MOSFET low-voltage power oscillator," in *Proceedings of the IEEE ISCAS*, Phoenix, AZ, May 2002, vol. 5, pp. 509–512.

[18] M. K. Kazimierczuk, V. G. Krizhanovski, J. V. Rassokhina, and D. V. Chernov, "Class-E MOSFET tuned power oscillator design procedure," *IEEE Trans. Circuits Syst. I*, vol. 52, no. 6, pp. 1138–1147, Jun. 2005.

[19] H. Hase, H. Sekiya, J. Lu, and T. Yahagi, "Novel design procedure for MOSFET class-E oscillator," *IEICE Trans. Fund.*, vol. E87-A, no. 9, pp. 2241–2247, Sep. 2004.

[20] V. G. Krizhanovski, D. V. Chernov, and M. K. Kazimierczuk, "Low-voltage electronic ballast based on class E oscillator," *IEEE Trans. Power Electron.*, vol. 22, no. 3, pp. 863–870, May 2007.

[21] L. R. Nerone, "Novel self-oscillating class E ballast for compact fluorescent lamps," *IEEE Trans. Power Electron.*, vol. 16, no. 2, pp. 175–183, Mar. 2001.

[22] H. Hase, H. Sekiya, J. Lu, and T. Yahagi, "Resonant dc/dc converter with class E oscillator," *IEEE Trans. Circuits Syst. I*, vol. 53, no. 9, pp. 2025–2035, Sep. 2006.

[23] T. Andersen, S. K. Christensen, A. Knott, and M. A. E. Andersen, "A VHF class E DC–DC converter with self-oscillating gate driver," in *Proceedings of the IEEE APEC*, Fort Worth, TX, March 2011, pp. 885–891.

[24] C. M. Zierhofer and E. S. Hochmair, "High-efficiency coupling-insensitive transcutaneous power and data transmission via an inductive link," *IEEE Trans. Biomed. Eng.*, vol. 37, no. 7, pp. 716–722, Jul. 1990.

[25] M. Qingyun, M. R. Haider, Y. Song, and S. K. Islam, "Power-oscillator based high efficiency inductive power-link for transcutaneous power transmission," in *Proceedings of the IEEE MWSCAS*, Seattle, WA, August 2010, pp. 537–540.

[26] F. Ellinger, U. Lott, and W. Bachtold, "Design of a low-supply-voltage high-efficiency class-E voltage-controlled MMIC oscillator at C-band," *IEEE Trans. Microw. Theory Tech.*, vol. 49, no. 1, pp. 203–206, Jan. 2001.

[27] M. K. Kazimierczuk, V. G. Krizhanovski, J. V. Rassokhina, and D. V. Chernov, "Injection-locked class-E oscillator," *IEEE Trans. Circuits Syst. I*, vol. 53, no. 6, pp. 1214–1222, Jun. 2006.

[28] T. Nagashima, X. Wei, H. Tanaka, and H. Sekiya, "Locking range derivations for injection-locked class-E oscillator applying phase reduction theory," *IEEE Trans. Circuits Syst. I*, vol. 61, no. 10, pp. 2904–2911, Oct. 2014.

[29] R. Miyahara, X. Wei, T. Nagashima, T. Kousaka, and H. Sekiya, "Design of class-E_M oscillator with second harmonic injection," *IEEE Trans. Circuits Syst. I*, vol. 59, no. 10, pp. 2456–2467, Oct. 2012.

[30] R. E. Zulinski and J. W. Steadman, "Idealized operation of class E frequency multipliers," *IEEE Trans. Circuits Syst. I*, vol. 33, no. 12, pp. 1209–1218, Dec. 1986.

[31] M. Albulet, "Analysis and design of the class E frequency multipliers with RF choke," *IEEE Trans. Circuits Syst. I*, vol. 42, no. 2, pp. 95–104, Feb. 1995.

[32] A. Telegdy, B. Molnár, and N. O. Sokal, "Class-E_M switching-mode tuned power amplifier – High efficiency with slow-switching transistor," *IEEE Trans. Microw. Theory Tech.*, vol. 51, no. 6, pp. 1662–1676, Jun. 2003.
[33] R. Miyahara, H. Sekiya, and M. K. Kazimierczuk, "Novel design procedure for class-E_M power amplifiers," *IEEE Trans. Microw. Theory Tech.*, vol. 58, no. 12, pp. 3607–3616, Dec. 2010.
[34] X. Wei, T. Nagashima, M. K. Kazimierczuk, H. Sekiya, and T. Suetsugu, "Analysis and design of class-E_M power amplifier," *IEEE Trans. Circuits Syst. I*, vol. 61, no. 4, pp. 976–986, Apr. 2014.

Chapter 13
Ring oscillators and self-timed rings in true random number generators

Viktor Fischer[1], *Patrick Haddad*[2] *and Abdelkarim Cherkaoui*[3]

In this chapter, we explain important role of ring oscillators in generating random bit streams in logic devices. We consider three types of ring oscillators used as sources of jittery clock signals: single-event ring oscillators, multi-event ring oscillators with collision and multi-event ring oscillator without collision, i.e. self-timed rings. We present some representative examples of oscillator-based true random number generators and their characteristics.

13.1 Introduction

Random numbers and random bit streams are unavoidable in data security systems. They are used as confidential keys and as other critical security parameters including masks in side channel attack countermeasures, initialization vectors, padding values, nonces in challenge/response protocols. Random numbers aimed at cryptographic applications must fulfil two basic security requirements: they must have good statistical properties and they must be unpredictable, i.e. their output should not be distinguishable from that of an ideal random number generator (RNG) and its past and future values must not be guessable.

Random numbers can be generated by deterministic random number generators (DRNGs) or physical true random number generators (PTRNG, for short TRNG). DRNGs are usually faster and they have better statistical properties which are ensured mathematically by the underlying algorithm. Compared to DRNGs, TRNGs have suboptimal statistical characteristics and are slower.

[1]Laboratoire Hubert Curien UMR 5516, Université de Lyon, Saint-Étienne, France
[2]Agence nationale de la sécurité des systèmes d'information (ANSSI), Paris, France
[3]Laboratoire TIMA UMR 5159, Grenoble Institute of Technology, Grenoble, France

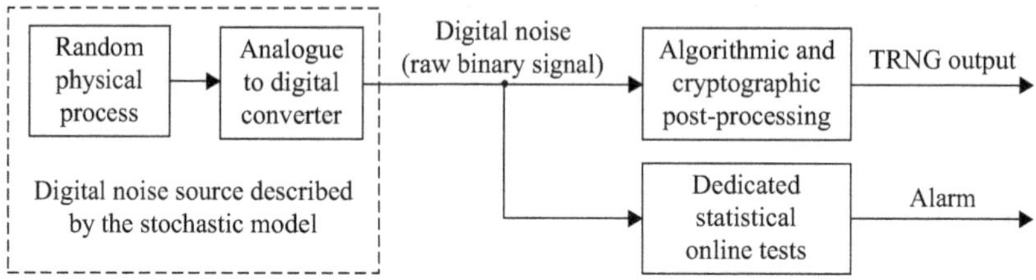

Figure 13.1 General structure of a TRNG

Unpredictability can be guaranteed by a computational difficulty to predict numbers generated in pseudo-random number generators, but it is preferable if unpredictability is ensured by a robust non-manipulable and stationary random physical process, which is used as a source of randomness.

For security reasons, critical security data like confidential keys must never leave a security area in clear. Therefore, RNGs must be implemented within a protected security zone, e.g. inside a cryptographic system on chip.

While the preferable security solution, TRNGs, exploits analogue physical processes, cryptographic systems on chip are mostly implemented using digital technologies. For that reason, recent TRNGs mainly exploit randomness in the timing of events of logic signals, e.g. the rising and falling edges of the clock signals generated in free-running ring oscillators, transition effect ring oscillators (TEROs), and self-timed rings (STRs).

13.2 Design of TRNGs

A TRNGs is composed of several basic and optional blocks (Figure 13.1). The digital noise source exploits an analogue physical process, e.g. an analogue noisy signal, as a source of randomness. The analogue signal exploited must first be converted into a digital form. If necessary, an entropy harvesting technique is applied to obtain a good-quality raw binary signal: a digital noise. If the statistical quality of the digital noise is not suitable for the application concerned, it can be enhanced using an optional post-processing algorithm and/or a cryptographic algorithm.

The main role of the post-processing block is to render the distribution of the generated numbers close to uniform, to reduce the correlation between subsequent numbers, and to cryptographically isolate the output of the generator from its core, e.g. through a one-way cryptographic function.

The unpredictability of the PTRNG is guaranteed by the entropy rate at the output of the generator. This entropy rate can only be specified using a stochastic model, i.e. the entropy cannot be measured or obtained from observation of the values of the generated numbers.

The stochastic model represents a set of distributions of generated numbers, which contains all possible distributions depending on the operating conditions concerned. The stochastic model should be based on verified physical assumptions and must include the statistical behaviour of the source of randomness and the entropy harvesting mechanism. It must also make it possible to specify dependences between generated numbers: only short-term dependences are allowed (i.e. small dependences between two successive values).

The physical parameters of the generator and the environmental parameters can change during the lifetime of the generator because of the effects of ageing or due to malicious attacks. It is thus important to check the statistical parameters of generated numbers online using dedicated online statistical tests based on the existing stochastic model. These tests must be robust and fast enough to avoid false alarms and/or outputting of low-quality random numbers.

The detailed security requirements of modern PTRNG designs are given in European and American standards [1,2].

13.3 Electric noise and clock jitter as a source of randomness

Electric noise represents random fluctuation of the amplitude and/or of the phase of an electrical signal. Because of its random character, it is often used as a source of randomness in TRNGs.

13.3.1 Electric noise in clock generators

Electric noises from various physical sources are inherently present in all electronic circuits. They can be divided into two categories: frequency-independent noises (e.g. white noise) and frequency-dependent noises (e.g. low-frequency pink noise).

The electric noise sources are characterized by the power spectral density (PSD) of the generated noisy electric current expressed as follows:

$$S_I(f) = \lim_{\Delta f \to 0} \frac{\overline{i_n^2(t)}}{\Delta f}, \tag{13.1}$$

where $\overline{i_n^2(t)}$ is the mean square value of the noisy current.

In logic devices, the thermal noise is widely used as a source of randomness. Its PSD is constant to frequencies as high as 10^{14} Hz [3]. It is thus often considered to be frequency independent (white noise).

The thermal noise is generated by the random thermal motion of charge carriers inside an electrical conductor and it is independent from the voltage applied. It was described for the first time by Nyquist in 1928 [4]. Thermal noise is unavoidable in all types of conductors at non-zero temperatures (over −273.15°C). It is characterized by the PSD of the random thermal current defined as follows:

$$S_{I_{th}}(f) = 4 \cdot k_B \cdot T \cdot g, \tag{13.2}$$

where k_B is the Boltzmann constant, T is the absolute temperature and g is the electrical conductance of the conductor.

In a transistor, the charge carriers fluctuate between the drain and the source. The current coming from the thermal noise source is added to the current flowing between the two transistor electrodes. As follows from (13.2), its variance increases with an increase in temperature and with an increase with the conductance between transistor electrodes.

Unlike thermal noise, flicker noise, another noise which has a major impact on oscillators in logic devices, is a frequency-dependent noise. Flicker noise depends on the current, the homogeneity of the material, its volume, and on the frequency. It is characterized by a PSD which is proportional to $\frac{1}{f^\alpha}$, where $\alpha \approx 1$. It is therefore also called the $1/f$ noise. This noise was first reported by McWhorter in 1957 [5].

The University of Berkeley developed accurate models of transistors which include noise sources. These models are used in many electrical simulators. In the Berkeley Short-Channel IGFET Model BSIM4 presented in Reference 6, the authors express the PSD of the flicker noise current as

$$S_{I_{fl}}(f) = \frac{KF \cdot I_{DS}}{C_{oxe} \cdot L_{eff}^2 \cdot f}, \tag{13.3}$$

where L_{eff} is the length of the transistor channel, C_{oxe} is silicon oxide capacitance, I_{DS} is the average value of the current between transistor's drain and source and KF is a constant which varies with the material.

As thermal noise and flicker noise originate from different physical phenomena, they are assumed to be mutually independent. Consequently, the PSD of the noise current between the drain and source (denoted $S_{I_n}(f)$), which is composed of both the thermal noise and the flicker noise component, is given as

$$S_{I_n}(f) = \frac{KF \cdot I_{DS}}{C_{oxe} \cdot L_{eff}^2 \cdot f} + 4 \cdot k_B \cdot T \cdot g = \frac{a_{fl}}{f} + a_{th}, \tag{13.4}$$

where a_{fl} and a_{th} are positive constants, which correspond to the impact of the flicker and thermal noises on the frequency/phase difference at the output of the transistor. The two noises have very different impacts on the stability of the generated clock period over time.

13.3.2 Jitter of the generated clock signal

Oscillator-based TRNGs in logic devices use the fact that the analogue random noise signals are transformed into the variation of the phase of the generated clock signal over time in the oscillator circuitry. In the frequency domain, this variation is considered to be a phase noise and in the time domain, a phase jitter.

To formally define the difference between phase noise and phase jitter, let us express the value of the clock signal V_{out} as a function of time t:

$$V_{out}(t) = P\left(\frac{2\pi}{T} \cdot t + \phi_0 + \phi(t)\right), \tag{13.5}$$

where P is a 2π-periodic square function, T is the average period of V_{out}, ϕ_0 is its initial phase and $\phi(t)$ is the *phase noise* (in radians).

Let t_n be the time instant corresponding to the nth rising edge of the signal V_{out}, then:

$$t_n = \left[n - \frac{\phi_0 + \phi(t_n)}{2\pi}\right]T. \tag{13.6}$$

The *phase jitter* (expressed in seconds) is then equal to $\frac{\phi(t_n)}{2\pi}T$. For clock applications such as random number generators in digital systems, time-domain measurements are preferable, since most specifications which are of concern involve time-domain values.

In addition to the absolute phase jitter which quantifies overall phase instability in the time domain, there are two other widely used time-domain measurements of jitter: period jitter and cycle-to-cycle jitter. Both are somehow related to phase jitter.

The *period jitter* represents variations in the clock period values over time. Consequently, the nth realization of the period jitter can be expressed as:

$$J_n = (t_{n+1} - (n+1)\cdot T) - (t_n - n \cdot T) = t_{n+1} - t_n - T,$$

and from (13.6), the period jitter J_n is defined as follows:

$$J_n = \frac{\phi(t_{n+1}) - \phi(t_n)}{2\pi}T. \tag{13.7}$$

The period jitter can thus be obtained as a first difference in the phase jitter.

Similarly, it can be shown that the *cycle-to-cycle jitter*, which represents a variation in the time interval between two successive rising edges of V_{out}, can be expressed as:

$$C_n = (t_{n+2} - t_{n+1}) - (t_{n+1} - t_n) = J_{n+1} - J_n.$$

Consequently, the cycle-to-cycle jitter can be seen as a first difference in the period jitter.

In order to understand the use of the jittery clock signal as a source of randomness, it is important to analyse how jitter behaves over time. First, let us consider the time interval between two rising edges appearing at time t_1 and t_2 separated by N clock periods.

The delay between two rising edges of the ideal clock signal with period T separated by N clock periods would be $\tau = N \cdot T$. The time interval $(t_2 - t_1)$ separating two actual clock edges is affected by the phase jitter $\delta(t) = \frac{\phi(t)}{2\pi}T$. It can be obtained as follows:

$$\begin{aligned} t_1 &= iT + \delta(iT), \\ t_2 &= (i+N)T + \delta((i+N)T) = (iT+\tau) + \delta(iT+\tau), \\ t_2 - t_1 &= \tau + (\delta(iT+\tau) - \delta(iT)). \end{aligned} \tag{13.8}$$

The accumulated error between the two clock edges is known as the time interval error (TIE). The TIE for an interval of N clock periods is defined as:

$$TIE(t,\tau) = \delta(t+\tau) - \delta(t) = \frac{1}{2\pi}(\phi(t+\tau) - \phi(t))\,T. \tag{13.9}$$

In other words, TIE is a measure of the accumulated timing error between endpoints t and $t+\tau$ over the time interval $\tau = N \cdot T_0$. If the measurement is initialized at $t=0$ and $\delta(t)=0$, the TIE is identical to the phase jitter $\delta_n = \delta(N \cdot T_0)$. TIE and phase jitter are thus sometimes used interchangeably. Taking the mean square average of the TIE gives:

$$\overline{TIE(t,\tau)^2} = \frac{1}{4\pi^2}\left[\overline{\phi(t+\tau)^2} - 2\overline{\phi(t+\tau)\phi(t)} + \overline{\phi(t)^2}\right]T^2. \tag{13.10}$$

This jitter measurement shows the cumulative effect of the jitter over time if one assumes ideal clock locations are known. The TIE can be obtained by integrating the period jitter after subtracting the ideal period from each measured period. This metric diverges over time.

Figure 13.2 illustrates statistical characteristics of a ring oscillator-based clock generator and shows changes in the clock period over time, TIE (obtained by integrating the period jitter in time and by subtracting the result from the mean value of the period) and the cycle-to-cycle jitter of a nine-element ring oscillator. The output

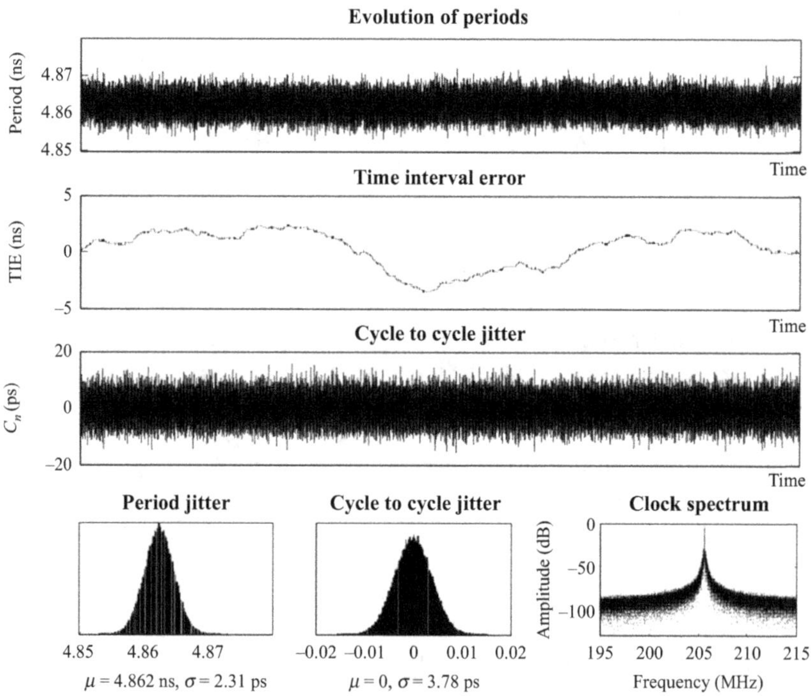

Figure 13.2 Clock generated by a ring oscillator implemented in an FPGA powered by linear power supplies: evolution of the period, time interval error (TIE) and cycle-to-cycle jitter (top panel) and histograms of period jitter and cycle-to-cycle jitter, and the spectrum of the generated clock signal (bottom panel)

signal of the generator was acquired using a Lecroy WavePro 735Zi oscilloscope with a 40 GS/s sampling rate over a low-voltage differential signalling (LVDS) output of the field-programmable gate array (FPGA) device and a differential probe D420 PS featuring 4 GHz bandwidth.

The differential signal transmission was used to increase the precision of the jitter measurement. The bottom panel in Figure 13.2 presents histograms of the period and of the cycle-to-cycle jitter and also the spectrum of the generated clock. The signals were obtained from an Altera Cyclone III family device EP3C25 placed on a hardware evaluation board dedicated to TRNG testing, the Evariste modular system [7]. To reduce the impact of external noise sources, the system was powered by linear power supplies.

In great majority of cases, clock jitter has two components: *random jitter*, which is caused by some non-deterministic phenomena like thermal or flicker noise, and *deterministic jitter*, which is caused by a deterministic process.

As the name of the random jitter suggests, its behaviour is random and statistical tools (e.g. mean value, variance and standard deviation) are therefore often used to quantify it. In most cases, it obeys the central limit theorem and has a Gaussian probability distribution function (PDF).

Deterministic jitter is typically caused by variations in the power supply (e.g. from switching power supplies or other periodical signal sources), crosstalk, electromagnetic interference (EMI), simultaneous switching outputs and other regularly occurring interference signals.

The main confusion when one is confronted with jitter measurement and quantification is that some jitter metrics apply only to random jitter and not to deterministic jitter, whereas deterministic jitter is almost always present.

Figure 13.3 presents the same clock characteristics obtained with the same measuring equipment as that presented in Figure 13.2, but this time using a commercial evaluation board featuring an Altera Stratix II EP2S60 FPGA device, which was powered from original switching power supplies. It is clear that, in this case, the deterministic jitter originating from the switching power supply largely dominated the random clock jitter.

Clock jitter can be caused by local and global sources, as shown in Figure 13.4. The global jitter sources, which can be random (e.g. the noise of the power supply) or deterministic (e.g. external electromagnetic emanations), affect all transistors in the device in the same way and are thus easier to reduce. The local jitter sources, also random (e.g. thermal noise or flicker noise) or deterministic (e.g. data-dependent crosstalks), affect all the transistors differently and are thus more difficult to eliminate.

Global jitter sources are very dangerous, because they are easy to manipulate. Fortunately, the impact of the global jitter sources on the generated random numbers can be reduced using the fact that identical blocks will be affected in the same way. Therefore, by using a differential TRNG design, e.g. exploiting pairs of identical clock generators and differentiating their output, the effect of global jitter sources can be significantly reduced [8].

The impact of the local deterministic sources on the generated numbers can be reduced only by carefully designed oscillators. The oscillators must be physically

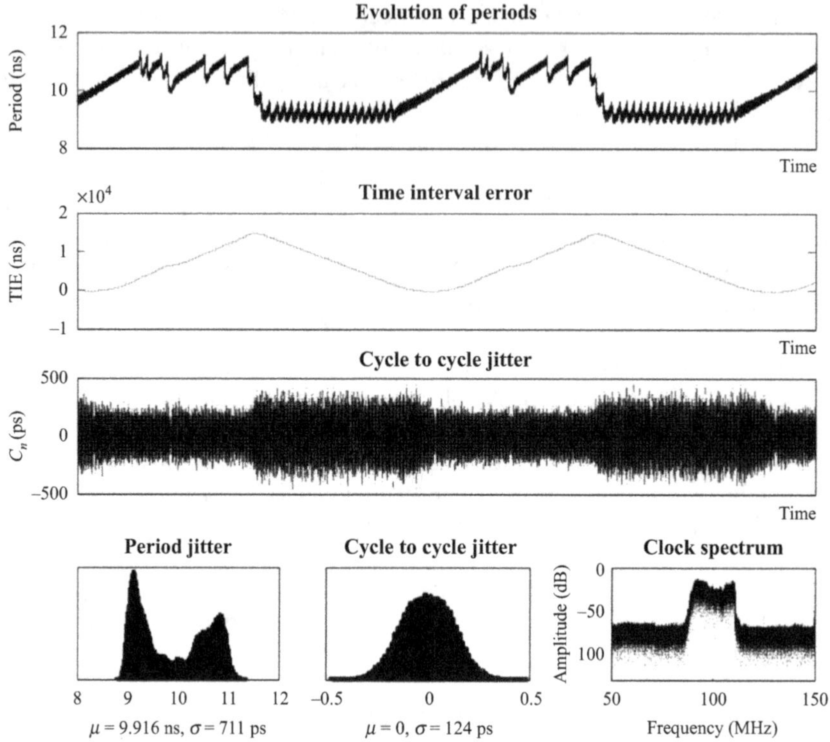

Figure 13.3 Clock generated by a ring oscillator implemented in an FPGA powered by switching power supplies: evolution of the period, TIE and cycle-to-cycle jitter (top panel) and histograms of period jitter and cycle-to-cycle jitter, and the spectrum of the generated clock signal (bottom panel)

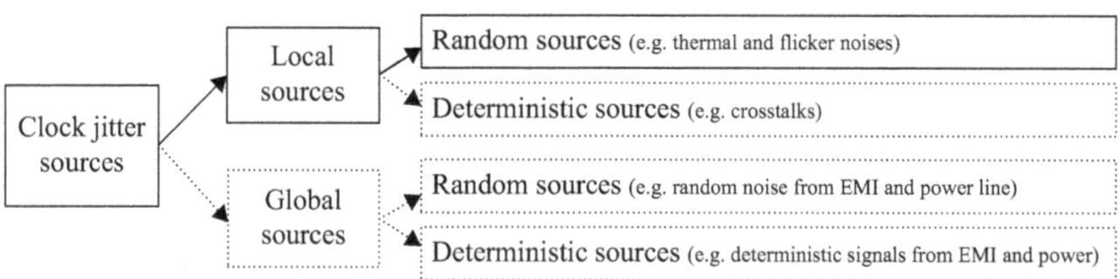

Figure 13.4 Sources of clock jitter in logic devices, the sources in the dashed boxes should be avoided as they can be used to attack generators

isolated from the rest of the device as well as from each other, the interconnection wires must be as thin and as short as possible in order to avoid capacitive and inductive coupling. Unfortunately, this is very difficult to achieve in FPGAs and sometimes also in application-specific integrated circuits (ASICs).

13.4 Harvesting the entropy from jittery clock signals

Entropy harvesting methods are used to obtain digital noise from the clock signals. In general, two principles can be used: *sampling* of a jittery clock signal and *counting* the edges of the jittery clock signal during a fixed time interval. Although the two principles may seem similar, their stochastic models can differ considerably.

The principle of the entropy harvesting method based on sampling of jittery clock signals is presented in Figure 13.5. It uses M jittery clock signals as a source of randomness. The clock signals are added modulo 2 in a XOR gate before being sampled at the reference clock in a D flip-flop. Although the phases of the clock signals are uniformly distributed, the frequency of the reference clock signal and hence the output bit rate of the generator can be very high comparing to the agility (spectrum) of the jitter. Another practical limit of the output bit rate is the bandwidth of the XOR gate (many high-frequency signals have to be XOR-ed). However, this problem can be solved by synchronizing the input clock signals on the reference clock using additional D flip-flops (not depicted in Figure 13.5).

The entropy harvesting method based on counting the jittery clock periods is presented in Figure 13.6. This method uses two clock signals: a reference clock signal and a jittery clock signal. In practice, both signals include some jitter, but if they come from independent generators, the jitter of the reference clock can be added to the jitter of the jittery clock signal and the reference clock can then be assumed to be jitter free. The time base generator generates the time interval τ which enables the m-bit counter. Because of period instability of the jittery clock signal, the counter values vary after each measurement interval τ as depicted in the right panel in Figure 13.6.

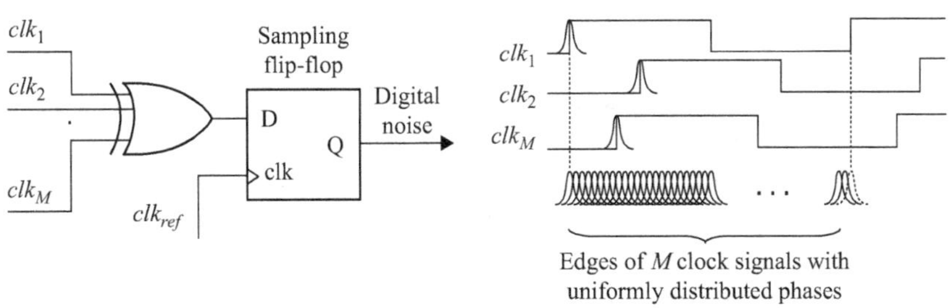

Figure 13.5 Entropy harvesting by sampling M jittery clock signals

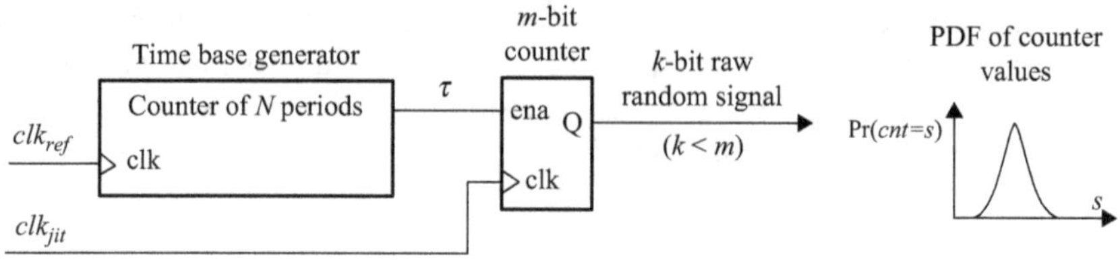

Figure 13.6 Entropy harvesting by counting the periods of the jittery clock signal

13.5 Single-event ring oscillators as sources of jittery clocks

A single-event ring oscillator is a loop built of logic gates containing an odd number of inverting gates, mostly inverters, and any number of non-inverting gates. Two kinds of ring oscillators are used as sources of randomness (see Figure 13.7): the conventional free-running ring oscillator in the left panel is composed of an odd number of inverters; the ring oscillator with a control input (in the right panel) is composed of a NAND gate and a sufficient number of non-inverting logic gates or buffers used as delay elements.

When the ring is oscillating (oscillations of the controlled ring from the left panel in Figure 13.7 must be enabled), the output of each inverting gate (the inverter on the left or the NAND gate output on the right) toggles periodically between two logic states. The main feature of the single-event ring oscillator is that only one event (i.e. a rising or falling edge of the generated periodic signal) is present in the ring at any given time.

The period of a single-event ring oscillator is equal to the sum of delays the event needs to pass across the ring twice (first as the rising and then as the falling edge of the generated signal or vice versa).

In the simplified noise-free model with ideal interconnections (noise free, zero impedance connections) and constant delay of all delay elements, the period of the generated signal is:

$$T = 2 \cdot \sum_{i=1}^{n} d_i,$$

where d_i is the delay of the ith element of the ring and n is the number of elements.

In the real world, elements of the ring do not have constant delays, their delays vary individually depending on local and global jitter sources causing the clock jitter.

Today, conventional ring oscillators are the most commonly used source of randomness. Their success is certainly due to the simplicity of their implementation in logic devices. Their main disadvantage is strong frequency dependence on temperature and on power supply voltage, which makes them vulnerable to external manipulation. Moreover, it was shown in Reference 9 that in some circumstances, ring oscillators can lock to each other due to crosstalk. This mutual locking can be fatal for the entropy harvesting mechanism. Jitter sources in ring oscillators are well

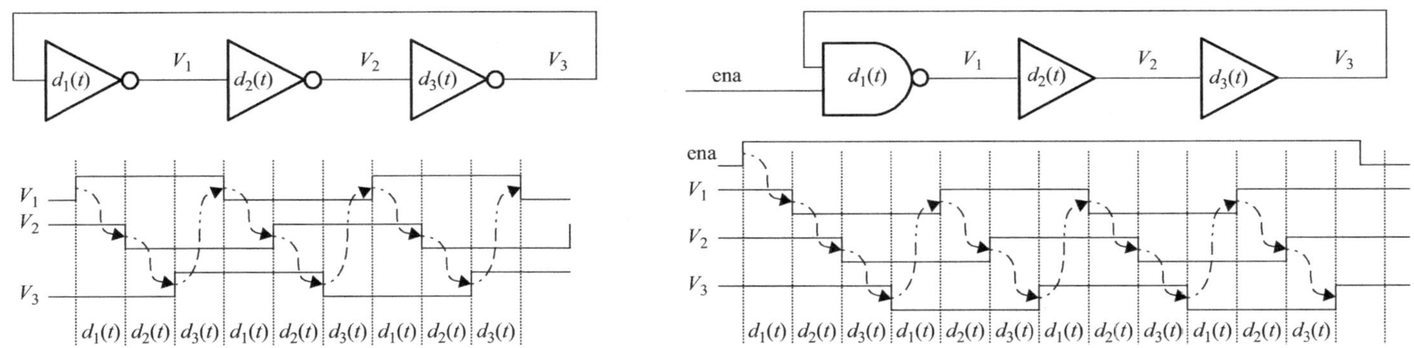

Figure 13.7 Free running, single-event ring oscillator (left panel) and single-event ring oscillator with a control input (right panel): only one event crosses the ring

studied and modelled, but because of shrinking submicron technologies, the models have to continuously evolve.

13.5.1 Modelling the jitter of clocks generated in ring oscillators

In the past decades, the authors in numerous articles including References 10–13 studied the origin of clock jitter in ring oscillators. They linked the properties of clock jitter with the electrical characteristics of ring oscillators, particularly with electric noises.

The works of Hajimiri and Lee [12] are a reference in this field. Hajimiri and Lee's approach is based on a simple experiment similar to the one shown in Figure 13.8. In their simulations, Hajimiri and Lee injected one current pulse at the time t_{inj} into one stage of a ring oscillator featuring average oscillation frequency f_0, and a load capacitance at the inverter outputs C_{node}.

The left panel in Figure 13.9 corresponds to when the current is injected far from the clock edge. As can be seen, in this case, the injected current has no impact on the phase jitter. The right panel in Figure 13.9 corresponds to when the pulse appears during the edge of the signal V_1. In this case, the current pulse does have an impact on the phase jitter – the clock edge is delayed and the delay is propagated over the ring.

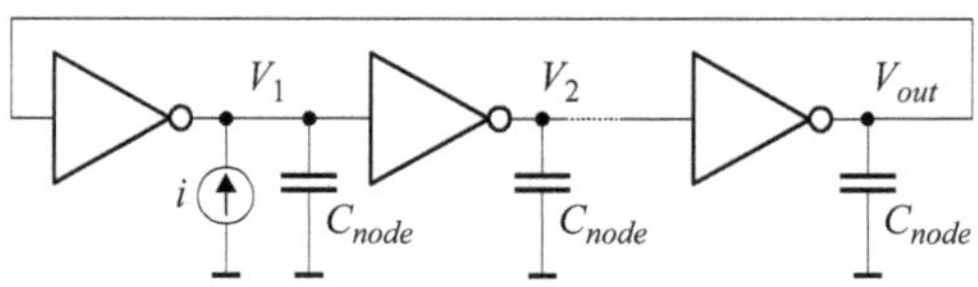

Figure 13.8 Injection of a current pulse into one stage of the ring oscillator in the experiment of Hajimiri and Lee

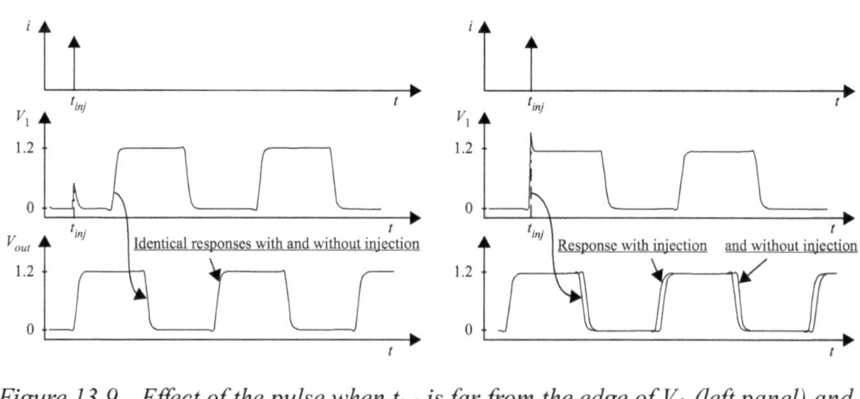

Figure 13.9 Effect of the pulse when t_{inj} is far from the edge of V_1 (left panel) and during its edge (right panel)

Hajimiri and Lee thus showed that the impact of the noisy current source (denoted i_n) on the phase noise ϕ can be approximated by a linear system with parameters which vary over time. They also showed that the infinite impulse response of such a system is a periodic function.

Let d_m be the mth component of Fourier series of the infinite impulse response. The model proposed in Reference 12 shows that if i_n is a sinusoidal signal of frequency f and amplitude I, it will cause a sinusoidal phase noise having the frequency $\nu = f$ mod f_0 and the amplitude $\frac{I \cdot d_m}{2 \cdot C_{node} \cdot V_{DD} \cdot \nu}$.

In the presence of both flicker and thermal noises, the PSD of the phase noise can thus be expressed as

$$S_\phi(f) = \frac{b_{fl}}{f^3} + \frac{b_{th}}{f^2}, \tag{13.11}$$

where b_{fl} and b_{th} are positive constants corresponding to the impact of flicker noise and thermal noise on ϕ.

As can be seen, the impact of the thermal and flicker noises depends on frequency: under the corner frequency f_c, the impact of flicker noise dominates over that of thermal noise. Using (13.4), the corner frequency is given by

$$f_c = \frac{KF \cdot I_{DS}}{C_{oxe} \cdot L_{eff}^2} \cdot \frac{1}{4 \cdot k_B \cdot T \cdot g} \tag{13.12}$$

In the time domain, for short accumulation times (short sampling periods), the contribution of thermal noise to the accumulated phase jitter will dominate. However, the jitter caused by the thermal noise is usually very small and consequently has to accumulate over longer periods of time. But for long accumulation times, the jitter from the flicker noise will accumulate faster and its contribution will dominate the jitter accumulated from the thermal noise.

Unfortunately, because of this $1/f$ dependence, the flicker noise samples are autocorrelated and their contribution to entropy increase is low. The autocorrelation of the flicker noise and its contribution to the phase jitter in ring oscillators after long accumulation times is often ignored by TRNG designers.

13.6 Multi-event ring oscillators with signal collisions

A multi-event ring oscillator with collisions is a loop built of logic gates, which contains an even number of inverting gates and any number of non-inverting gates. Because of the even number of inverting gates, the oscillator must be restarted regularly – the multiple events created after each restart circulate inside the loop until collision occurs, during which the edge which moves faster reaches the slower one. The difference in speed of circulating events is caused by differences in delays in loop branches between inverters and by analogue phenomena in the inverter circuitry. The circulating events create temporary oscillations which disappear after the last collision.

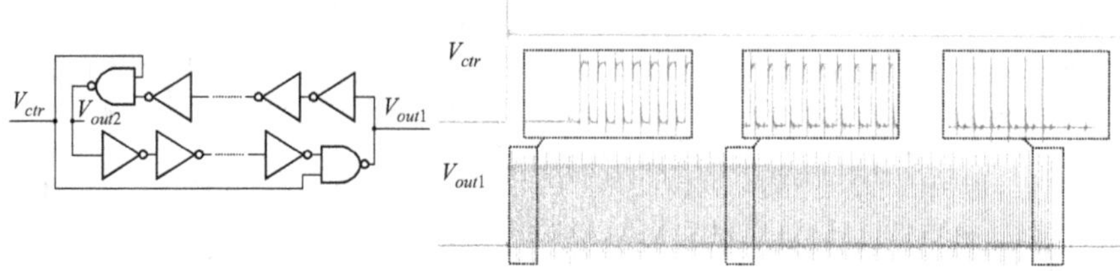

Figure 13.10 Circuit diagram of a typical TERO structure and its input/output waveforms

The most common configuration of the multi-event ring oscillator with collisions is the TERO, which features so-called oscillatory metastability, in which oscillations last random time intervals, theoretically even infinitely [14].

The TERO is a loop composed of an even number of inverters and a couple of gates which restart temporary oscillations (e.g. two NAND or two XOR gates). A typical TERO configuration is presented in the left panel in Figure 13.10: it is composed of two NAND gates and two inverter chains. The TERO can be seen as an RS latch with two inputs controlled by the same signal V_{ctr} and two outputs V_{out1} and V_{out2}.

Following the rising edge of the V_{ctr} input, the outputs V_{out1} and V_{out2} start to oscillate. The oscillations have a constant mean frequency, but their duty cycle varies over time: it changes monotonously and after a certain number of oscillations, it reaches the rate of either 0% or 100%. At this point, outputs V_{out1} and V_{out2} stop oscillating and remain stable at two opposite logic values. The right panel in Figure 13.10 presents traces of the V_{ctr} input and V_{out1} output signals captured from the oscilloscope.

The three zooms presented in this panel reveal the varying duty cycle: immediately after the rising edge of the V_{ctr} signal, it is close to 50%, after which it decreases (as the faster edge catches up with the slower one) until it reaches 0%. Consequently, the signal V_{out1} stabilizes at logic level 0. Of course, as far as the duty cycle is concerned, the signal V_{out2} behaves in the opposite way and stabilizes at logic level 1.

The number of oscillations before the outputs stabilize is not constant but variable because it is affected by the electronic noises which disturb the normal behaviour of transistors in the TERO structure, which in turn dynamically alters the delays of individual delay elements.

13.6.1 Modelling number of oscillations in TERO

Recently, Haddad *et al.* [15] proposed a stochastic model which specifies the number of temporary oscillations in the TERO structure. The model is based on the work of Reyneri *et al.* [14], in which the authors studied oscillatory metastability in an ideal noiseless circuitry.

Haddad *et al.* completed the physical model of Reyneri *et al.* by including the noisy behaviour of transistors in their noise-free model. Haddad *et al.* showed that the PDF of counter values has the following form:

$$Pr(N_{OSC} = q) = \frac{1}{2}\left[erf\left(K\frac{1 - R^{q-q_0-1}}{\sqrt{R^{2q} - 1}}\right) - erf\left(K\frac{1 - R^{q-q_0}}{\sqrt{R^{2q+2} - 1}}\right)\right], \quad (13.13)$$

where K and q_0 are equal to:

$$K = \frac{\sqrt{R^2 - 1}}{2\sqrt{2}\sigma_r},$$

$$q_0 = -\frac{\log(\Delta_r)}{\log(R)},$$

and σ_r and Δ_r are equal to:

$$\sigma_r = \sigma_{out}\sqrt{\frac{R^2 - 1}{(1 + H_d)^2 - 1}} \Big/ (\tau_1 + \tau_2),$$

$$\Delta_r = (\tau_1 - \tau_2)/(\tau_1 + \tau_2).$$

Equation (13.13) is very important, because it can be used to model the distribution of the number of temporary oscillations. Its main advantage is that the parameters of the model (R, σ_r and Δ_r) are easy to quantify. Parameter R is the ratio of the geometric series [14], σ_r is the relative jitter and Δ_r is the relative difference between TERO branches (τ_1 and τ_2 correspond to the delays of the two TERO branches). The model characterizing the number of temporary oscillations can serve as a basis for the TERO TRNG stochastic model.

13.7 STR oscillators

STRs are oscillators in which several events (electrical transitions) can propagate evenly spaced in time thanks to analogue mechanisms which are characteristic of the architecture of their stage. The STR structure is depicted in Figure 13.11. It corresponds to the control circuit of an asynchronous micropipeline as proposed by Sutherland in Reference 16, which has been closed to form a ring.

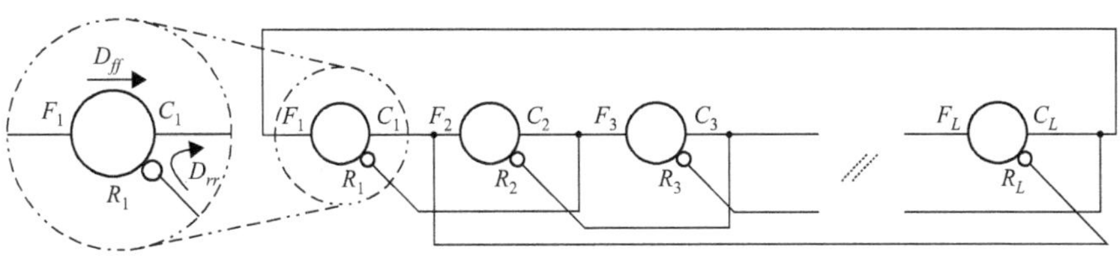

Figure 13.11 Architecture of a self-timed ring

Table 13.1 Truth table of a stage of the self-timed ring

F_i	R_i	C_i
0	0	Unchanged
0	1	0
1	0	1
1	1	Unchanged

The STR is composed of L stages, each consisting of a Muller gate and an inverter. F_i is the forward input of the ith stage, associated with a forward static delay D_{ff}, R_i is the reverse input of the same stage, associated with a reverse static delay D_{rr}, and C_i is the output of the stage. As it is obvious from the truth table in Table 13.1, the forward input value is written to the stage output if the forward and reverse input values are different. Otherwise, the previous output is maintained.

The STR stages communicate using the two-phase handshake protocol described in Reference 16. Each request and acknowledgement signifies the transfer of an event between interconnected stages. In contrast to inverter ring oscillators, several events can propagate without colliding thanks to this handshake protocol, which enables precise, built-in frequency and phase control of the internal clock signals by setting up the appropriate number of propagating events when the ring is initialized.

In practice, the ring is initialized with N events which start propagating during a transient state. Independently of their initial positions and thanks to two analogue mechanisms inferred in the ring (the Charlie and the drafting effect), they end up in a steady state in which they arrange themselves in one of two ways: they either form a cluster which propagates in the ring (burst oscillation mode), or they spread out around the ring and propagate with a constant temporal spacing (evenly spaced oscillation mode). Both these oscillation modes are stable and depend on the static parameters of the ring (mainly the ring occupancy $\frac{N}{L-N}$ with respect to the ratio of static delays $\frac{D_{ff}}{D_{rr}}$). Figure 13.12 shows the waveforms of the evenly spaced propagation of two events in a five-stage STR.

Charlie diagrams were first used by Ebergen *et al.* to predict the timing behaviour of a Muller gate as a function of the separation time between the input events which drive the gate [17]. Winstanley and Greenstreet carried on the study by introducing another analogue effect which affects the timings of events in STR stages – the drafting effect [18].

The Charlie effect determines the impact of the separation time between input events on a Muller gate delay: the closer the arrival times of events at inputs, the longer the gate propagation delay. The drafting effect describes the impact of the elapsed time from the last output commutation on the stage propagation delay: the shorter this time, the shorter the stage propagation delay.

In Reference 19, Fairbanks showed that the Charlie effect promotes the evenly spaced propagation mode while the drafting effect promotes the burst propagation

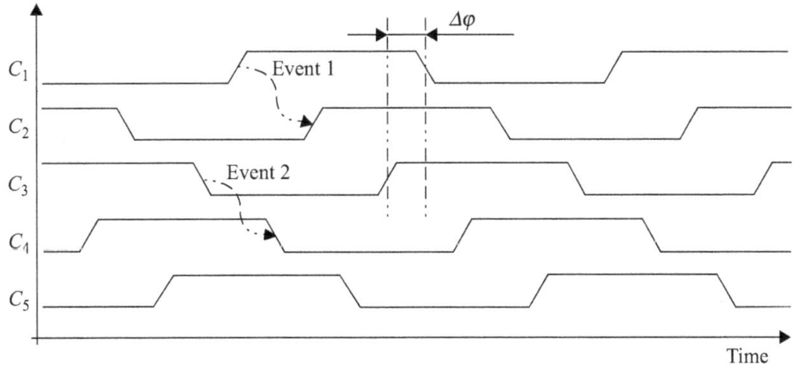

Figure 13.12 Illustration of the evenly spaced propagation of two events in a five-stage STR

mode. The Charlie effect causes two close events to push away from each other due to the increased delay experienced by a ring stage when driven by two events separated with a short time lapse.

If several events are constrained in a short ring structure, the Charlie effect is retroactive: events keep pushing away from each other until they are evenly spread out across the ring, which causes the evenly spaced propagation mode of the STR.

On the other hand, the drafting effect causes two events to approach each other because of the reduced propagation delay of a ring stage when it switches faster, thereby promoting the burst propagation mode of the STR.

In practice, the evenly spaced mode is obtained for a range of events centred around N_0 which fulfils the following relation [20]:

$$\frac{N_0}{L - N_0} = \frac{D_{ff}}{D_{rr}}. \tag{13.14}$$

The burst oscillation mode is obtained for corner values of the number of events. The stronger the Charlie effect, the longer the interval between events when the ring achieves the evenly spaced propagation mode. For example, a 64-stage STR in Altera Cyclone III with $\frac{D_{ff}}{D_{rr}} \simeq 1$ in Reference 21 used the evenly spaced mode for N varying between 22 and 42, while the same 64-stage configuration in Xilinx Virtex 5 achieved the evenly spaced mode for N between 28 and 38. This suggests the Charlie effect is stronger in the Cyclone III STR implementation.

In contrast to inverter ring oscillators, the frequency of an STR in the evenly spaced regime is a function of its occupancy and not of the number of its stages. The frequency of oscillations increases with number of events N, then starts to drop when the number of free stages is less than the number of events to be processed with respect to the asymmetry of the ring stages. The maximum frequency is reached when the number of events is equal to N_0 from (13.14).

Conversely with inverter ring oscillators, STRs can reach phase resolutions which are fractions of the propagation delay of a single logic gate. Figure 13.12 shows how a phase resolution below to the propagation delay of a single ring stage can be obtained using the STR. The propagation of an event in an STR causes a 180° phase shift of the oscillating signal. If N events are confined in L stages and spread evenly around the ring, the phase shift between two stages separated by n stages is [19]:

$$\varphi_n = n \times \frac{N}{L} \times 180^\circ. \tag{13.15}$$

Therefore, if the number of events is a factor of the number of stages, some stages may exhibit the same absolute phase. But if the number of events and the number of stages are co-prime, the STR exhibits as many different equidistant phases as the number of stages. In this case, if T is the oscillation period, the phase resolution can be expressed as follows [21]:

$$\Delta\varphi = \frac{T}{2L}. \tag{13.16}$$

On the other hand, the oscillation period of an STR is a function of its occupancy rather than of the number of its stages. This means that it is possible to increase the number of ring stages (L) while keeping a constant frequency. Consequently, the phase resolution of an STR can theoretically be set as finely as needed. Elissati *et al.* demonstrated the efficiency of the method in Reference 22 by implementing several designs and obtained phase resolutions in the order of picoseconds.

13.7.1 Clock jitter in STRs

Jitter in STRs was first discussed in Reference 23. In this work, the authors analysed and compared the period jitter of clocks generated using STRs and conventional free-running ring oscillators. First, they reported that the distribution of the periods had a Gaussian profile with a small standard deviation even in the case of large STRs (about 5 ps for rings oscillating at frequencies ranging from 400 to 600 MHz).

They observed that, contrary to conventional ring oscillators, the jitter size did not increase with the number of ring stages and remained constant for a fixed ratio $\frac{N}{L-N}$. The main result of this work was the estimation of the size of the period jitter σ_{period} with respect to the size of the local jitter generated in a single ring stage σ_{stage} (i.e. the standard deviation associated with the propagation delay of one ring stage):

$$\sigma_{period} \simeq \sqrt{2}\sigma_{stage}. \tag{13.17}$$

In Reference 21, the author studied the accumulation of the clock jitter in STRs implemented in ASICs and FPGAs. First, using a digital model, which took into account the analogue effects in STRs and the random jitter issued from white noise, the author showed that the variations in timing barely propagate from one stage to another due to the self-timed behaviour of the ring (Chapter 4.3.1 of Reference 21).

Moreover, this work showed that the higher the Charlie effect magnitude, and the closer N (the number of events) to N_0 defined in (13.14), the less variations in timing propagate from one stage to another. This explains why the period jitter does not increase with the number of stages, and why it features the same order of magnitude as the local jitter generated in one ring stage.

In conclusion, the name STRs originates from the fact that the timing of events in these rings is self-regulated thanks to inherent analogue mechanisms. This has a major consequence for how the locally generated jitter in one ring stage affects other stages: the propagation of events is constrained by the Charlie and drafting effects, which limit the propagation of variations in timing from one stage to another. All available studies indicate that jitter in STRs is mostly composed of the jitter generated locally in individual ring stages, with a notable white noise component and a negligible flicker component [21].

13.8 Examples of oscillator-based TRNGs

In the following paragraphs, we give some representative examples of oscillator-based TRNGs and their characteristics, which depend on the characteristics of the oscillating rings themselves and on the entropy harvesting algorithm used.

13.8.1 Elementary ring oscillator-based TRNG

The elementary ring oscillator-based TRNG (ERO TRNG) is a PTRNG similar to the one proposed by Fairfield *et al.* in 1985 [24]. Its structure is very simple (see Figure 13.13): the ERO TRNG consists of two free-running ring oscillators, a counter and a D flip-flop as a sampler.

The output of one ring oscillator is periodically sampled by a D flip-flop. The sampling period lasts K clock periods of the reference clock signal generated in the second ring oscillator.

The ERO TRNG was thoroughly analysed by Baudet *et al.* in Reference 25, and the authors proposed a stochastic model which can be used to determine the entropy rate at the ERO TRNG output. The model relies on exploitation of thermal noise. It is based on observations and on characterization of the thermal noise in the frequency domain and it is thus easy to understand and simple to use.

The main advantage of the ERO TRNG is that it is very simple to implement in logic devices. Since it only contains two ring oscillators, their placement and routing is easier and can be done in such a way that the mutual influence of the rings, which

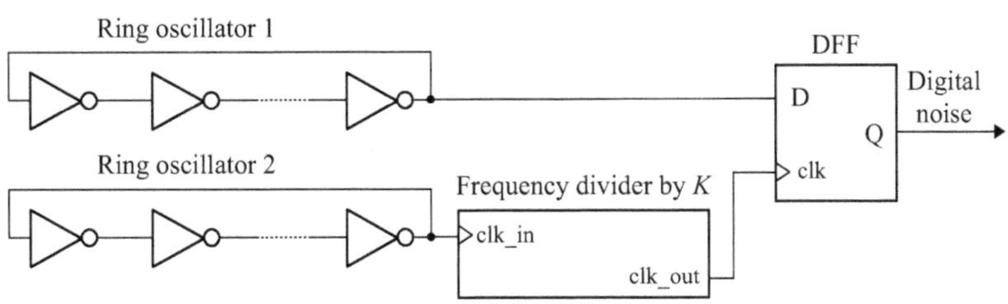

Figure 13.13 Elementary ring oscillator-based TRNG

creates autocorrelations in the RNG output signal, is reduced to a minimum. It is also easy to detect mutual locking of the two rings, which would have catastrophic consequences for security, as the entropy rate at the output would drop to zero.

The main disadvantage of the ERO TRNG is its low-output bit rate. Depending on the model, the entropy rate at the output of the generator required by security standards comes from the thermal noise. Because the jitter originating from the thermal noise is very small (only few picoseconds), it has to accumulate for a very long time: several hundred thousand periods of the jittery clock signal are necessary [26]. The output bit rate is thus in the order of few kilobits per second.

13.8.2 Multiple ring oscillator-based TRNG

The multiple ring oscillator-based TRNG (MURO TRNG) shown in Figure 13.14 was proposed by Sunar *et al.* in 2007 [27]. In the original design, outputs of M ring oscillators were added modulo 2 in a XOR gate. The resulting logic signal was sampled in a D flip-flop at a reference clock frequency.

The main difference between the ERO and MURO TRNG is in the number of input ring oscillators and in the stochastic model exploited. To increase the output bit rate of the generator, Sunar *et al.* used M rings as M sources of randomness. The number was obtained from the model, in which the authors assumed that the free-running oscillators were independent and consequently, that their phase was uniformly distributed across the mean period of the rings (rings had the same topology and thus oscillated at a similar frequency). In practice, more than 100 rings were needed. In MURO TRNG, the uniform distribution of the phases required by the

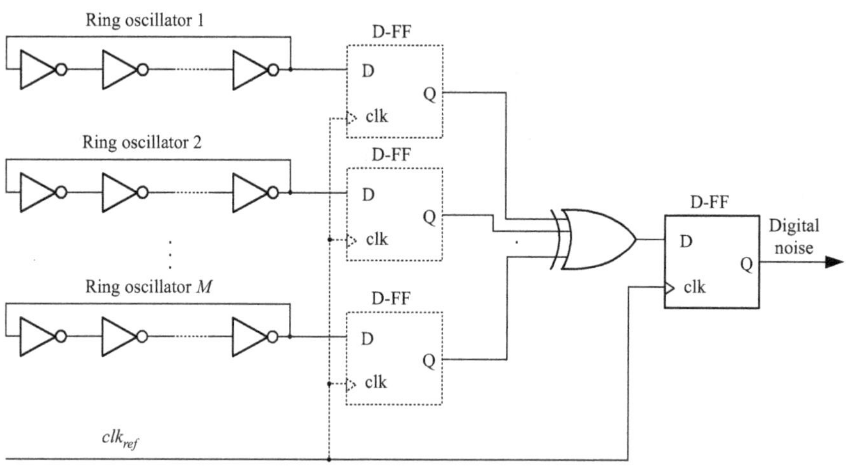

Figure 13.14 Multiple ring oscillator-based TRNG proposed by Sunar et al. and enhanced by Wold and Tan (modification of Wold and Tan in dashed lines)

principle of the entropy harvesting mechanism presented in Section 13.4 is guaranteed by the stochastic model, in which the rings are assumed to be independent.

The assumption of the independence of the rings was later criticized by Dichtl *et al.* in Reference 28, who also pointed out that the XOR tree is not fast enough to correctly process more than a hundred high-frequency signals. To overcome the problem, Wold and Tan in Reference 29 added a flip-flop at the output of each ring, in order to synchronize the outputs of the rings with the reference clock signal and hence to reduce the agility of inputs of the XOR gate. This improvement is shown by dashed lines in Figure 13.14.

Following this improvement, Wold and Tan suggested reducing the number of rings to 25, because using statistical tests, they estimated that this number of rings would suffice to obtain required entropy rate. However, this was in complete contradiction with the model of Sunar *et al.*, which was the most valuable part of Sunar *et al.*'s design (independently of their incorrect initial assumptions).

In References 9,30, Bochard *et al.* showed that the modification proposed by Wold and Tan (the use of additional flip-flops) did not affect the stochastic model proposed by Sunar *et al.* They showed in simulations, that even in the total absence of jitter, the sequence produced by 18 rings with a similar but not identical frequency, passed all statistical tests independently of the fact that its entropy rate was zero. This last point is crucial. Indeed, it shows that the criteria based on statistical testing, which were selected by Wold and Tan, do not evaluate the quality and security of the TRNG.

13.8.3 TERO-based TRNG

The PTRNG based on the TERO structure (TERO TRNG) proposed by Varchola and Drutarovsky in 2010 [31] is depicted in Figure 13.15. In this design, the TERO circuitry generating a random number of temporary oscillations is completed by an n-bit asynchronous counter which counts the rising edges of the oscillating signal. The counter output represents realizations of the random variable, i.e. the number of oscillations in subsequent control periods. Because of the temporary character of oscillations, TERO must be periodically restarted by the control signal. The control period defines the output bit rate of the generator.

The random binary sequence is obtained by successively concatenating the least significant bits of the counter, i.e. only one T flip-flop is needed in the counter

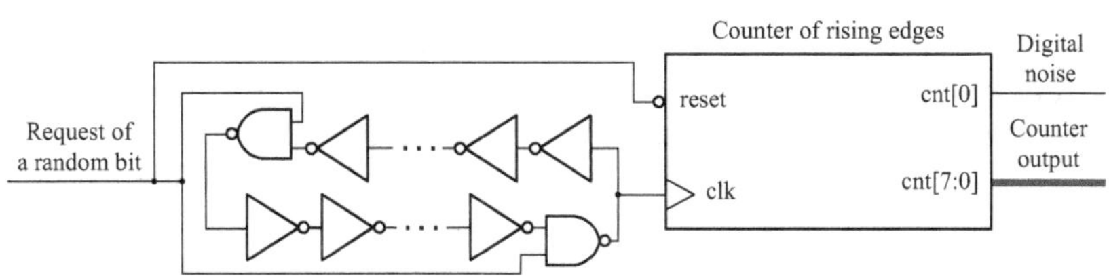

Figure 13.15 TRNG based on the TERO structure

(the n bit counter is only needed for the characterization of the generator, or alternatively, it can be used to build embedded randomness tests).

The main advantage of the TERO TRNG is its simple structure (especially if only a T flip-flop is used instead of an n bit counter) and its high-output bit rate – several megabits per second can be obtained. The main weakness of this design is related to its repeatability.

We stress that the mean number of oscillations is a function of the TERO internal structure and parameters. Unfortunately, TERO behaviour is very sensitive to variations in the production process and several TEROs with identical topology could feature a very different number of oscillations.

As shown by Haddad *et al.* in Reference 15, the entropy rate in the generated bit stream is closely linked to the number of oscillations. In order to accumulate enough entropy, a sufficient number of oscillations defined by the stochastic model must occur after each restart.

Two negative events can occur: (1) because of the internal parameters of the TERO, which are determined by variations in the production process, the oscillator does not oscillate for a sufficiently long time; (2) again, because of variation in the production process, the TERO is too symmetrical and the oscillations last longer than the control period and consequently, number of oscillation periods is not random. In both cases, the entropy rate at the output of a given TERO TRNG would be insufficient because of variations in the production process.

It thus seams that TERO TRNGs are better suited for ASICs, because their internal parameters can be better adjusted. Nevertheless, they will remain sensitive to variations in the production process. On the other hand, since the TERO circuitry is very small, many cores can be implemented in one device and the core with the highest entropy rate can be selected during the initialization phase of the device.

13.8.4 STR-based TRNG

As explained in Section 13.7, the STR is a multi-event ring oscillator without collision. In the evenly spaced mode, the events are uniformly distributed over half an oscillation period. It will also be recalled that the oscillation period in the steady state of STR is defined by the time interval between two subsequent events traversing the stage and not by the number of stages. A STR thus provides L jittery signals C_i, $1 \leq i \leq L$, with the same period T and a constant mean phase difference between them.

In contrast to MURO TRNG, in which the uniform distribution of the phases of multiple clock signals originates from the independence of rings, the uniformity of distribution of phases in STR is guaranteed by the STR principle, which is based on two analogue phenomena – the Charlie and the drafting effect.

Nevertheless, because of the presence of many jittery clock signals with uniformly distributed phases, the entropy harvesting principle can be the same. Left panel in Figure 13.16 depicts the STR-based TRNG published in Reference 32: the outputs of all the STR stages C_i are first sampled at reference clock frequency to obtain samples s_i, $1 \leq i \leq L$, which are then added modulo 2 using an XOR gate to obtain the digital noise signal

$$\psi = s_1 \oplus s_2 \oplus \cdots \oplus s_L.$$

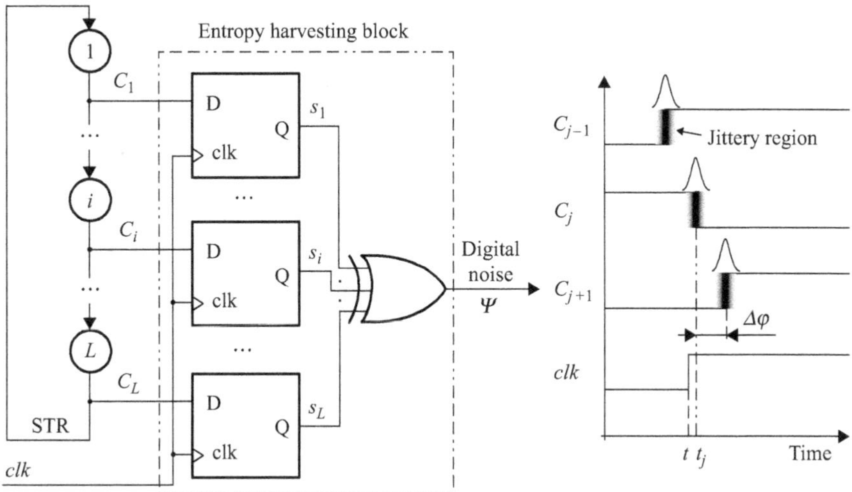

Figure 13.16 STR TRNG core architecture and working principle

The entropy harvesting principle is illustrated in the right panel in Figure 13.16. Since each signal C_i is sampled at the same frequency (reference clock *clk*), for any sampling instant t, there exists j such that $|t - t_j| \leq \frac{\Delta\varphi}{2}$, where t_j is the switching time of the signal C_j.

If the jittery region is larger than the phase difference $\Delta\varphi$, the signal C_j is sampled in this region, as shown in the right panel in Figure 13.16. The resulting sample s_j then has a random value, and hence the output of the XOR gate is also random.

The entropy rate of the output bit ψ is at least equal to the entropy of the sample s_j. The higher the magnitude of the jitter and the lower the phase difference $\Delta\varphi$, the higher the entropy of the sample s_j and at the output of the TRNG.

The main advantage of using the STR for the generation of random numbers is that it makes it possible to adjust the mean time elapsed between successive events precisely. This time lapse can be set as short as needed, and can thus be adjusted to the jitter magnitude of a STR stage.

Since the phase resolution and thus the number of phases is linked to the number of stages, the STR with a high resolution (required to capture small jitters) needs many stages. This can be easily implemented in ASICs, but is more difficult in FPGAs. This is apparently the biggest disadvantage of the STR TRNG and the price to pay for obtaining a very high-output bit rate of several hundred megabits per second [33].

Another advantage of the STR is that the entropy rate at the output of the generator is practically independent of the sampling clock frequency. It is sufficient that this frequency is smaller than or equal to the generated clock frequency. Last but not least, the locking of the ring clock to the reference clock thus will not have catastrophic consequences.

On the other hand, an attacker could reduce the number of events in the ring by fault injection and thereby reducing the phase resolution. However, this is quite easy to detect and the ring can be restarted with a right number of events.

13.9 Conclusions

In this chapter, we have explained important role of ring oscillators in generating random bit streams in logic devices. We analysed and characterized sources of noise, the transformation of electric noise into the phase noise or into the phase jitter of the generated clock signal. We specified the local and global jitter sources and analysed their role in random number generation for cryptography. We then explained the basic principles of the entropy harvesting mechanisms used in ring oscillator-based TRNGs.

The core of this chapter dealt with three types of ring oscillators used as sources of jittery clock signals: single-event ring oscillators, multi-event ring oscillators with collision and multi-event ring oscillator without collision, i.e. STRs. We analysed the principle of each of the three types of oscillators and studied the characteristics of the jitter of the generated clock signals.

Finally, at the end of the chapter, we presented some representative examples of oscillator-based TRNGs and their characteristics, which depend on the parameters of the oscillating rings themselves as well as on the entropy harvesting algorithm used. In conclusion, each oscillator, each TRNG has its strong and weak points and there is – as yet – no universally acceptable generator. This opens the way for further research in the field.

References

[1] Killmann W., Schindler W. *A Proposal For: Functionality Classes for Random Number Generators, Version 2.0*. Bundesamt fur Sicherheit in der Informationstechnik (BSI), Bonn [online]; 2011. BSI, Germany. Available from https://www.bsi.bund.de/SharedDocs/Downloads/DE/BSI/Zertifizierung/Interpretationen/AIS_31_Functionality_classes_for_random_number_generators_e.pdf [Accessed Sept 2016].

[2] Barker E., Kelsey J. *Recommendation for the Entropy Sources Used for Random Bit Generation, NIST Special Publication 800-90B*. Federal Information Processing Standards Publication [online]; 2012. Available from http://csrc.nist.gov/publications/drafts/800-90/draft-sp800-90b.pdf [Accessed Sept 2016].

[3] McShane E.A., Shenai K. 'Noise in analog and digital systems'. In: Chen WK, editor. *The Electrical Engineering Handbook*. London: Elsevier Academic Press; 2005. pp. 101–108.

[4] Nyquist H. 'Thermal agitation of electric charge in conductors'. *Physical Review*. 1928;32(1):110–113.

[5] McWhorter A.L. '1/*f* noise and germanium surface properties'. *Semiconductor Surface Physics*, University of Pennsylvania Press, 1957; pp. 207–228.

[6] Dunga M.V., Xi X., He J., *et al.* BSIM4.6.0 MOSFET model [online]; 2006. Available from http://www-device.eecs.berkeley.edu/bsim/Files/BSIM4/BSIM460/doc/BSIM460_Manual.pdf [Accessed Sept 2016].

[7] Fischer V., Bernard F., Haddad P. 'An open-source multi-FPGA modular system for fair benchmarking of true random number generators'. In *International Conference on Field Programmable Logic and Applications (FPL 2013)*; Porto, Portugal, 2013. pp. 1–4.

[8] Fischer V., Bernard F., Bochard N., Varchola M. 'Enhancing security of ring oscillator-based RNG implemented in FPGA'. In *International Conference on Field Programmable Logic and Applications (FPL 2008)*; Heidelberg, Germany, 2008. pp. 245–250.

[9] Bochard N., Bernard F., Fischer V., Valtchanov B. 'True-randomness and pseudo-randomness in ring oscillator-based true random number generators'. *International Journal of Reconfigurable Computing*. 2010;879281(2010): 1–13.

[10] Abidi A.A. 'Phase noise and jitter in CMOS ring oscillators'. *IEEE Journal of Solid-State Circuits*. 2006;41(8):1803–1816.

[11] Hajimiri A., Limotyrakis S., Lee T.H. 'Jitter and phase noise in ring oscillators'. *IEEE Journal of Solid-State Circuits*. 1999;34(6): 790–804.

[12] Hajimiri A., Lee T.H. 'A general theory of phase noise in electrical oscillators'. *IEEE Journal of Solid-State Circuits*. 1998;33(2):179–194.

[13] McNeill J.A. 'Jitter in ring oscillators'. *IEEE Journal of Solid-State Circuits*. 1997;32(6):870–879.

[14] Reyneri L.M., Del Corso D., Sacco B. 'Oscillatory metastability in homogeneous and inhomogeneous flip-flops'. *IEEE Journal of Solid-State Circuits*. 1990;25(1):254–264.

[15] Haddad P., Fischer V., Bernard F., Nicolai J. 'A physical approach for stochastic modeling of TERO-based TRNG'. *Cryptographic Hardware and Embedded Systems (CHES 2015)*, vol. 9293 of LNCS. Berlin: Springer; 2015. pp. 357–372.

[16] Sutherland I.E. 'Micropipelines'. *Communications of the ACM*. 1989;32(6): 720–738.

[17] Ebergen J.C., Fairbanks S., Sutherland I.E. 'Predicting performance of micropipelines using Charlie diagrams'. In *IEEE International Symposium on Asynchronous Circuits and Systems (ASYNC 1998)*; San Diego, CA, USA, 1998. pp. 238–246.

[18] Winstanley A., Greenstreet M. 'Temporal properties of self-timed rings'. *Correct Hardware Design and Verification Methods*, vol. 2144 of LNCS. Berlin: Springer; 2001. pp. 140–154.

[19] Fairbanks S. *High Precision Timing using Self-Timed Circuits*. University of Cambridge, Computer Laboratory; 2009. Technical Report No. UCAM-CL-TR-738.

[20] Hamon J., Fesquet L., Miscopein B., Renaudin M. 'High-level time-accurate model for the design of self-timed ring oscillators'. In *IEEE International Symposium on Asynchronous Circuits and Systems (ASYNC 2008)*; New Castle, UK, 2008. pp. 29–38.

[21] Cherkaoui A. 'Générateurs de nombres véritablement aléatoires à base d'anneaux asynchrones: conception, caractérisation et sécurisation'; 2014. Université Jean Monnet de Saint Etienne.

[22] Elissati O., Yahya E., Rieubon S., Fesquet L. 'A novel high-speed multi-phase oscillator using self-timed rings'. In *International Conference on Microelectronics (ICM 2010)*; 2010. pp. 204–207.

[23] Cherkaoui A., Fischer V., Aubert A., Fesquet L. 'Comparison of self-timed ring and inverter ring oscillators as entropy sources in FPGAs'. In *Design, Automation Test in Europe, Conference and Exhibition (DATE 2012)*; Dresden, Germany, 2012. pp. 1325–1330.

[24] Fairfield R.C., Mortenson R.L., Coulthart K.B. 'An LSI random number generator (RNG)'. *Advances in Cryptology*, vol. 196 of LNCS. Berlin: Springer; 1985. pp. 203–230.

[25] Baudet M., Lubicz D., Micolod J., Tassiaux A. 'On the security of oscillator-based random number generators'. *Journal of Cryptology*. 2011;24(2): 398–425.

[26] Fischer V., Lubicz D. 'Embedded evaluation of randomness in oscillator based elementary TRNG'. *Cryptographic Hardware and Embedded Systems (CHES 2014)*, vol. 8731 of LNCS. Berlin: Springer; 2014. pp. 527–543.

[27] Sunar B., Martin W.J., Stinson D.R. 'A provably secure true random number generator with built-in tolerance to active attacks'. *IEEE Transactions on Computers*. 2007;56(1):109–119.

[28] Dichtl M., Meyer B., Seuschek H. 'SPICE simulation of a 'provably secure' true random number generator"; 2008. *IACR Cryptology ePrint Archive* [online]. Available from https://eprint.iacr.org/2008/403.pdf [Accessed Sept 2016].

[29] Wold K., Tan C.H. 'Analysis and enhancement of random number generator in FPGA based on oscillator rings'. In *Proceedings of International Conference on Reconfigurable Computing and FPGAs (ReConFig 2008)*; Cancun, Mexico, 2008. pp. 385–390.

[30] Bochard N., Bernard F., Fischer V. 'Observing the randomness in RO-based TRNG'. In *Proceedings of International Conference on Reconfigurable Computing and FPGAs (ReConFig 2009)*; Cancun, Mexico, 2009. pp. 237–242.

[31] Varchola M., Drutarovsky M. 'New high entropy element for FPGA based true random number generators'. *Cryptographic Hardware and Embedded Systems (CHES 2010)*, vol. 6225 of LNCS. Berlin: Springer; 2010. pp. 351–365.

[32] Cherkaoui A., Fischer V., Aubert A., Fesquet L. 'A self-timed ring based true random number generator'. In *IEEE International Symposium on Asynchronous Circuits and Systems (ASYNC 2013)*; Santa Monica, CA, USA, 2013. pp. 99–106.

[33] Cherkaoui A., Fischer V., Fesquet L., Aubert A. 'A very high speed true random number generator with entropy assessment'. *Cryptographic Hardware and Embedded Systems (CHES 2013)*, vol. 8086 of LNCS. Berlin: Springer; 2013. pp. 179–196.

Chapter 14
Attacking on-chip oscillators in cryptographic applications

Lilian Bossuet[1], Pierre Bayon[2] and Viktor Fischer[1]

In this chapter, we describe a complete attack on a ring oscillator (RO) based true random number generator using the electromagnetic side channel. First, we analyze the frequency of the oscillators and their placement in the device. This analysis leads to the active attack that modifies the behavior of the random number generator using a strong electromagnetic field. We show that it is possible to dynamically control the bias of a RO based random number generator implemented in logic devices.

14.1 Introduction

Although true random number generators (TRNGs) based on ring oscillators (ROs) are the most widely expanded, they seem to be easier to attack. As presented in Chapter 13, the most cited TRNG principle is that proposed in 2007 [1] and in 2008 [2]. In order to guarantee sufficient amount of entropy, both designs use many ROs and the generated jittery clock signals are used as independent sources of randomness. However, high number of oscillators guarantees sufficient amount of entropy only if the oscillators are independent. This constraint represents a weakness that can be used in side channel analysis [3]. Indeed, an adversary could reduce entropy and introduce a bias on the TRNG output by locking the ROs between them, for example by manipulating the supply voltage [4].

Similarly to manipulations presented in Reference 4, an active attack targeting RO-TRNG was published in 2009 by Markettos and Moore [5]. It used the injection of a harmonic signal through the power supply channel in order to lock the ROs. However, the targeted RO-TRNG was rudimentary since it was made up of two ROs only. Following this work, Wold and Petrović studied in detail the robustness of RO-TRNG against attacks in which a locking signal was superimposed on the supply network [6]. They concluded that RO-TRNGs were in most cases robust against this

[1]Laboratoire Hubert Curien UMR 5516, Université de Lyon, Saint-Étienne, France
[2]Brightsight, Delft, The Netherlands

attack, and additionally that the attack can be detected by embedded statistical tests or it can be detected by checking the correlation between the oscillator frequency and the sampling clock [7]. However, we will show in this chapter that RO-TRNGs are susceptible to active electromagnetic attacks injecting harmonic signals to the generator circuitry [8].

14.2 Background

This section discusses the TRNG threat model, describes briefly the generator adopted as a design under test (DUT) in the rest of this chapter, and explains how the two attacks are working together.

14.2.1 TRNG model and implementation

The general structure of a TRNG is depicted in Figure 14.1. The generator is composed of:

- A digital noise source (randomness source + entropy extractor) that should give as much entropy per bit as possible, enable a sufficient bit-rate and be robust to environmental (voltage, temperature) variations.
- An algorithmic post-processing could be added at the output of the TRNG to enhance statistical properties without reducing the entropy.
- In some cases, the designer could add some embedded tests in order to evaluate on-chip and in real time the quality of the randomness source or to detect online the generator's permanent or temporal security failure. However, advanced and complex statistical tests are time and energy consuming. Commonly, the functionality and the quality of a TRNG are only periodically tested on-chip.

Two types of attacks on TRNGs can be considered: passive and active attacks. Passive attacks collect some information about the generator in order to predict future values with a non-negligible probability or to easily tune an upcoming active attack (attacks 1, 2, 4, and 5 in Figure 14.1—see arrow orientation). Active attacks tend to modify the behavior of the generator in order to control somehow its output (attacks 1, 3, 4, and 5 in Figure 14.1). According to Figure 14.1, the adversary may target different parts of the TRNG in different ways. We could expect that the statistical tests (simple embedded tests or complex external tests) could detect the attack. One could also argue that the algorithmic post-processing would reduce the impact of the attack. However, algorithmic post-processing is missing in some generators [2] or embedded tests are not used, because the generator is considered to be "provably secure" [1]. Nevertheless, it is a common practice in applied cryptography that the security of all building elements is evaluated separately. For this reason, evaluation of the robustness of the generator and all its parts is of great interest.

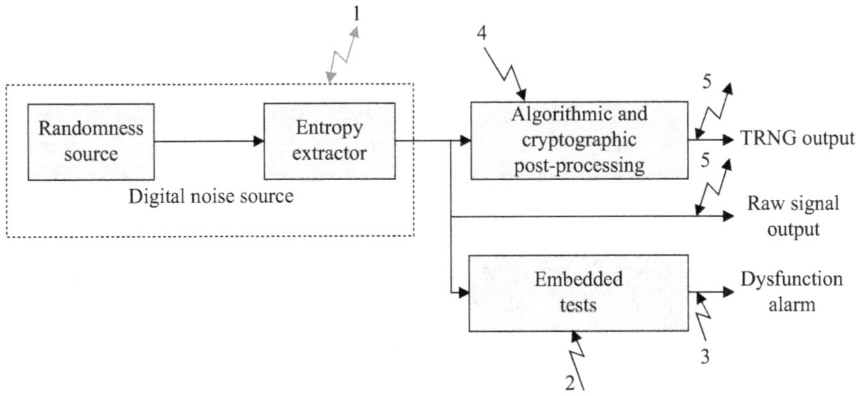

Figure 14.1 Passive (1, 5) and active (1–4) attacks on a TRNG general structure

14.3 Figures

Many sources of randomness such as thermal noise, $1/f$ noise, shot noise or metastability can be used in TRNGs. A good source of randomness should not be manipulable (and therefore not attackable) or the manipulation should be prevented. For example, the thermal noise quality can be guaranteed by controlling the temperature. It is thus reasonable to expect that attacks will not target the source of randomness.

In this chapter, we consider both passive and active attacks on entropy extraction (1). The passive attack is used to retrieve information on the TRNG itself (involved frequencies, position, sampling frequency, etc.). On the other hand, the objective of the active attacks can be to bias the generator output or to reduce the digital noise entropy, since both bias and entropy reduction can simplify the subsequent attack on the cryptographic system, since the exhaustive key search can be significantly shortened. We do not consider other attacks from Figure 14.1, such as attacks on tests (2 and 3) and post-processing (4), because of huge number of methods and cases that should be considered. It is up to the designer, to adapt post-processing and embedded tests to the weaknesses of the generator.

As it will be shown in the chapter, the electromagnetic waves can be considered to be good candidates for performing both passive and active attacks on RO-based TRNG. For this work, we have chosen the TRNG proposed in Reference 2 (this TRNG was described in Section 13.8.2 of Chapter 13). It needs only inverters (for implementing ROs), flip-flops (as samplers), and a large XOR gate (as an entropy collector). In Reference 1, authors proposed a mathematical model of the RO-TRNG that guarantees enough entropy in the output bit and thus the robustness and security. In their model, ROs are assumed to be independent.

The generator has several parameters that can be tuned: number of inverters composing ROs, number of ROs, and the sampling frequency. Modifying these parameters, the designer can change the statistical properties of the random bitstream produced by the TRNG. For example, according to Wold and Tan [2], for a sampling frequency of 100 MHz, the generator composed of 25 ROs, each using three inverters, generates the bitstream passing common statistical tests even without post-processing.

All electromagnetic attacks (both active and passive) presented in this chapter were realized on a board featuring Microsemi Fusion AFS600 Field Programmable Gate Array (FPGA). The TRNG was built using 50 ROs (that is more than the recommendation given in Reference 2), each composed of three inverters. We recall that two attacks (passive and active) are not invasive and so the FPGA package could remain intact.

14.3.1 Attack scenario

When an adversary wants to attack a secured device containing a cryptographic system, he usually does not possess much information about the system. The device, which includes the system, can thus be seen as a black box.

Figure 14.2 shows the proposed scenario of performing attack on a RO-TRNG. The first thing the attacker has to do is to gather information on the TRNG using the electromagnetic analysis during the passive attack. Namely, he will try to retrieve some information about working frequencies of the ROs and to locate the TRNG inside the device in order to tune his active attack. After having successfully retrieved the information needed, the attacker can proceed to the active attack. The use of the electromagnetic field as an attack channel is motivated by the contactless property of this channel and the sensitivity of most of electronic devices to the electromagnetic waves. That explain why the electromagnetic channel is more and more used as a channel for fault injection [9].

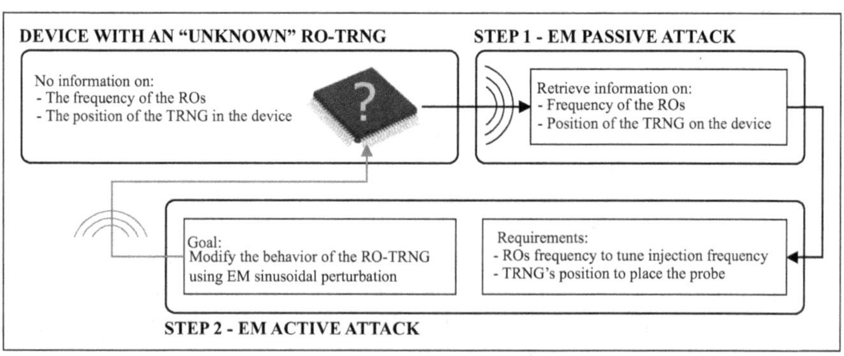

Figure 14.2 Scenario of coupled electromagnetic attacks on RO-TRNGs

14.4 Retrieving information on the RO-TRNG: passive electromagnetic attack

This section explains how to gather the information about the unknown device containing a RO-TRNG by exploiting electromagnetic field emanated by the device and namely by ROs [10,11].

14.4.1 Electromagnetic analysis platform

The electromagnetic radiation of the device is evaluated using a near-field (the limit between the far field and the near field can be considered to be located roughly to a distance of 23 mm from the device, according to the considered hardware) electromagnetic analysis test bench from Figure 14.3. The most important part of the test bench is the acquisition chain. It determines the signal-to-noise ratio and precision of the measurement. It is composed of:

- A Langer magnetic probe with the frequency range from 30 MHz to 3 GHz and the spatial resolution of approximately 500 μm.
- A Miteq low-noise amplifier with the frequency range from 100 MHz to 1 GHz.
- A Lecroy oscilloscope with the frequency range up to 3.5 GHz and a sampling rate up to 40 GS/s.

The device under test (the board) is fixed on a XYZ table with repeatability of movement of 1 μm. The test bench, including acquisition chain, XYZ table,

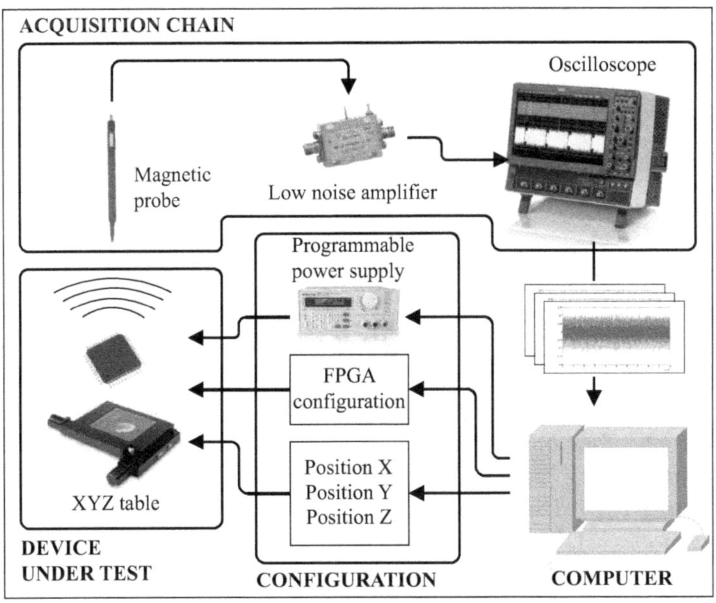

Figure 14.3 Electromagnetic analysis test bench

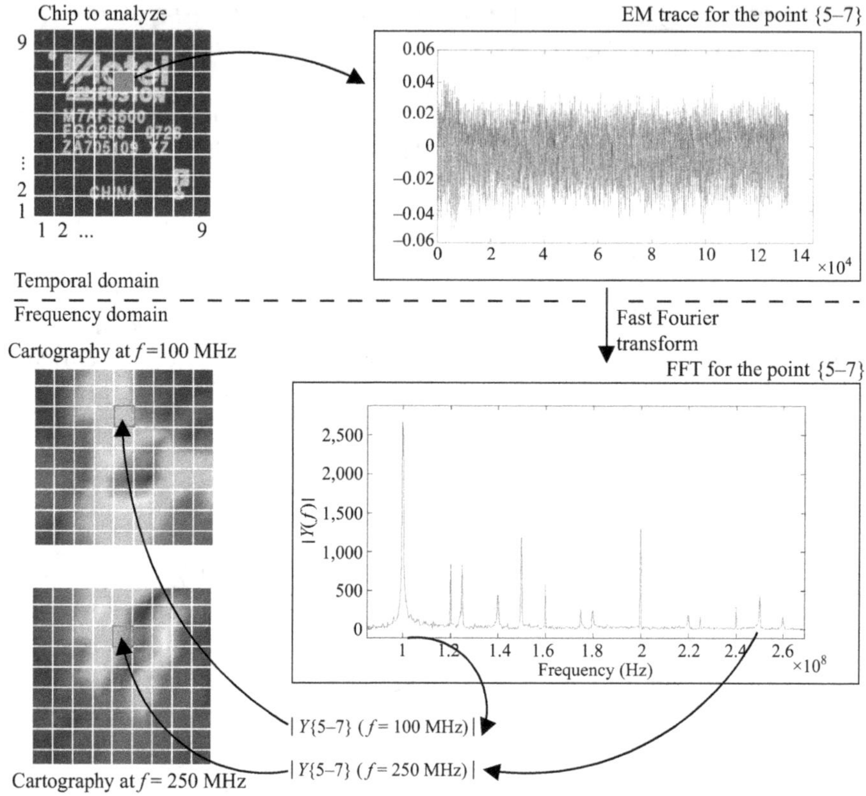

Figure 14.4 Electromagnetic cartography using frequency analysis

FPGA configuration, and power supply variations, is controlled by a personal computer (PC).

14.4.2 Frequency analysis

The goal of the passive attack is to retrieve as much information as possible on the cryptographic system and namely on the TRNG. Considering RO-TRNG, we suppose that the attacker will tend to determine the working frequencies of the ROs, their locations on the chip, and if possible, the sampling frequency of the flip-flops. To extract such information from the electromagnetic radiations, we have used a so-called electromagnetic cartography method presented in Figure 14.4. In this method, the electromagnetic radiations of the device are analyzed point-by-point using a frequency analysis [12]. The aim of the electromagnetic cartography is to obtain an electromagnetic map of the device at certain frequency (e.g., see two maps on the left side of Figure 14.4, one obtained at 100 MHz and the second one at 250 MHz). For each point (i,j), an electromagnetic trace is acquired and the power spectral

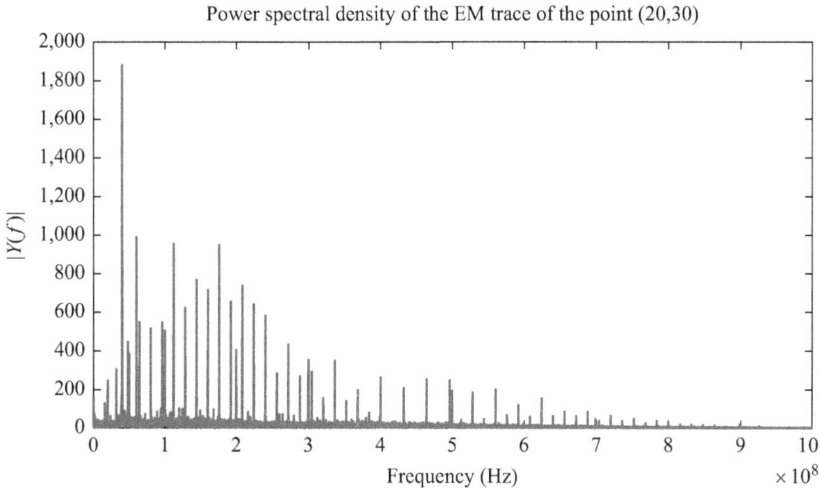

Figure 14.5 Typical power spectral density of the electromagnetic traces at a random location in the electromagnetic device map

density (PSD) of this trace denoted $|Y\{i,j\}(f)|$ is computed using fast Fourier transform (FFT). Figure 14.5 presents a typical PSD of the electromagnetic radiation of a cryptographic circuit including a RO-TRNG embedded in an FPGA. Notice that it is not possible to directly guess the ROs working frequencies. Indeed, the RO-TRNG represents a small part of the cryptographic system and its contribution to the electromagnetic radiation is insignificant compared to the contribution of large synchronous blocks such as the cipher. As a consequence, the attacker will need to discern the frequency contributions of the ROs from those of the other modules contributing to the PSD.

14.4.3 Differential frequency analysis

It is commonly known that the RO frequency depends dynamically on the power supply voltage and temperature. In synchronous systems clocked by an external quartz oscillator (most of today's systems), the frequency of RO is more sensitive to environmental variations than that of the system clock. This RO characteristic can be used to discern ROs frequency contributions to the PSD by performing a differential frequency analysis. This analysis consists in performing two successive electromagnetic frequency analyses for two distinct working conditions {Cond #1, Cond #2} (i.e., two distinct power supply voltages or two distinct temperatures). A differential electromagnetic radiation of PSD is computed by subtracting the two PSDs at Cond #1 and Cond #2. This differential PSD contains two symmetrical spectral contributions with large frequency shifting, which highlights the presence of condition-sensitive block such as RO in the targeted chip. The final electromagnetic device maps at Cond #1

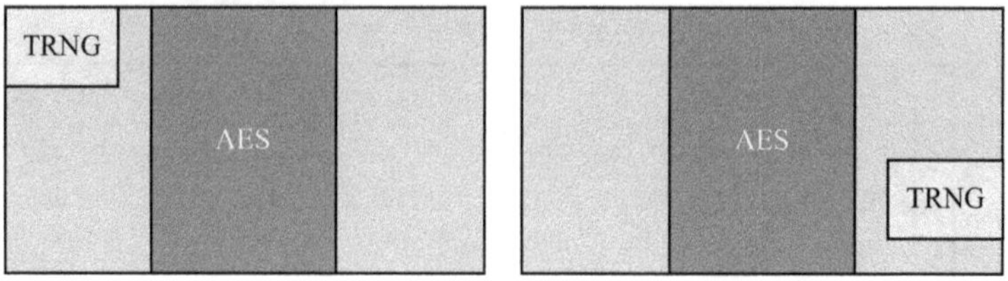

Figure 14.6 From left to right, floor plan of Experiment #1 and Experiment #2

and Cond #2 permit to detect the presence of ROs and determine their location and frequency. Experimental results presented in Section 14.4.5 illustrate this method.

14.4.4 Experiments

In order to demonstrate the efficiency of the proposed analysis, three experiments were realized:

- The first two experiments aimed at showing that the proposed electromagnetic analysis is efficient in retrieving the information about the RO-TRNG, and this independently from the activity of the circuitry surrounding the RO-TRNG. For this reason, besides RO-TRNG placed in the upper left corner in Experiment #1 and in the lower right corner in Experiment #2, as shown in Figure 14.6, the device contained also an Advanced Encryption Standard (AES) cipher encrypting continuously random data and perturbing thus the analysis (moving our attack closer to real operating conditions).
- The third experiment aimed at demonstrating that it is possible to find the sampling frequency of the RO-TRNG. It consisted of using the hardware design from Experiment #2 but using a different sampling frequency for the TRNG for each measurement.

14.4.4.1 Design setup

We have implemented the two designs in the Microsemi Fusion AFS600 FPGA device. The TRNG was built using 50 ROs, each composed of three inverters. For the first two experiments, the sampling frequency was 6.9 MHz. For the third one, we used two sampling frequencies: 6.9 and 8.9 MHz. In both cases, the 20 MHz clock used by the AES cipher came from an embedded RC oscillator. We recall that the electromagnetic analysis is not invasive and so the FPGA package could remain intact.

14.4.4.2 Acquisition setup

We decided to perform the electromagnetic cartography of the whole FPGA package (1.7 cm $\times$ 1.7 cm). In order to get a sufficient resolution and to have a reasonable amount of points, we have chosen a resolution of 90 $\times$ 90 points ($\sim$200 μm resolution). For each point of the electromagnetic device map, ten traces (used to average the PSD of each point) of 200,000 points (with a sampling frequency of 20 GHz) were acquired successively. In order to simplify the analysis (obtention of the RO's frequency range), the points were gathered in nine larger "areas" (30 $\times$ 30), for which the PSDs were averaged. Note that for localizing ROs, the map resolution remained unchanged.

As stated previously, in our differential frequency analysis, the attacker needs to obtain two electromagnetic device maps at different voltage or temperature conditions. It is important to note that the working conditions (voltage or temperature) must remain sufficiently stable during the construction of each of two electromagnetic device maps. Notice that in our experiments, the acquisition took at least 6 h. Since it would be more difficult to control the temperature, we decided to modify the power supply voltage of the device. The electromagnetic device maps are performed at 1.5 V (nominal V_{dd}) and 1.7 V for Experiments #1 and #2.

14.4.5 Experimental results

14.4.5.1 Results of Experiments #1 and #2: retrieval of the RO frequency and localization of RO

The differential PSDs of the first two experiments are depicted in Figures 14.7 and 14.8. We recall that the aim of the first two experiments was to show that it is possible to find the position and frequencies of the ROs composing a RO-TRNG while the AES cipher is running in the same device. In Section 14.4.3, we expected that only the frequencies corresponding to the ROs would clearly appear in the differential PSDs. As it can be seen in Figures 14.7 and 14.8, other frequencies are not completely suppressed by the differential frequency analysis. This is mainly due to the fact that, for a given system frequency, the amplitude of this frequency in the PSD varies slightly from one measurement to another. Nevertheless, this effect remains negligible and the proposed method is still very useful for locating frequencies of interest in the PSDs.

In Figures 14.7 and 14.8, the positive part of the differential PSDs corresponds to the electromagnetic device map at 1.5 V and the negative part to the electromagnetic device map at 1.7 V.

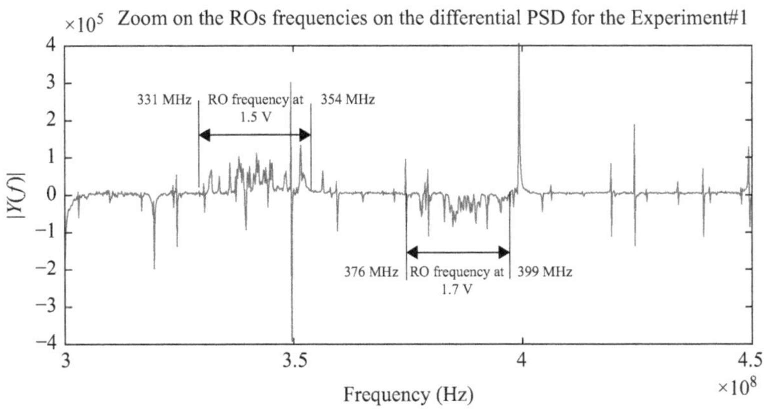

Figure 14.7 Differential PSD of Experiment #1 at the center of the chip

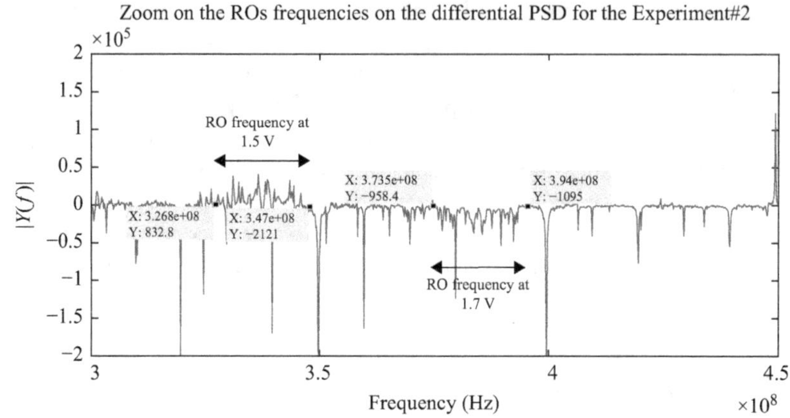

Figure 14.8 Differential PSD of Experiment #2 at the center of the chip

It can be seen in Figure 14.7 that for Experiment #1:

- At 1.5 V, ROs frequencies range from 331 to 354 MHz.
- At 1.7 V, ROs frequencies range from 376 to 399 MHz.

From Figure 14.7, we can expect that the RO-TRNG radiation will be visible in the upper left corner of the die only for the couple of conditions (1.5 V and $f = 331$–354 MHz) and (1.7 V and $f = 376$–399 MHz).

It can be seen in Figure 14.8 that for Experiment #2:

- At 1.5 V, ROs frequencies range from 327 to 347 MHz.
- At 1.7 V, ROs frequencies range from 374 to 394 MHz.

From Figure 14.8, we can expect that the RO-TRNG radiation will be visible in the bottom right corner of the DIE only for the couple of conditions (1.5 V and $f = 327$–347 MHz) and (1.7 V and $f = 374$–394 MHz).

Figures 14.9 and 14.10 show electromagnetic device maps of Experiments #1 and #2 for the two power supply voltage conditions (1.5 and 1.7 V). The white dotted rectangle represents the FPGA's die and the solid rectangle represents the location of the ROs. These electromagnetic device maps are plotted for the frequencies ranges obtained by the differential frequency analysis. We can conclude that we are able to locate the RO-TRNG on the chip by finding the RO's working frequencies. Moreover, we can note that the activity of the AES block inside the FPGA has no influence on the result of the electromagnetic differential frequency analysis.

14.4.5.2 Results of Experiment #3: retrieval of the sampling frequency

The aim of this experiment was to retrieve the sampling frequency of the RO-TRNG. Figure 14.11 depicts the PSDs for the whole circuit (all the PSDs were averaged) at two different sampling frequencies, one at 6.9 MHz and the other at 8.9 MHz. As it

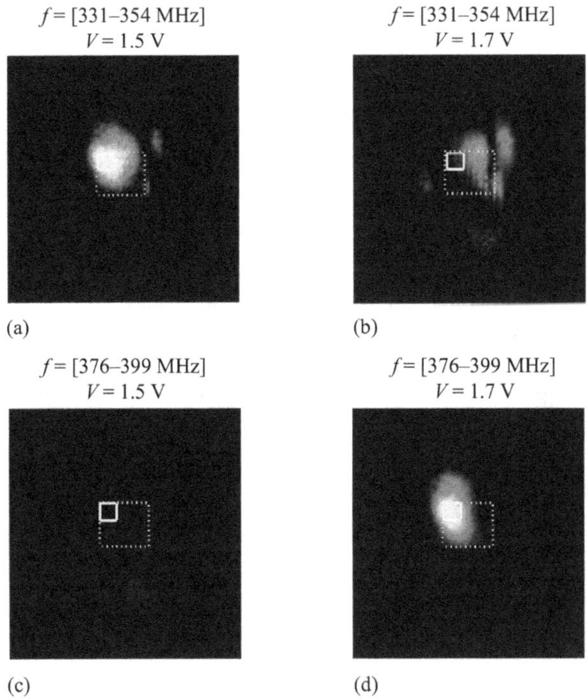

Figure 14.9 Electromagnetic device maps for Experiment #1

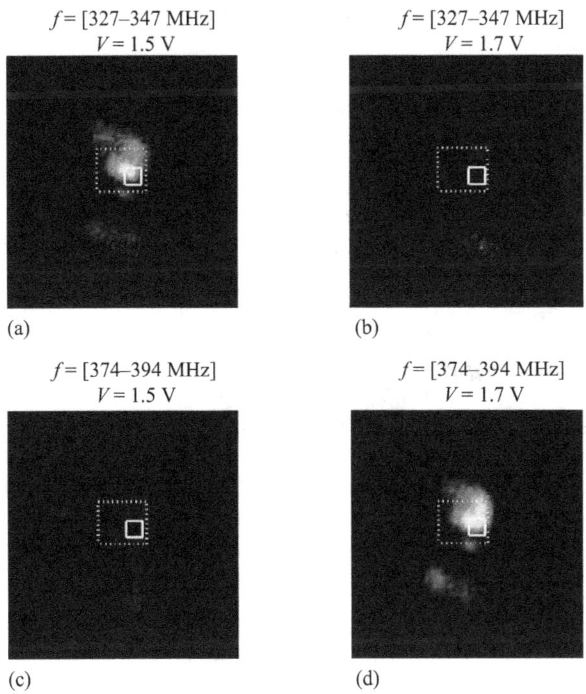

Figure 14.10 Electromagnetic device maps for Experiment #2

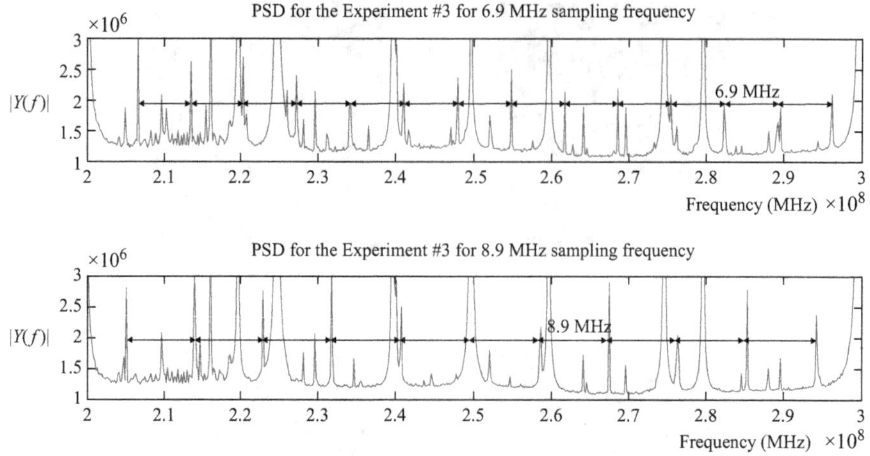

Figure 14.11 PSD of Experiment #3 for the whole circuit—upper figure for a sampling frequency of 6.9 MHz and bottom figure for a sampling frequency of 8.9 MHz

can be seen, it is easy to determine the sampling frequency of the RO-TRNG: the attacker has just to find frequency peaks that are repeated with the same spacing all over the PSD trace. The sampling frequency is determined by this spacing.

14.5 Modifying the RO-TRNG behavior: active electromagnetic attack

Having retrieved the information about location of the RO-TRNG and their working frequencies, we can now tune properly the active electromagnetic attack. This section will describe how the attack works and it will then give results of injection of electromagnetic signals on both RO-TRNG configurations presented before.

14.5.1 Injection platform

The electromagnetic signal injection platform is presented in Figure 14.12. The platform embeds a power injection chain supplying the micro-antenna, but also two other chains: one for controlling the whole platform and the other one for data acquisition and storage.

The main element of both control and data acquisition chains is a PC, which:

- controls the amplitude and the frequency of the sine waveform signal provided by the signal generator to the input of the power amplifier,
- positions the micro-antenna above the integrated circuit surface thanks to the XYZ motorized stage,

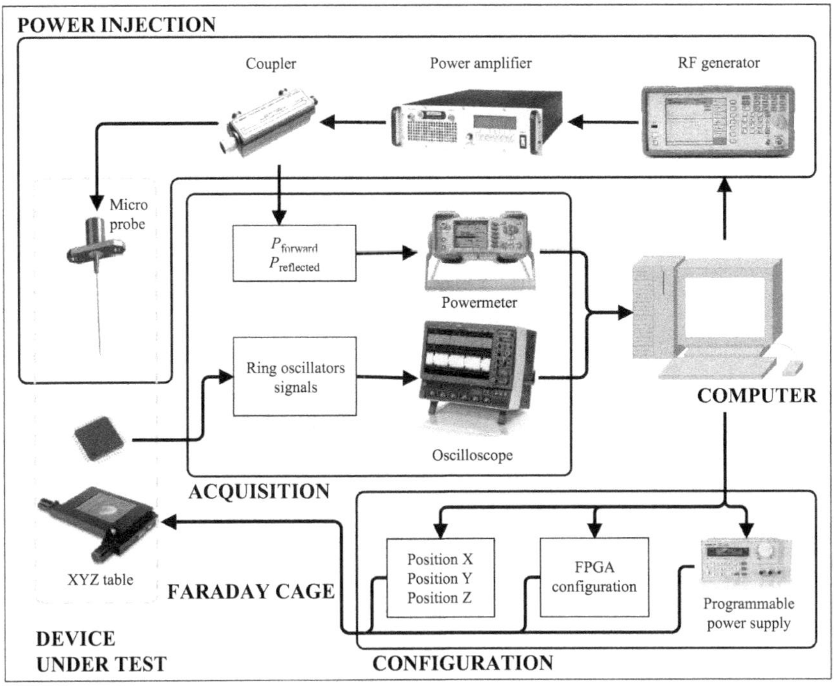

Figure 14.12 Direct power injection platform

- collects data provided by the power meter, connected to a bidirectional coupler, in order to monitor the forward ($P_{forward}$) and reflected ($P_{reflected}$) powers,
- sends configuration data to the Microsemi Fusion FPGA and supplies target boards via USB,
- stores the time domain traces of all signals of interest acquired using the oscilloscope.

Note that according to safety standards, also in order to limit the noise during acquisitions, the whole electromagnetic injection platform is placed in a Faraday cage with a 120 dB isolation at 1 GHz.

A key element of this platform is the probe that converts electric energy in a powerful electromagnetic field (active attacks). Most of micrometric electromagnetic probes used generally to characterize the susceptibility of IC are inductive, composed of a single coil in which a high amplitude and thus a sudden current variation is injected. These probes cannot be used in our context. Indeed, reducing the coil diameter to micrometric dimensions (200–20 μm) implies reducing the coil wire diameter too. As a result, the amplitude of the current injected into the probe must be reduced to avoid any deterioration of the coil. Consequently, the power that can be effectively injected into such probes was experimentally found too small for disturbing significantly the behavior of the logic device. After several attempts and prototype versions,

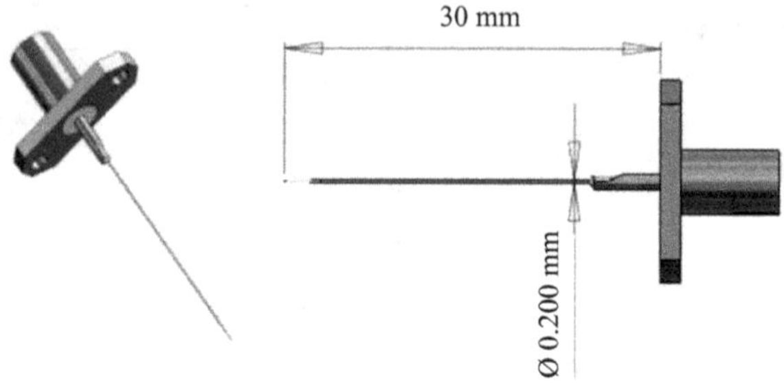

Figure 14.13 Unipole microprobe

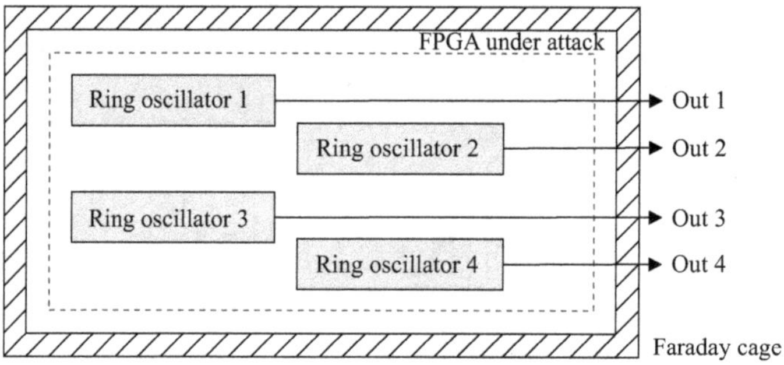

Figure 14.14 Implementation of ROs in Target #1

we adopted a probe shown in Figure 14.13. It is constituted of a thin tungsten rod. Its main characteristics are: a length of 30 mm and a diameter of 200 μm at one end and 10 μm at the other end.

This probe involves predominantly electric field, and we can assume that only this component, at the tip end, can couple with the metal tracks inside the IC. Further information about the platform and the effects of electromagnetic injection are available in Reference 13.

14.5.2 Experiments

In order to demonstrate that the electromagnetic injection can disturb both RO and TRNG behavior, we performed attacks on two kinds of implementations:

- The first one was composed of four three-element ROs. It was aimed at the measurement of the phase difference between four generated clocks (see Figure 14.14). This implementation is called Target #1.
- In the second implementation depicted in Figure 14.15, the TRNG core was implemented in an FPGA board under attack. Another board that was protected from electromagnetic emanations generated reference clock signals read data from TRNG and communicated with the computer. We decided to separate the

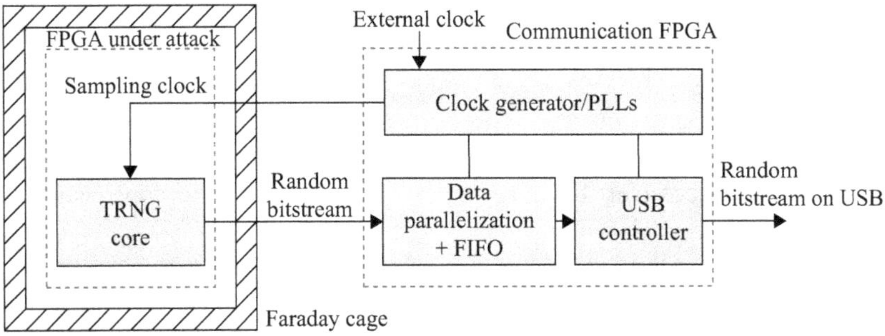

Figure 14.15 TRNG testing architecture—Target #2

communication from random data generation in order to be sure that it was the TRNG that was disturbed by the attack and not the communication. The communication module is composed of a serial to parallel converter, a First In-First Out (FIFO) memory, and a USB controller. USB interface throughput (up to 20 MB/s) was sufficient to handle the bit rate of the TRNG. The FIFO guarantees that no data are lost during the transfer. Two signals were exchanged between the boards: a clock signal coming from the communication board and the random bitstream produced by the TRNG inside the FPGA under attack. These two signals were monitored with an oscilloscope during the attack in order to ensure that their integrities were untouched. This implementation is called Target #2.

We ensured that the ROs were not initially locked due to their placement. In the rest of the chapter, the acronym "locked" will refer to the phenomenon described in Reference 4.

In both cases, ROs were composed of three inverters (NOT gates), giving the working frequencies of about 330 MHz. For Target #2, the TRNG was composed of 50 ROs of three inverters. A sampling clock of 24 KHz was generated in an embedded phase-locked loop (PLL). This sampling frequency was chosen in order to make the RO-TRNG output pass the National Institute of Standards and Technology (NIST) statistical tests. In general, decreasing the speed of the sampling clock improves the behavior of the TRNG (the jitter accumulation time is longer), while increasing it will potentially reduce the entropy extracted by the whole TRNG. Moreover, we used more ROs than Wold and Tan in Reference 2 (50 in our case versus 25 in the case of Wold and Tan). We stress that the TRNG featuring 50 ROs should pass Federal Information Processing Standard (FIPS) and NIST statistical tests under normal conditions without any problems (giving that the sampling frequency is properly tuned).

14.5.3 Attack description

Inside the Faraday cage the probe was located in the close vicinity of the FPGA plastic seal (the FPGA packaging was left intact), i.e., at a distance of roughly 100 μm from the DUT packaging. In order to maximize the impact of electromagnetic injections, the tip of the probe was placed near ROs implemented inside the FPGA.

- The first set of experiments, realized on Target #1, was aimed at analyzing the influence of electromagnetic injections on the ROs. The electromagnetic signals power level P_{forward} was set successively to 34 nW; 340 μW; 1 mW; 3 mW, in a frequency range of 300–325 MHz. As stated before, this frequency range was chosen according to the results in Section 14.4. With a sampling rate of 20 MS/s, we acquired successively ten traces on each of the four oscilloscope channels, in order to record:
  - Out_1, the signal provided by the RO#1 used as a trigger to synchronize the oscilloscope.
  - Out_2 to Out_4, the signals provided by RO#2, RO#3, and RO#4.

  Finally, all acquired data were analyzed off-line according to several criteria. Among them, one is the mutual information (MI). This point is detailed in Section 14.5.4.2. Another one (detailed in Section 14.5.4.3) is the phase difference between the oscillating signals Out_1 and Out_3 during the electromagnetic injection.
- The second set of experiments was aimed at studying the behavior of a complete TRNG (Target #2) under electromagnetic emanation attacks. For each configuration, the TRNG output bitstream was stored and analyzed with and without electromagnetic injections. This latter set of experiments was conducted with a periodic signal of 309.7 MHz. This frequency corresponded to the value maximizing the coupling between the probe and the IC. It was found by analyzing the results of a discrete Fourier transform (DFT) applied on the simple electromagnetic analysis (SEMA) signal that was obtained at different electromagnetic emanation frequencies. This point is further explained in the next section.

14.5.4 Effect of the electromagnetic waves on the ROs—Target #1

14.5.4.1 Choice of the injected signal frequency

The frequency of the injected signal determines success of the attack. Indeed, the coupling between the IC and the probe tip end depends strongly on this parameter.

Our first aim was to find the frequency that will impact a maximum number of ROs. For this reason, the electromagnetic injections were realized at different frequencies. More precisely, the frequency was swept over the range of 300–325 MHz by steps of 50 kHz. This range was chosen because the oscillating frequencies f_{RO_i} of all ROs were measured and found to be spread between 325 and 330 MHz (using the electromagnetic differential analysis method presented in Section 14.3). During frequency sweeping, we analyzed the evolution of the ratio $DFTRi = Y_{f_{inj}}/Y_{f_{RO_i}}$, where $Y_{f_{inj}}$ is the amplitude of the spectral decomposition of Out_i at the injected frequency and $Y_{f_{RO_i}}$ is the amplitude at f_{RO_i}. As shown in Figure 14.16, within given frequency range, all the *DFTRi* ratio values reach their maximum at around 309.7 MHz. For this reason and also because this frequency maximizes the electromagnetic injection effects on all ROs, it was selected for realizing all the following experiments. Figure 14.17(b) illustrates the effect of the electromagnetic injection at this frequency. It can be seen that the spectral decomposition of Out_1 and Out_3 shows a maximum at 309.7 MHz during injection of the perturbing signal. This maximum value is 15 times higher than the amplitude at f_{RO_1} and f_{RO_3}, because ROs oscillate at the injected frequency.

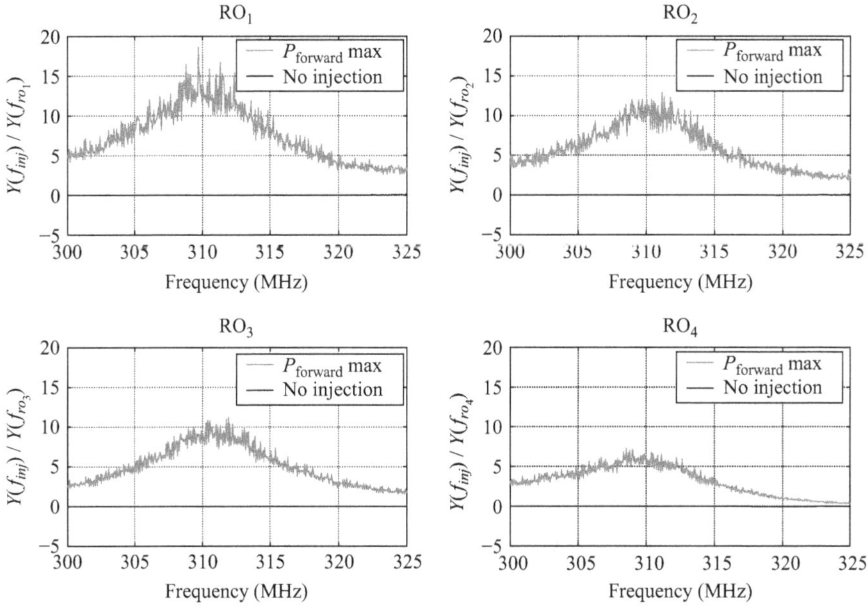

Figure 14.16 Discrete Fourier transforms (DFTs) factor Y_{finj}/Y_{fRO_i} versus injection frequency, after analyzing signals Out_1, Out_2, Out_3, and Out_4

However, this also means that all ROs (or at least most of them) are mutually locked. The selected frequency was kept unchanged during the rest of experiments and also during the specific attacks on TRNGs.

When the RO was not perturbed by an electromagnetic injection, the signal was composed only of the fundamental frequency. Its magnitude was equal to 0.25 (Figure 14.17(a)). As a result, the DFT factor was near to 0. Then, the electromagnetic sinusoidal signal of 309.7 MHz was injected. The 309.7 MHz harmonic signal was so strong that it appeared on the DFT and its amplitude became 15 times higher than that of the fundamental frequency (Figure 14.17(b)). This injected harmonic signal took the control of ROs and generated signals.

14.5.4.2 Proof of effectiveness

In order to verify that all ROs were effectively locked, we analyzed the evolution of the MI between the four ROs output voltages $(V_i(t), V_j(t))$ for the injected power $P_{forward}$. The MI is a general measure of the dependence between two random variables, and this parameter is often used as a generic side-channel distinguisher [14]. Concerning our experiments, we expected to observe:

- low MI values between $V_i(t)$ and $V_j(t)$ when for $P_{forward} = 340$ nW, meaning that the ROs were not locked,
- increased MI values when $P_{forward}$ was higher, meaning that electromagnetic injections effectively lock the ROs.

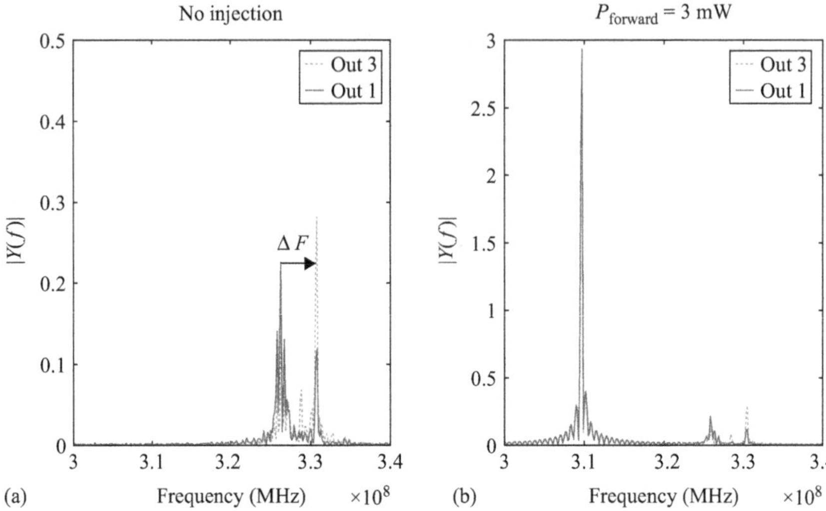

Figure 14.17 Discrete Fourier transform of the signals Out_1 and Out_3 under: (a) normal conditions and (b) electromagnetic injection at $F_{inj} = 309.7$ MHz and $P_{forward} = 3$ mW

Table 14.1 MI values for selected RO couples obtained at different injection powers

P_{Forward} **at 309.7 MHz**	**340 nW**	**34 μW**	**1 mW**	**3 mW**
MI(RO#1,RO#2)	0.0267	0.1746	0.5478	1.5729
MI(RO#1,RO#3)	0.0305	0.7697	0.7889	1.1029
MI(RO#1,RO#4)	0.0135	0.2838	0.6747	0.8221
MI(RO#2,RO#3)	0.1055	0.1086	0.3872	0.8379
MI(RO#2,RO#4)	0.0245	0.1332	0.2247	0.6477
MI(RO#3,RO#4)	0.0383	0.3196	0.8053	0.9382
MI average	0.0398	0.2983	0.5715	0.9870

Table 14.1 shows MI values at different levels of injection. As expected, the MI values were really low (0.02 bit) when $P_{\text{forward}} = 340$ nW. On the other side, when $P_{\text{forward}} = 3$ mW, the MI average increased up to 0.99 bits at $f = 309.7$ MHz. This demonstrated clearly that the ROs were locked or at least interdependent. This interdependence was also visible on the oscilloscope using the persistent screen mode. Figure 14.18 shows signals Out_1 and Out_3 obtained without (Figure 14.18(a)) and with (Figure 14.18(b)) signal injection. As it can be seen, if under attack, the two ROs were synchronized and operated at the same frequency (note that for other ROs, we observed the same behavior).

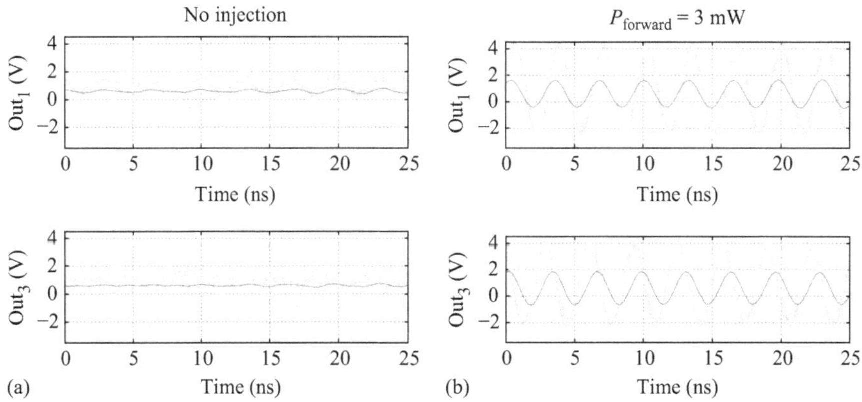

Figure 14.18 Subsequent traces in persistent display mode (bold) and mean traces (fine) of Out_1 and Out_3 corresponding to RO's outputs during (a) normal conditions and (b) submitted to $P_{forward} = 3\ mW$ of 309.7 MHz electromagnetic injections

14.5.4.3 Phase reduction

Under normal conditions, the ROs have different operating frequencies due to different interconnection delays. This is visible in Figure 14.17(a). The difference $\Delta F = f_{RO_1} - f_{RO_3}$ produces a linear drift between the rising edges of the RO signals (their positions will also depend on the jitter, but compared to ΔF, the jitter impact is smaller).

In the case of strong electromagnetic harmonic injections, the two ROs are locked on the injection frequency. This is clearly visible in Figure 14.17(b), where the biggest harmonic is one of the injected frequency. Next, we propose to evaluate the phase difference between output signals of the two ROs. The evolution of the phase differences between signals Out_1 and Out_3 is plotted in Figure 14.19(a). According to the histogram from Figure 14.19(b), the phase is distributed between 222° and 252° and centered around 237°. This gives a range of variation for the phase of 30°. If we look at the phase evolution over the time, it is following an almost sinusoidal tendency. As said before, during the harmonic injection, Out_1 and Out_3 are mainly composed of two frequencies, one coming from the injection itself (f_{inj}) and the working frequency of the ring (f_{RO_1} and f_{RO_3}). These two frequencies in the frequency spectrum of each RO produce a beat phenomenon (as it is defined in acoustics). This beat phenomenon explains the sinusoidal tendency of the phase.

14.5.5 Effect of the electromagnetic waves on the TRNG—Target #2

14.5.5.1 Impact of the dependence between ROs on the random bitstream

The TRNG output bitstream produced for several levels of electromagnetic injection power is shown in Figure 14.20. Each sample is composed of 120 successive

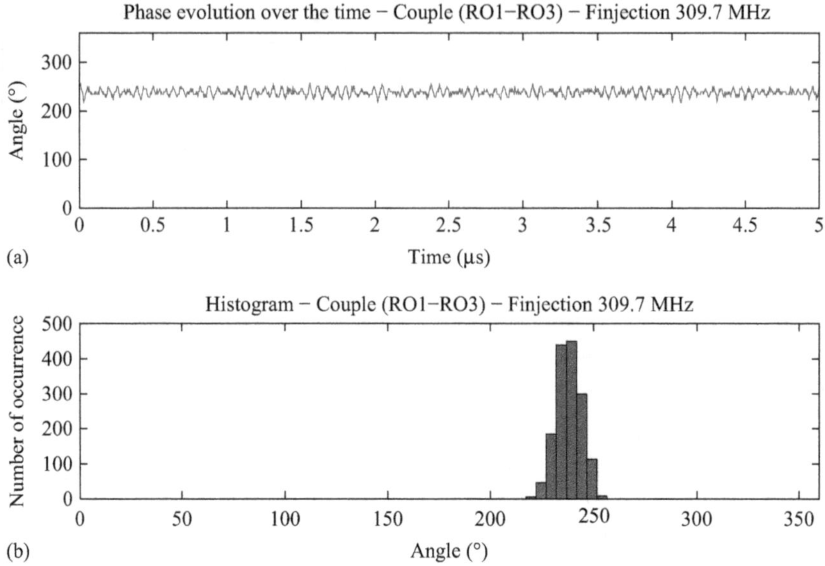

Figure 14.19 (a) Phase difference between Out_1 and Out_3 over the time and (b) phase distribution

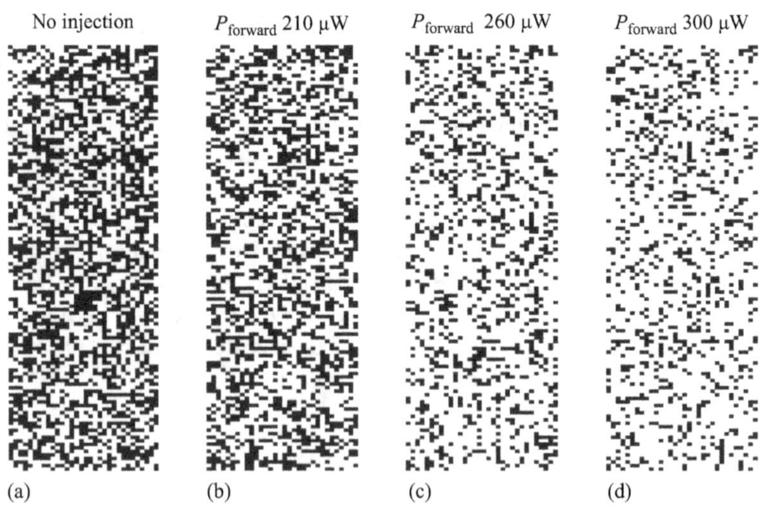

Figure 14.20 Bitstream produced by the TRNG under different attack powers at 309.7 MHz using electric probe (120 × 32)—starting from left to right: (a) no injection; (b) $P_{forward} = 210$ μW; (c) $P_{forward} = 260$ μW; and (d) $P_{forward} = 300$ μW

Table 14.2 Statistical parameters of the TRNG output bitstream

$P_{forward}$	**No injection**	**210 μW**	**260 μW**	**300 μW**
Bias (%)	0.1	15.87	51.57	55
NIST tests	Success	Fail	Fail	Fail

32-bit frames (black and white squares correspond to ones and zeros, respectively). Notice that under normal conditions (Figure 14.20(a)), the TRNG bitstream passed NIST statistical tests with confidence level $\alpha = 0.01$, applied on 1,000 sequences of 1 Mbit. In order to speed-up testing, we used only the frequency test (also called monobit test) in subsequent experiments. The frequency test evaluates the balance between number of ones and zeros in the bitstream. If the generated data do not pass this test, it is not necessary to continue with more complex tests—the data would not pass the complete test suite.

In Table 14.2, the bias is defined as $Bias = abs(0.5 - P(0)) = abs(0.5 - P(1))$, where $P(x)$ is the probability of the element x. The bias can vary between 0 and 0.5. The bias is usually reported in percentage, after extrapolation between 0% and 100% corresponding to 0 and 0.5 bias values. We will use this bias representation in the rest of the chapter. A good TRNG must have a bias close to 0%.

According to Figure 14.20 and Table 14.2, the electromagnetic injection effect on the bias is clear. For example, for a power of 210 μW (Figure 14.20(b)), the bias reaches 15% (15 bits out of 100 bits of the bitstream are impacted by signal injection). Increasing the injected power up to 260 μW, the bias raises up to 50% (Figure 14.20(b) and (c)).

14.5.5.2 Control of the bias

Previous experiments confirmed that it is possible to control statically the bias of a RO-TRNG. In the next experiments, we wanted to observe the dynamic behavior of the TRNG under attack. In the power injection platform (Figure 14.12), we added an amplitude modulator (AM) between the RF generator and the input of the power amplifier. This system achieves the analog multiplication between the injection signal—a sine waveform signal fixed at 309.7 MHz (the active harmonic needed to perform the attacks)—and a square waveform signal (the control signal), which controls accurately the beginning and the end of the electromagnetic injection. The control signal is provided by an external FPGA in order to deliver the desired timing of the injection sequence. Figure 14.21(a) represents the timing evolution of the AM signal in volts. Figure 14.21(b) shows the effect on the TRNG output bitstream. Finally, Figure 14.21(c) represents the evolution of the bias in time. It was computed using a sliding window of 10,000 bits. The sliding step was 32 bits.

Looking at the bitstream or the bias, it is clear that the behavior of the TRNG is quickly (in less than 1 ms) impacted by the electromagnetic perturbation and it returns to its initial state with the same speed. In fact, we observed that the bias was

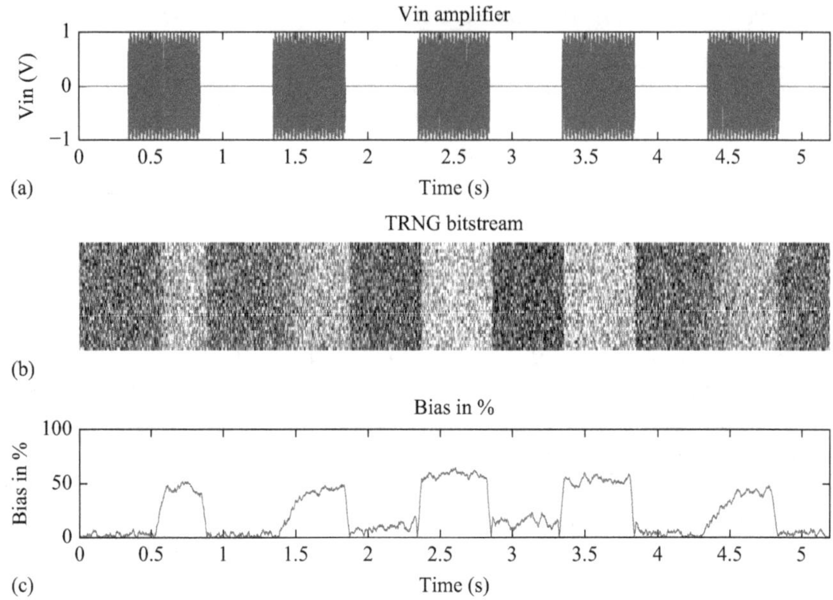

Figure 14.21 (a) AM signal; (b) TRNG stream of bits (raster scanning from bottom to top and left to right); and (c) bias in % for the TRNG stream of bits

changing according to the dynamics of the power amplification chain. In our case, it has a time response of roughly 1 ms. The difference in the bias for the different periods of attack is due to the fact that the response time of the power amplifier is not adapted to operate in an AM mode. This experiment shows clearly that the dynamic electromagnetic harmonic injection is feasible and that it can be very powerful and able to control the behavior of a RO-based TRNG even if it is composed of a big number of ROs. The dynamic control of the electromagnetic harmonic injection is of a paramount importance, because it can be used in order to bypass statistical-embedded tests launched periodically.

14.5.6 Discussion

As stated in Section 14.3, the RO-TRNG has several parameters that can be tuned in order to guarantee sufficient entropy amount and/or output bit rate:

- Number of ROs composing the TRNG.
- Number of elements composing the ROs (or in another words the frequency of the ROs).
- Sampling frequency of the flip-flops.

For example, by increasing the number of ROs and/or their frequency, the designer can increase the output entropy rate. On the other hand, by increasing the sampling frequency, subsequent bit may become correlated and the entropy reduced.

In our case, the number of ROs and their frequencies are fixed (the frequencies of the RO are slightly changing due to the injection as seen in Section 14.5.4). Having a flip-flop sampling on both falling and rising edges of the clock signal is equivalent to doubling the frequency of the clock for a regular flip-flop. If the sampling frequency of the TRNG is set with the aim of having the best ratio between good statistical property of the output and the highest throughput possible, doubling the value of the sampling frequency will further reduce the output entropy rate (we suppose that the successful attack has already reduced the entropy because of locking of ROs to the injected signal). We can therefore conclude that by setting suitable parameters of the injected signal (i.e., its amplitude and frequency), the attacker can further increase efficiency of the attack.

As far as vulnerability to the proposed electromagnetic attack is concerned, the main weakness of the entropy extraction in RO-TRNG is the use of an external clock as the sampling clock for our TRNG. Indeed, this clock has inconveniently long rising and falling times, making the generation of fake clock edges easier. We believe that the best way to reduce this weakness is to use an internal PLL to generate a proper clock signal. This will at least help to prevent the flip-flops to sample on fake edges and will definitely make the attack harder to perform. Removing the fault on the flip-flops will not remove completely the effects of the attack. The locking phenomenon of the ROs induced by the electromagnetic injection remains possible and it should be detected by some lock detection circuitry.

14.6 Conclusion

In this chapter, we described a complete attack on a RO-TRNG using the electromagnetic side channel, starting from an analysis on the TRNG to retrieve crucial information on it (the frequency of the ROs and their localization in the device). This analysis leads to the active attack that modifies the behavior of the TRNG using a strong electromagnetic field. According to the experiments presented in this chapter, we demonstrated that it is possible to dynamically control the bias of a RO-TRNG embedded in an FPGA. The attacker can thus manipulate statistically the output bitstream at his good will. This work clearly challenge the use of ROs for both TRNG and physical unclonable function (PUF) [15]. Nevertheless, it opens a new opportunity of using ROs to contactless, rapid, and lightweight transmission of hidden data (such as intellectual property data) from an integrated circuit on the electromagnetic channel [16], but it is a new story.

References

[1] Sunar B., Martin W.J. and Stinson D.R. 'A provably secure true random number generator with built-in tolerance to active attacks'. *IEEE Transactions On Computers*. 2007;56(1):109–119.

[2] Wold K. and Tan C.H. 'Analysis and enhancement of random number generator in FPGA based on oscillator rings'. In *Proceedings of International Conference*

on Reconfigurable Computing and FPGAs, ReConFig 2008; Cancún, Mexico, Dec. 2008. IEEE: Piscataway, NJ; 2008. pp. 385–390.

[3] Kocher P., Jaffe J. and Jun B. 'Differential power analysis'. In *Proceedings of the 19th Annual International Cryptology Conference, CRYPTO 1999*; Santa Barbara, CA, Aug. 1999. Berlin: Springer; 1999; LNCS1666, pp. 388–397.

[4] Bochard N., Bernard F., Fischer V. and Valtchanov B. 'True-randomness and pseudo-randomness in ring oscillator-based true random number generators'. *International Journal of Reconfigurable Computing*. vol. 2010, Article ID 879281, 13 pages, 2010. doi:10.1155/2010/879281.

[5] Markettos A.T. and Moore S.W. 'The frequency injection attack on ring-oscillator-based true random number generators'. In *Proceedings of Workshop on Cryptographic Hardware and Embedded Systems, CHES 2009*; Lausanne, Switzerland, Sep. 2009. Berlin: Springer; 2009; LNCS5747, pp. 317–331.

[6] Wold K. and Petrović S. 'Robustness of TRNG against attacks that employ superimposing signal on FPGA supply voltage'. In *Proceedings of the Norwegian Information Security Conference, NISK 2010*; Universitetsforlaget AS, Gjovik, Norway, Nov. 2010. pp. 81–92.

[7] Böhl E. and Ihle M. 'A fault attack robust TRNG'. In *Proceedings of 18th International On-Line Testing Symposium, IOLTS 2012*; Sitges, Spain, Jun. 2012. IEEE: Piscataway, NJ; 2012. pp. 114–117.

[8] Bayon P., Bossuet L., Aubert A., *et al.* 'Contactless electromagnetic active attack on ring oscillator based true random number generator'. In *Proceedings of Constructive Side-Channel Analysis and Secure Design, COSADE 2012*; Darmstadt, Germany, May 2012. Berlin: Springer; 2012; LNCS7275, pp. 151–166.

[9] Kune D.F., Backes J., Clark S.S., *et al.* 'Ghost talk: Mitigating EMI signal injection attacks against analog sensors'. In *Proceedings of the 34th Annual Symposium on Security and Privacy*; Berkeley, CA, May 2013. IEEE: Piscataway, NJ; 2013. pp. 145–159.

[10] Bayon P., Bossuet L., Aubert A. and Fischer V. 'EM radiation analysis on true random number generators: Frequency and localization retrieval method'. In *Proceedings of the Asia-Pacific International Symposium and Exhibition on Electromagnetic Compatibility, APEMC 2013*; Melbourne, Australia, May 2013. IEEE: Piscataway, NJ; 2013. pp. 1–4.

[11] Bayon P., Bossuet L., Aubert A. and Fischer V. 'Electromagnetic analysis on ring oscillator-based true random number generators'. In *Proceedings of the International Symposium on Circuits and Systems, ISCAS 2013*; Beijing, China, May 2013. IEEE: Piscataway, NJ; 2013. pp. 1954–1957.

[12] Sauvage L., Guilley S. and Mathieu Y. 'Electromagnetic radiations of FPGAs: High spatial resolution cartography and attack on a cryptographic module'. *ACM Transactions on Reconfigurable Technology and Systems*. 2009;2(1):4.1–4.24.

[13] Poucheret F., Robisson B., Chusseau L. and Maurine P. 'Local electromagnetic coupling with CMOS integrated circuits'. In *Proceedings of*

Eighth International Workshop on Electromagnetic Compatibility of Integrated Circuits; Dubrovnik, Croatia, Nov. 2011. IEEE: Piscataway, NJ; 2011. pp. 137–141.

[14] Batina L., Gierlichs B., Prouff E., Rivain M., Standaert F.X. and Veyrat-Charvillon N. 'Mutual information analysis: A comprehensive study'. *Journal of Cryptology*. 2010;2(2):269–291.

[15] Bossuet L., Ngo X.T., Cherif Z. and Fischer V. 'A PUF based on a transient effect ring oscillator and insensitive to locking phenomenon'. *IEEE Transactions on Emerging Topics in Computing*. 2014;2(1):30–36.

[16] Bossuet L., Fischer V. and Bayon P. 'Contactless transmission of intellectual property data to protect FPGA designs'. In *Proceedings of the 23rd IFIP/IEEE International Conference on Very Large Scale Integration, VLSI-SoC*; Daejeon, Korea, Sep. 2015. IEEE; Piscataway, NJ, 2015. pp. 19–24.

Index

www.ingramcontent.com/pod-product-compliance
Lightning Source LLC
LaVergne TN
LVHW011913200926
848484LV00002B/606
* 9 7 8 1 7 8 5 6 1 0 5 7 8 *